COMPLETE GENTLEMEN

A British Academy Monograph

The British Academy has a scheme for the selective publication of monographs arising from its British Academy Postdoctoral Fellowships, British Academy Newton International Fellowships and British Academy/Wolfson Fellowships. Its purpose is to assist individual scholars by providing a prestigious publishing opportunity to showcase works of excellence.

Richard Ansell is a historian of seventeenth- and eighteenth-century Britain and Ireland, focusing on travel to continental Europe. He studied at Selwyn College, Cambridge, Brown University and Hertford College, Oxford. He has held the Irish Government Senior Scholarship at Oxford and a British Academy Postdoctoral Fellowship at the University of Leicester, where he is currently a Research Associate. Alongside educational travel, he is now working on travelling servants and British encounters with Spain and Portugal. His research has appeared in the *Historical Journal*, *English Historical Review* and several edited collections.

COMPLETE GENTLEMEN
Educational Travel and Family Strategy, 1650–1750

Richard Ansell

Published for THE BRITISH ACADEMY
by OXFORD UNIVERSITY PRESS

Published for THE BRITISH ACADEMY
by OXFORD UNIVERSITY PRESS

Oxford University Press, Great Clarendon Street, Oxford OX2 6DP

First edition published in 2022

British Library Cataloguing in Publication Data
Data available

Library of Congress Cataloging in Publication Data
Data available

Typeset by
Marie Doherty

Printed and bound by CPI Group (UK) Ltd, Croydon, CR0 4YY

ISBN 978–0–19–726727–1

Contents

List of Illustrations

Acknowledgements

This book presents research undertaken during a British Academy Postdoctoral Fellowship at the University of Leicester and builds on work done at Hertford College, Oxford, for an AHRC-funded D.Phil. and the Irish Government Senior Scholarship. Over these periods, I also received a postgraduate bursary from the British Association for Irish Studies and fellowships at the Huntington Library and the Clark Library, UCLA. Towards the end of the project, I benefited from a visiting bye-fellowship at Selwyn College, Cambridge. I thank all these institutions for supporting my research.

The book developed in conversation with generous scholars, including Toby Barnard, Gabriel Glickman, Sarah Goldsmith, David Hayton, Roey Sweet, Gerrit Verhoeven and Mark Williams. Sarah and Mark, in particular, have offered friendship and historical discussion from Oxford and Leicester to Louisiana. Toby, Gabriel, Sarah, David, Roey and Mark have also shared unpublished work or archival notes that have contributed to this book, as have Jean Boutier, John Gallagher, Hugh Mayo, William Molesworth and Michael Page. Jason Kelly, Brendan Rooney and especially David White helped me to track down illustrations.

I am particularly grateful to Toby Barnard and Roey Sweet for their mentorship. Toby was a great source of guidance and archival knowledge as the themes of this project first developed at Oxford. At Leicester, Roey has offered invaluable judgement, support and friendship as it became a book. I am indebted to her for reading and discussing the whole manuscript, though any errors are mine. I am also thankful to Michèle Cohen, Simon Ditchfield, Tim Harris, John Morrill and David Smith for supporting my work.

The highlight of this project has been its many research trips. I am grateful to the staff of all the libraries and archives mentioned in my bibliography, but particularly to those of Blackburn Central Library; Cambridge University Library; Christ Church, Oxford; Christ's College, Cambridge; the Clark Library; the Huntington Library; Trinity College Dublin; and West Yorkshire Archive Service, Leeds, for help with specific questions.

As the book came together, I relied on Portia Taylor at the British Academy and Elizabeth Stone of Bourchier for help with the publication process. Tony Claydon and John Gallagher offered constructive comments on my proposal and, later, the anonymous reader and the publications committee engaged carefully with my manuscript.

Away from academia, Mark Bosley, Adam Clifton, Sophie Topham, Giles Ekblom and Camille Langlois-Bertrand provided hospitality and distractions on research trips to London, as did Audrey, George and Siân Kearns and Sinéad Gannon in Ireland, while Gabrielle Narcy's encouragement made sure that my project became a book. I would not have got near that point without the love and support of my parents, Chris and Dolores Ansell, to whom I am immensely grateful.

List of Abbreviations

BL	British Library, London
Blackburn CL	Blackburn Central Library
BNCF	Biblioteca Nazionale Centrale di Firenze, Florence
BNF	Bibliothèque Nationale de France, Paris
CSPD	*Calendar of State Papers, Domestic Series*
DIB	*Dictionary of Irish Biography*
HL	Huntington Library, San Marino, California
HMC	Historical Manuscripts Commission
LLRRO	Leicester, Leicestershire and Rutland Record Office, Wigston
NAL	National Art Library, Victoria and Albert Museum, London
NLI	National Library of Ireland, Dublin
Northants RO	Northamptonshire Record Office, Northampton
NYCRO	North Yorkshire County Record Office, Northallerton
ODNB	*Oxford Dictionary of National Biography*
PRONI	Public Record Office of Northern Ireland, Belfast
SBT	Shakespeare Birthplace Trust, Stratford-upon-Avon
SHC	Surrey History Centre, Woking
TCD	Trinity College Dublin
TNA	The National Archives, Kew
WYAS	West Yorkshire Archive Service, Leeds
WYAS, Bradford	West Yorkshire Archive Service, Bradford
YCBA	Yale Center for British Art, New Haven, CT

Note on the Text

In a book that deals with social mobility among the landed elite, changing titles can lead to confusion. James Brydges became Baron Chandos (1714), Earl of Carnarvon (1714) and Duke of Chandos (1719). Sir John Perceval, a baronet from 1691, was in turn Baron Perceval (1715), Viscount Perceval (1722) and Earl of Egmont (1733). Robert Molesworth became Viscount Molesworth (1716) and Sir Alan Brodrick progressively Baron Brodrick (1715) and Viscount Midleton (1717). I stick with the name or title used for someone's first chronological appearance and acknowledge a contemporary title where relevant. As one exception, I refer to Robert Southwell (1635–1702) as Sir Robert after his 1665 knighthood to help in distinguishing him from his father, Robert (1607–77) and to emphasise the advancement that followed his travels of 1659–62.

Working across generations means dealing with different people under the same names and titles. Regarding first names, I have tried to avoid confusion in the text and have distinguished namesakes with II and III in footnotes. Sons shadowed their fathers' advances, such that the same dependent titles also recur. For the one major instance—that John Brydges, Henry Brydges and James Brydges II were all Marquis of Carnarvon—I have omitted the title unless relevant to the discussion.

Introduction

Visitors to the public galleries and country houses of Britain invariably come across at least one portrait of an aristocratic young man on the Grand Tour. Classical props or a background of ruins indicate that he has reached Rome, the most important stage in a journey of several years through France, Italy, Germany and the Low Countries, and he has usually crossed the Channel in the mid-eighteenth century, the supposed heyday of this rite of passage. Such pictures are even more recognisable thanks to academic and popular books on the Grand Tour, which cast early modern travel as an artistic and literary phenomenon centred on Italy. In this widespread telling, wealthy young men travelled abroad in order to illustrate their education in classical history and poetry, to develop an appreciation of Renaissance masterpieces and to come of age among the courtesans of Venice. As a result, Grand Tour portraits have come to stand for all adolescent or early adult travel in the period, sometimes for all encounters with the Continent.

These familiar depictions offer only a partial view of seventeenth- and eighteenth-century educational travel. They reveal nothing of the family decisions that sent young men abroad or the preparations that they made at home, and they give little indication of what rested upon their successful returns. Although some sitters appear with fellow travellers, meanwhile, they are always isolated from their servants and from the people they met abroad, whether Anglophone or local. Artists based in Italy tended instead to emphasise the classical learning that drew many to that country, but they cannot capture the fact that most educational journeys between the mid-seventeenth and mid-eighteenth centuries were motivated by the pursuit of French culture, whether in France itself or in nearby territories. Visual evidence of the Grand Tour encourages the idea that travel was only for those who could afford a lengthy period in Italy, and indeed a Roman or Venetian portrait, but such souvenirs are only a snapshot of the most eye-catching voyages. Not only were Paris and the French-influenced Low Countries attractive destinations for cultural reasons, but they also formed a well-connected cluster of north-western European cities that offered practical options to those with less time or money to invest in a foreign voyage. Recognising that several forms of travel could coexist in the same period, this book offers a comprehensive view of how landed families of varying means participated in a wider culture of educational mobility between 1650 and 1750.[1]

In order to do so, it needs to play down the Grand Tour. This means working against a steady stream of publications over the past century, becoming a flood from the 1990s onwards.[2] Even as historians of art, architecture and culture have come together for interdisciplinary initiatives in recent decades, few have challenged the basis of their efforts.[3] Most still make Italy central to their definitions of the Grand Tour, which in turn often stands for early modern travel in general.[4] Literary scholars usually work separately on published travel writing, but many also operate on the assumption that the Italian Grand Tour framed encounters with the Continent.[5] As a result, in both historical and literary approaches, other European destinations receive far less treatment. These priorities also warp chronologies, with the mid-eighteenth century a 'golden age' mainly because it was the high point of Italian travel and collecting. Accordingly, when historians of the Grand Tour have strayed before 1700, it has been to establish the origins of a set phenomenon.[6] Across a

1 For 'coexisting and competitive' styles of travel, see Judith Adler, 'Travel as Performed Art', *American Journal of Sociology*, 94:6 (1989), 1366–91, at 1372–4, 1380.

2 For overviews, see Edward Chaney, *The Evolution of the Grand Tour: Anglo-Italian Cultural Relations since the Renaissance* (London: Routledge, 2000), pp. 383–404; John Wilton-Ely, '"Classic Ground": Britain, Italy, and the Grand Tour', *Eighteenth-Century Life*, 28:1 (2004), 136–65.

3 Andrew Wilton and Ilaria Bignamini, eds, *Grand Tour: The Lure of Italy in the Eighteenth Century* (London: Tate Gallery, 1996); Clare Hornsby, ed., *The Impact of Italy: The Grand Tour and Beyond* (London: British School at Rome, 2000); María Dolores Sánchez-Jáuregui and Scott Wilcox, eds, *The English Prize: The Capture of the Westmorland, An Episode of the Grand Tour* (New Haven, CT, and London: Yale University Press, 2012).

4 Chaney, for instance, treats the Grand Tour and 'Anglo-Italian cultural relations' as synonymous: *Evolution of the Grand Tour*. For the Grand Tour as early modern travel, including destinations outside Italy, see Jeremy Black, *The British Abroad: The Grand Tour in the Eighteenth Century* (Stroud: Sutton Publishing, 2003). For the enduring focus on southern Europe, see Robert Holland, *The Warm South: How the Mediterranean Shaped the British Imagination* (New Haven, CT: Yale University Press, 2018).

5 Chloe Chard, *Pleasure and Guilt on the Grand Tour: Travel Writing and Imaginative Geography, 1600–1830* (Manchester: Manchester University Press, 1999); Chard, *A Critical Reader of the Romantic Grand Tour: Tristes Plaisirs* (Manchester: Manchester University Press, 2014). For an attempt to unite literary and historical approaches, see Richard Ansell, 'Reading and Writing Travels: Maximilien Misson, Samuel Waring and the Afterlives of European Voyages, *c.*1687–1714', *English Historical Review*, 133:565 (2018), 1446–77.

6 Michael G. Brennan, ed., *The Origins of the Grand Tour: The Travels of Robert Montagu, Lord Mandeville (1649–1654), William Hammond (1655–1658) and Banaster Maynard (1660–1663)* (London: Hakluyt Society, 2004); Edward Chaney and Timothy Wilks, *The Jacobean Grand Tour: Early Stuart Travellers in Europe* (London: I.B. Tauris, 2014).

lengthening eighteenth century, moreover, scholars of travel have rarely related it to other aspects of social and cultural history. In various ways, 'the lure of Italy' continues to distort the study of early modern travel.

The last few years have brought promising exceptions, as recent work on Britain has integrated travel with discussions of language-learning, confessional identity, elite masculinity, danger and urbanism.[7] These books often still refer to the Grand Tour in their titles or their text, but they increasingly see the need to examine more diverse encounters with the Continent. Where interdisciplinarity in practice cemented the Grand Tour as a shared object of discussion, this more promising departure emerges from international collaboration among historians.[8] After all, foreign travel was widespread among European elites throughout the period. Continental historiography initially adopted the British Grand Tour framework with enthusiasm, but scholars now pay closer attention to the differences as well as the similarities between national cultures. Alongside lengthy voyages to Italy, more common from some countries than others, travellers across Europe pursued shorter, more northerly journeys for health and pleasure or longer stays at one educational institution.[9] Beyond the Grand Tour, international comparisons are helping to lay out the wider possibilities of early modern travel.

The decline of the Grand Tour may be welcome, but it brings new challenges. Several of the families who appear in this book sent eldest sons to Italy, and sometimes second sons too, while younger brothers made less ambitious travels. Other parents or guardians preferred French, Dutch or German destinations for a host of reasons, including but surpassing the financial. Still more young men undertook domestic tours or attended French institutions in London. Outside the familiar but restrictive framework

[7] Rosemary Sweet, *Cities and the Grand Tour: The British in Italy, c.1690–1820* (Cambridge: Cambridge University Press, 2012); Henry French and Mark Rothery, *Man's Estate: Landed Gentry Masculinities, c.1660–1900* (Oxford: Oxford University Press, 2012), pp. 137–83; Liesbeth Corens, *Confessional Mobility and English Catholics in Counter-Reformation Europe* (Oxford: Oxford University Press, 2019), pp. 79–104; John Gallagher, *Learning Languages in Early Modern England* (Oxford: Oxford University Press, 2019), pp. 157–207; Sarah Goldsmith, *Masculinity and Danger on the Eighteenth-Century Grand Tour* (London: University of London Press, 2020).

[8] Rosemary Sweet, Gerrit Verhoeven and Sarah Goldsmith, eds, *Beyond the Grand Tour: Northern Metropolises and Early Modern Travel Behaviour* (London: Routledge, 2017).

[9] Jean Boutier, 'Le grand tour: une pratique d'éducation des noblesses européennes (XVIe–XVIIIe siècles', in *Le voyage à l'époque moderne* (Paris: Presses de l'université de Paris-Sorbonne, 2004), pp. 7–21; Mathis Leibetseder, 'Across Europe: Educational Travelling of German Noblemen in a Comparative Perspective', *Journal of Early Modern History*, 14 (2010), 417–49; Gerrit Verhoeven, *Europe within Reach: Netherlandish Travellers on the Grand Tour and Beyond (1585–1750)*, trans. Diane Webb (Leiden: Brill, 2015).

of the Grand Tour, how can historians explain this diversity? How can they keep Italy as one expensive option while recognising that younger brothers and families of lesser means pursued other routes to continental experience? And, without the Grand Tour, how can scholars keep sight of the didactic purpose that so many ascribed to travel in adolescence or early adulthood? This book finds answers in a broader understanding of educational travel and through attention to its place in family strategy, where time abroad emerged from trade-offs between means, ambitions and obligations. Rather than starting with preconceived notions of the Italian Grand Tour, it looks at what families thought and did about travel.

The term 'educational travel' captures a wider range of continental experience than the Grand Tour. This book defines educational travel loosely, as a first independent journey to continental Europe made before or during a young man's mid-twenties, to allow for the diversity of family practice. Historians have variously characterised the Grand Tour as travel by aristocrats, travel with a tutor or travel to Italy, but each decision closes off routes through which families of lesser means or different priorities secured time abroad for their sons.[10] As a subject, educational travel has several other virtues. It allows for a connection between foreign and domestic travel in a culture of educational mobility, outlined in Chapter 1. As explored in Chapter 3, it recognises that contemporaries wrote of 'travels' far more often than 'tours' when they wanted to refer to their whole time abroad. And finally, it forces journeys that culminated in Italy to take their place among several options through which families provided sons with continental experience.

This definition does not include girls, although their education was a contemporary priority. When so much depended on marrying well, daughters were afforded a good share of family budgets.[11] Consequently, despite easy contrasts between female domestic tuition and male travel, many gentlemen's daughters in fact undertook some educational mobility. Girls' schools arose in the suburbs of London in the late seventeenth century, while parents continued the old custom of placing daughters with friends or relations, a practice that also allowed those from the provinces to appear in the capital.[12]

[10] Bruce Redford's oft-cited definition blends all three: *Venice and the Grand Tour* (New Haven, CT, and London: Yale University Press, 1996), p. 14.

[11] Linda Pollock, '"Teach Her to Live under Obedience": The Making of Women in the Upper Ranks of Early Modern England', *Continuity and Change*, 4, special issue 2 (1989), 231–58, at 238–9.

[12] Rosemary O'Day, *Education and Society, 1500–1800: The Social Foundations of Education in Early Modern Britain* (London: Longman, 1982), pp. 179–95; Ralph A. Houlbrooke, *The English Family 1450–1700* (London: Longman, 1984), pp. 176–7.

Foreign travel was usually a different matter, however. When one nobleman thought of sending his daughters abroad in the 1660s, his son's tutor convinced him that 'the coming of English ladies into France for an education is a thing singular'.[13] He evidently ignored Catholic girls, who studied at convents there and in Flanders throughout the seventeenth and eighteenth centuries, but Protestant daughters rarely went abroad for their education.[14] One family in this study, the Percevals, nevertheless offers two instances of adolescent female travel. Kate (15) and Helena (12) went to France in 1677–9 with their mother, who had left home on medical advice, and, on another journey prompted by maternal illness, their great-niece, Kitty (13), saw Paris and the Low Countries with her parents and brother in 1725–6. The Percevals reflect a trend examined in Chapter 4, in which families increasingly made short trips abroad for health and leisure, but their daughters did not cross the Channel with the same objectives as their sons. Kate and Helena left London to help their ailing mother, to profit from the air of France and only incidentally, as their uncle put it, for 'the improvement of their education'. Since they had come to Saumur and Paris, however, their mother sent them for tuition in French, dancing and music, hoping to keep her girls abroad 'till thay had what were nessery for them to enable them to come into the best company'.[15] Such concerns shaped the marriageability of daughters, so central to family strategy, but there is little wider sense that it improved with travel. Women and girls from Protestant families in Britain and Ireland usually went abroad as wives and mothers, or as daughters with parents, rather than as educational travellers. They had earlier, more widespread and more frequent continental experience than would appear from work on Italy, which finds them in significant numbers only by the mid-eighteenth century.[16] Throughout the period, however, travel was never a central part of women's upbringing.

This book establishes the position that male educational travel held in family strategy. The concept of family strategy usually features in studies of the industrial working class, helping to reconstruct the lives of those for

[13] Theophilus Gale to Lord Wharton, quoted in John Stoye, *English Travellers Abroad, 1604–1667: Their Influence in English Society and Politics* (New Haven, CT, and London: Yale University Press, 1989), p. 304.

[14] Caroline Bowden, 'Convent Schooling for English Girls in the "Exile" Period, 1600–1800', *Studies in Church History*, 55 (2019), 177–204.

[15] BL, Add MS 46954A, fo. 167: Sir Robert Southwell to Katherine, Lady Perceval, 19 May 1677; 46955B, fo. 128: Lady Perceval to Southwell, 8 Oct 1678.

[16] Brian Dolan, *Ladies of the Grand Tour* (London: Flamingo, 2002); Sweet, *Cities and the Grand Tour*, p. 27.

whom few documents remain.[17] It views family decisions, such as those over the upbringing of children, as trade-offs between means and ambitions; this book adds obligations, in recognition of the performative nature of early modern gentility and nobility. Scholars apply the term 'family strategy' with little consistency, however, sparking debates over its purpose and utility. Does it reflect behaviour or is it only a metaphor? How thought-out were actions? What power did actors have to shape their circumstances? And, amid varying accounts of family dynamics, whose strategies were they anyway?[18] Most issues come down to a lack of sources, since historians usually turn to such arguments to fill documentary gaps. By contrast, confronted with the ample archives of the early modern gentry and nobility, what this study requires is less a rigorous approach than a set of questions to ask of abundant material. What options were available when families planned to educate their sons? How did continental travel fit into that range? What possibilities existed for those who decided to send young men abroad? How did they use their varying means and connections to do so? How were resources allocated among brothers? Among parents, guardians, relations and travellers, how were decisions over educational journeys made? What did families hope to achieve through time abroad? And did it help them to meet these ends? Answers to these questions will establish what travel meant to the families who pursued it.

In turn, this book examines what family decision-making meant for educational travel. First, in the sense that family strategy shaped individual encounters with the Continent: trade-offs between means, ambitions and obligations informed the timing, destinations and duration of journeys, the activities that young men undertook, the company in which they moved and the ways in which they remembered their experiences. Secondly, this study places family strategy at the centre of social and cultural change, recognising that each decision to send a son abroad contributed to the rise of travel as an educational phenomenon. Families may have been conservative institutions but, over generations, the cumulative weight of their choices lay at the forefront of change.[19] Travel became an integral part of many family

[17] Tamara K. Hareven, 'The History of the Family and the Complexity of Social Change', *American Historical Review*, 96:1 (1991), 95–124.

[18] Pier Paolo Viazzo and Katherine A. Lynch, 'Anthropology, Family History, and the Concept of Strategy', *International Review of Social History*, 47:3 (2002), 423–52; Hannah Barker, *Family and Business during the Industrial Revolution* (Oxford: Oxford University Press, 2017), pp. 78–81.

[19] Keith Wrightson, 'The Family in Early Modern England: Continuity and Change', in Stephen Taylor, Richard Connors and Clyve Jones, eds, *Hanoverian Britain and Empire: Essays in Memory of Philip Lawson* (Woodbridge: Boydell, 1998), pp. 1–22, at pp. 18–19.

identities but also, in a period when landowners performed for their peers, an expected part of gentlemanly education. Historians of Britain and Ireland recognise an increasingly homogeneous elite during the seventeenth and eighteenth centuries, but the role of education is more often assumed than explored.[20] Educational travel was clearly an important factor, however. Throughout this book, parents, relations, guardians, trustees, tutors and young men will argue that it was necessary to what many called the 'compleat gentleman', an ideal pursued by the gentry and nobility alike. As such, shared time abroad was not only a way for families to secure or advance their standing, but also a crucial part of elite formation. Travel is too often a stand-alone subject, reduced to anecdotal treatment, but much of its contemporary meaning derived from preparations, alternatives and afterlives that were interwoven with wider social, cultural and economic themes. Attention to family strategy reconnects travel to these broader historical currents.

This book covers the period 1650–1750 so that it can include generations of decision-making, encompassing the earliest and latest examples of educational travel in most of the families under discussion. It also cuts across standard chronologies of the Grand Tour, which supposedly began in the early seventeenth century, spread to the gentry during the 1640s and 1650s, flourished in the mid-eighteenth century and declined, with the rise of middle-class travel, in the few decades either side of 1800. Within this timeline, the eighteenth century was a long period of interchangeable engagement with the history and culture of Italy. This study finds the origins of the Grand Tour—in practice the Italian experiences and artistic collecting of a handful of noblemen—less interesting than the popularisation of educational travel in its several forms.[21] As the following chapters will show, travel became an expected part of a gentleman's education between the mid-seventeenth and mid-eighteenth centuries not because of a few trailblazers but on account of the repeated decisions of unremarkable landowning families. The individual journeys in this book claim little lasting cultural significance, but they contributed to a profound social change. Equally, where most historians introduce less affluent travellers only in the late eighteenth century, this book sees no need to await the decline of the

[20] For example, James M. Rosenheim, *The Emergence of a Ruling Order: English Landed Society, 1650–1750* (London: Longman, 1998).
[21] Chaney and Wilks, *Jacobean Grand Tour*.

Grand Tour before finding other forms of continental travel.[22] There had always been options to suit different purses. The book ends in around 1750 because, with a couple of exceptions, the evidence of educational travel then fades in the archives under discussion. Although gentry and noble families still crossed the Channel for many reasons, more patriotic forms of English manliness left less room for educational options on the Continent.[23]

If change depended on innumerable decisions to pursue educational travel, what were families hoping to achieve? In broad terms, the gentry and nobility aimed to maintain social standing while entertaining hopes of advancement.[24] Newcomers to the landed elite were apparently scarce, at least from business, trade and the professions, but landmark studies have defined that elite as the peerage and, sometimes, the 'greater gentry'.[25] Far less attention has been paid to movement between the middling sort and the 'lesser gentry' or within the landed elite itself, from 'lesser' to 'greater' gentry. These omissions are understandable, given how difficult it can be to ascertain status below the titled nobility, but this very fuzziness, in which identification as a gentleman depended upon the acknowledgement of one's peers, in fact opened the ranks of the gentry to wealthy newcomers.[26] In turn, families could contemplate ascending some of the ill-defined rungs between provincial gentleman and peer; in this book, several who started the period among the gentry of Yorkshire or Ireland would end it with titles. Whether or not such rises were unusual, they were achievable, and belief in the possibility of upward mobility could motivate action. At the same time, though, the fear of falling was usually a greater incentive at all levels of early modern society.[27] The social hierarchy was one of consumption rather than

22 Black, *British Abroad*, pp. 9–10; John Towner, 'The Grand Tour: A Key Phase in the History of Tourism', *Annals of Tourism Research*, 12 (1985), 297–333, at 310; Stephen Conway, *Britain, Ireland and Continental Europe in the Eighteenth Century: Similarities, Connections, Identities* (Oxford: Oxford University Press, 2011), pp. 189–91.

23 Michèle Cohen, 'Manliness, Effeminacy and the French: Gender and the Construction of National Character in Eighteenth-Century England', in Tim Hitchcock and Michèle Cohen, eds, *English Masculinities, 1660–1800* (London: Longman, 1999), pp. 44–61.

24 French and Rothery, *Man's Estate*, p. 231.

25 Lawrence Stone and Jeanne C. Fawtier Stone, *An Open Elite? England 1540–1880* (Oxford: Clarendon, 1984); John Cannon, *Aristocratic Century: The Peerage of Eighteenth-Century England* (Cambridge: Cambridge University Press, 1984).

26 Henry French, 'Gentlemen', in Keith Wrightson, ed., *A Social History of England, 1500–1750* (Cambridge: Cambridge University Press, 2017), pp. 269–89, at p. 272.

27 Michael Mascuch, 'Social Mobility and Middling Self-Identity: The Ethos of British Autobiographers, 1600–1750', *Social History*, 20:1 (1995), 45–61; Keith Thomas, *The Ends of Life: Roads to Fulfilment in Early Modern England* (Oxford: Oxford University Press, 2009), pp. 110–11.

wealth, with members of the gentry and nobility obliged to maintain lifestyles appropriate to their condition. This necessity, rather than any fall in rental income, tends to explain the financial difficulties that many elite families continued to face. The pressure to perform was especially strong in Britain and Ireland, where those below the peerage lacked the legal recognition afforded to counterparts elsewhere in Europe, and questions of standing demanded the resources of landed families throughout the period.[28]

Childhood and adolescence were key moments for family standing, and education a route to securing or advancing one's position.[29] The right upbringing not only displayed wealth and kept up with neighbours, but it also imparted the skills and connections with which young men could maintain or improve upon the rank of their parents. When one father worried that the cost of educating his sons would 'diminish' their inheritance, his brother-in-law urged him to consider 'that Education is the greatest Estate one can give'. Men of 'parts and education, seldome want Estate', he continued, 'and oft times arrive at the highest'.[30] Elites and their observers have stressed breeding as well as birth in many times and places, but in England the balance had increasingly tipped towards the former since the sixteenth century, and especially after 1660.[31] Jean Gailhard, tutor to Kate and Helena Perceval's brother, Sir Philip, went as far as to argue in print that 'nature makes us all alike ... but Education makes a difference, and sets a mark of distinction'.[32] Such ideas fitted well with the understanding that children's minds were malleable, a commonplace before John Locke's famous description of 'white Paper, or Wax, to be moulded and fashioned as one pleases'.[33] Education

[28] Thomas, *Ends of Life*, pp. 114–15; French, 'Gentlemen', pp. 274, 281–3; Ronald G. Asch, *Nobilities in Transition 1550–1700: Courtiers and Rebels in Britain and Europe* (London: Arnold, 2003), pp. 25–8. For a contrasting emphasis on prudence, see Mark Rothery and Jon Stobart, 'Inheritance Events and Spending Patterns in the English Country House: The Leigh Family of Stoneleigh Abbey, 1736–1806', *Continuity and Change*, 27:3 (2012), 379–407.

[29] Peter Borsay, 'Children, Adolescents and Fashionable Urban Society in Eighteenth-Century England', in Anja Müller, ed., *Fashioning Childhood in the Eighteenth Century: Age and Identity* (Aldershot: Ashgate, 2006), pp. 53–62; Thomas, *Ends of Life*, p. 30. Stone and Stone, *Open Elite?*, pp. 249–51.

[30] HL, mssHA 14482: Earl of Conway to Sir George Rawdon, 28 Mar 1671.

[31] Felicity Heal and Clive Holmes, *The Gentry in England and Wales, 1500–1700* (Basingstoke: Macmillan, 1994), pp. 38–42, 244–7, 277; Keith Thomas, *In Pursuit of Civility: Manners and Civilization in Early Modern England* (New Haven, CT: Yale University Press, 2018), pp. 31–2.

[32] Jean Gailhard, *The Compleat Gentleman: or, Directions for the Education of Youth as to their Breeding at Home and Travelling Abroad*, 2 vols (London: 1678), i, pp. 2–3.

[33] John Locke, *Some Thoughts Concerning Education* (London: 1693), p. 261. Few of his views on the topic were new: Margaret J. M. Ezell, 'John Locke's Images of Childhood: Early Eighteenth-Century Response to *Some Thoughts Concerning Education*', *Eighteenth-Century Studies*, 17:2 (1983–4), 139–55.

was, therefore, an investment in family standing, both advertised at the time and secured for the future.[34] By extension, it was a crucial point at which the landed elite, as an agglomeration of families, actively reproduced itself. This book shows that, for sons, educational travel was an important part of that process. Surprisingly little is clear about how average landowners used their wealth, but time abroad was one way, alongside hospitality, office-holding, collecting, building and dress, for families to maintain or acquire pre-eminence.[35] Indeed, travel was arguably the most accessible of these methods, requiring little more than time and money.[36] At the same time, it provided skills and contacts for the renewed performance of status at home. Rather than simply seeing the Grand Tour as something that the elite did and the less fortunate could not afford, this book asks how travel helped to build those boundaries and the cluster of attributes within. What did it mean that travel made the 'compleat gentleman'?

If education and travel were investments in family standing, the urgent question was how best to lay out resources. Although this book covers a period of demographic crisis among landowners, in which mortality rose as nuptiality and fertility declined, sons aged between 5 and 9 were three times more likely to survive than those of 4 and under.[37] They became physically and mentally capable of mobile education, which they began in those years, but also more viable prospects for a course that would ideally culminate in foreign travel. Families divided funds among sons in various ways, a dynamic explored in Chapter 2, but heirs tended to receive the greatest investments. The downward trajectory of younger brothers was accordingly a major aspect of landed society, so the degree to which they should spend time abroad was

[34] J. H. Plumb waits for birth and survival rates to rise in the mid-eighteenth century before declaring that parents saw children as 'a less risky vehicle for capital investment', made in the hope of 'social emulation': 'The New World of Children', in Neil McKendrick, John Brewer and J. H. Plumb, eds, *The Birth of a Consumer Society: The Commercialisation of Eighteenth-Century England* (London: Europa, 1982), pp. 286–312, at pp. 287, 292. Others correctly find such investments throughout the early modern period: Houlbrooke, *English Family*, p. 194; Ilana Krausman Ben-Amos, 'Reciprocal Bonding: Parents and Their Offspring in Early Modern England', *Journal of Family History*, 25:3 (2000), 291–312. Since education only became mobile and expensive once sons were more likely to survive, the demographic crisis of 1650–1740 must rather have improved their prospects.

[35] Rothery and Stobart, 'Inheritance Events and Spending Patterns'.

[36] Gerald Newman, *The Rise of English Nationalism: A Cultural History, 1740–1830* (Basingstoke: Macmillan, 1997), pp. 42, 44.

[37] T. H. Hollingsworth, 'Mortality in the British Peerage Families since 1600', *Population*, 32:1 (1977), 323–52, at 327.

always open for debate.[38] Although new options for female education added to its cost, meanwhile, most money after early childhood went towards sons, since the skills required of daughters had changed comparatively little.[39] This disparity is even more apparent when one factors in the scarcity of female educational travel. Finally, the 'family' comprised all those who lived in the same household, including servants and other dependents, but contemporaries differentiated between this larger group and the dynastic unit on questions of education and travel.[40] When it came to it, mothers and fathers invested in their children.

The intellectual blank slate was a source of anxiety, putting pressure on parents to make the right decisions and on children to use their time well.[41] One father wanted 'to have every Moment imploy'd the right way' in the education of his 14-year-old boy, fearing 'any one false stepp', and others warned sons against wasting time at school and university.[42] After all, youth offered opportunities that would not come again. The strongest fears surrounded travel. Not only was it generally the most expensive period of education but, far from home, its success particularly depended on individual ability and application. Mothers, fathers and older relations warned travellers against wasting 'the time of Labor for the attaining all that will be usefull for you', 'this spare time of Life you have for improving your naturall Parts', and urged them 'to imploy this time with most benefitt'.[43] Parents encouraged a fear of time-wasting among sons as a pedagogical tool, with one young man speaking for many when he promised 'a proper Account of my Time at my

[38] Joan Thirsk, 'Younger Sons in the Seventeenth Century', *History*, 54:182 (1969), 358–77; Henry French and Mark Rothery, 'Male Anxiety among Younger Sons of the English Landed Gentry, 1700–1900', *Historical Journal*, 62:4 (2019), 967–95; Rory Muir, *Gentlemen of Uncertain Fortune: How Younger Sons Made Their Way in Jane Austen's England* (New Haven, CT: Yale University Press, 2019).

[39] O'Day, *Education and Society*, pp. 191, 194–5.

[40] Naomi Tadmor, *Family and Friends in Eighteenth-Century England: Household, Kinship and Patronage* (Cambridge: Cambridge University Press, 2001); French and Rothery, *Man's Estate*, p. 191.

[41] Anthony Fletcher, *Growing up in England: The Experience of Childhood, 1600–1914* (New Haven, CT, and London: Yale University Press, 2008), p. 7; French and Rothery, *Man's Estate*, pp. 65–6.

[42] Blackburn CL, Southwell Papers, vol. 1: Sir Robert Southwell to William Webster, 1 Sep 1686; SHC, 1248/4/150: Sir Alan Brodrick to Alan Brodrick II, 10 Apr 1718; HL, mssST57, vol. 15, fo. 206: James Brydges to Dr Hudson, 23 Apr 1718.

[43] BL, Add MS 46954A, fo. 83v: Lady Perceval to Sir Philip Perceval, 30 March 1677; 46954B, fo. 3v: Sir Robert Southwell to Alexandre de Rasigade, 9 Jul 1677; HL, mssST57, vol. 21, fo. 34: James Brydges to John Brydges, 22 Jul 1722.

Return'.[44] Travellers also played on these fears, however, as in the fourth Viscount Irwin's battles with the trustees of his father's will, during which he often called on them not to delay his progress by dragging their heels over funds or recalling him temporarily to England. The trustees grudgingly agreed that 'if he must Travell', the young viscount should take no longer than necessary. Irwin worked with a shared acceptance that, once embarked upon, time abroad was 'very pretious'.[45]

The fear of wasting time was only one manifestation of family anxiety over travel. Young men might also waste money abroad, and the other side of Irwin's case was the fear, on the part of the trustees, that the viscount would saddle his successors with debt. Such financial calculations are the subject of Chapter 2. Families also worried that greater independence would expose young travellers to the corrupting influences of Catholicism and absolutism, despite a 'slow and uneven transition' away from 'vehement anti-popery' after the mid-seventeenth century.[46] These threats often intertwined, in the minds of those at home, with the potential for moral and sexual corruption as young men headed southwards. One father worried that his son, then departing for Italy, was 'a sightly lad & must be well watched to keep him in the innocence he goes over in'.[47] If travel was central to family strategy, it also threatened marriageability and reproductive capacity. Finally, families feared for the lives of the young men they sent abroad, worrying about political instability, fragile health or the perils of sea travel, river crossings and the plague.[48] They thought carefully before sending 'an onely sonne to seek dangers abroad' and were reluctant to expose all sons at once 'to the many and great Hazards of travelling into Foreign parts'.[49] As Chapter 4 will show, a culture of duelling meant that one of the most serious threats

44 HL, mssSTB, Box 12, Folder 53: James Brydges II to Colonel Robinson, 4 Jan 1751; French and Rothery, 'Male Anxiety'.

45 WYAS, WYL100/C/9A: Edward Ingram, fourth Viscount Irwin to John Roades, 8 Aug 1704; Thomas Worsley to same, 16 Feb 1704[/5]; Irwin to same, 20 Feb 1705; same to same, [1705]; WYL100/C/20, fo. 158: same to Isabella, Dowager Viscountess Irwin, 8 Jan [1705].

46 Tony Claydon, *Europe and the Making of England, 1660–1760* (Cambridge: Cambridge University Press, 2007), p. 18; Conway, *Britain, Ireland and Continental Europe*, pp. 197–9; Ansell, 'Reading and Writing Travels', 1454.

47 NLI, microfilm P.3753: Robert Molesworth to [John Molesworth], 27 Feb 1721/2.

48 NYCRO, ZON13/1/259: Thomas Worsley to John Covel, 27 Nov 1703; SHC, 1248/6/44: Thomas Brodrick to Alan Brodrick II, 10 Jul 1724; fo. 205: Martha Courthope to same, 23 Apr 1725.

49 BL, Add MS 46962, fo. 196v: Sir John Perceval to Sir Robert Southwell, 2 Feb 1685/6; [Tobias Pullen], *A Vindication of Sr. Robert King's Designs and Actions, in Relation to the Late, and Present Lord Kingston* (Dublin: 1699), p. 31.

to a young man abroad in fact came from his countrymen. Each of these fears will recur in what follows, and each of them arose from the increased independence enjoyed by educational travellers.

In the middle of a demographic crisis, one might wonder why families took such risks at all.[50] Ultimately, they judged them worthwhile because they felt that travel would help sons to develop the authority, independence, manners, knowledge and connections expected of an adult male member of the landed elite. Abroad, they could become men, gentlemen and Englishmen.[51] The attendant risks were part of the point, promising high rewards to families and the chance to prove individual honour.[52] Educational travel was regularly viewed as an investment by those who questioned it in print, even before any boost to such language from the financial revolution of the 1690s. Travel could be 'profitable' to young men and their country if done with care, but it would otherwise 'wast[e] their Bodies and Estates'; it risked 'poor Returns', since French polish was 'not worth the Charge the Gentleman is at, that travels for it'.[53] Such criticism often appears in books on the Grand Tour.[54] It is never acknowledged, however, that families also saw time abroad as an investment, and that the popularisation of educational travel depended on their willingness to make it. There was a chance that young men might die abroad or come home corrupted, but travel would display family standing and, it was hoped, develop the skills and connections needed to maintain or advance it. As the following chapters will show, families came to different conclusions about whether travel was worth the outlay and, if it was, about appropriate levels of spending. If gentlemen were made rather than born, many tried to build with cheap materials.[55] Equally, it is now clear that just because parents and guardians assumed the risks did not mean that they were comfortable with them. Families nevertheless invested time, money and emotional energy in various forms of travel, which they hoped would bring social returns.[56]

[50] Black, *British Abroad*, p. 334; Michèle Cohen, 'The Grand Tour: Constructing the English Gentleman in Eighteenth-Century France', *History of Education*, 21:3 (1992), 241–57, at 248.

[51] Michèle Cohen, *Fashioning Masculinity: National Identity and Language in the Eighteenth Century* (London: Routledge, 1996), p. 58; French and Rothery, *Man's Estate*, pp. 141–2.

[52] Goldsmith, *Masculinity and Danger*. On understandings of risk, see pp. 37–73.

[53] Margaret Cavendish, Marchioness of Newcastle, *Orations of Divers Sorts Accommodated to Divers Places Written by the Lady Marchioness of Newcastle* (London: 1662), p. 74; Richard Lingard, *A Letter of Advice to a Young Gentleman Leaving the University* (London: 1671), pp. 30–1.

[54] For example, Black, *British Abroad*, pp. 315–16.

[55] Pollock, 'Teach Her to Live under Obedience', 248–9.

[56] On the emotional side of family strategy, see French and Rothery, 'Male Anxiety'.

Attention to family strategy demands immersion in generations of experience, which this book finds in four clusters from England and Ireland. First, the Ingrams (Viscounts Irwin), Worsleys, Robinsons and Franklands, from the North and West Ridings of Yorkshire, travelled together in the decades either side of 1700, offering a range of journeys by the gentry and nobility. Secondly, the Brydges family, Dukes of Chandos from 1719, came into phenomenal wealth in the early eighteenth century, ploughing some of it into the education of children and dependents. Their rapid rise means that their papers give access not only to aspiring nobles but also to relatively modest relations such as the Leighs of Gloucestershire. Thirdly, the Southwells, Percevals and Brodricks, from the Cork gentry in Ireland, invested heavily in travel over three or four generations, a period in which they rose socially and moved to England. The Molesworths, of County Dublin and later Yorkshire, had similar experiences over a shorter time frame. Finally, the Rawdons and Warings, from the Down/Antrim border, initially enjoyed more modest incomes and secured travel only though patronage, borrowing and limited ambitions. The Rawdons too had risen by the mid-eighteenth century, however, when they undertook more lavish Italian journeys. Taken together, these families offer generations of continental experience that included but went beyond recognisable Grand Tours.

Establishing the position of travel in family strategy requires evidence of thought and action before, during and after time abroad. Traditionally, though, histories of travel have more accurately been studies of travel literature, as scholars have used printed books to assemble accounts that must, given the filters of writing and publication, bear questionable relation to actual encounters with the Continent, let alone to the experiences of the average educational traveller. This study instead follows the archival approach pioneered by Jeremy Black, whose many books have encouraged others in the past couple of decades to root their discussions in personal experience and manuscript sources.[57] Black is among those for whom the Grand Tour refers to all encounters with the Continent, however, and he often prefers compiling evidence to analysing travel in relation to wider social and cultural themes. He has regularly regretted that 'Many tourist letters are scattered in general family correspondence and therefore difficult to find'.[58] By contrast, this book depends on the fact that such traces are embedded

[57] Black, *British Abroad*, pp. v–vi; Black, *Italy and the Grand Tour* (New Haven, CT, and London: Yale University Press, 2003), pp. 17–22; Black, *France and the Grand Tour* (Basingstoke: Palgrave Macmillan, 2003), p. ix. Archival studies include Sweet, *Cities and the Grand Tour*; French and Rothery, *Man's Estate*, pp. 137–83; Goldsmith, *Masculinity and Danger*.

[58] Black, *Italy and the Grand Tour*, p. ix; Black, *France and the Grand Tour*, p. ix.

within family archives, returning travel to the contexts within which it was understood and experienced.

Not all manuscripts are equally useful. It would be tempting to prioritise diaries and journals, seemingly the most coherent accounts of time abroad and certainly those most easily located by the historian, set apart from correspondence and individually catalogued in modern archives. Chapter 3 establishes that diaries and journals tend only to record mobile periods of travel, however, and deliberately omit long residences. This book leans more heavily on letters and financial accounts for a comprehensive picture of young men's encounters with the Continent. These two kinds of source also place travel within family strategy, providing evidence of decision-making and its constraints. Diaries and journals shed little light on such questions, although they remain important evidence of self-fashioning when restored to family contexts, and they feature here only when they can be set alongside other sources. Letters and ledgers are more painstaking to sift through, but they alone can reveal how families thought about, discussed and afforded educational travel.

Death and distance created the most voluminous correspondence and financial records. The untimely demise of fathers such as the third Viscount Irwin and Sir John Perceval was hardly unusual, in an age when perhaps half of elite men did not live to see their eldest son married, but it has made their families particularly good sources for educational strategy.[59] Widows, relations and friends set down in letters matters that would, in more fortunate circumstances, have been discussed in person between husband and wife. They also laid out finances for the scrutiny of trustees. Young boys have left few archival traces, meanwhile, but emerge precisely when their upbringing became more mobile, progressively leaving home for school, university and travel; the combination of lengthening distances and increasingly weighty decision-making required ever more copious letters.[60] In correspondence with adolescents, parents and guardians tended to lay out values and objectives that they stressed in person to younger offspring and safely assumed among adults.[61] 'Anglo-Irish' families such as the Southwells, Percevals, Brodricks, Molesworths and Rawdons discussed household questions and educational plans particularly frequently in letters, with the habitual separation of spouses, siblings, parents and children by the Irish

[59] Stone and Stone, *Open Elite?*, p. 260.

[60] Houlbrooke, *English Family*, p. 178.

[61] R. C. Richardson, 'The Generation Gap: Parental Advice in Early Modern England', *Clio*, 32:1 (2002), 1–25, at 23; French and Rothery, *Man's Estate*, pp. 67–8.

Sea.[62] These families also offer a valuable sidelight on the construction of the English gentleman abroad, a process into which they wholeheartedly bought.

Archival sources allow the book to explore what families thought and did before, during, after and instead of foreign travel. Chapter 1 places time abroad within longer courses of education, assembled from a range of options. It finds that institutions in southern England and domestic tours offered preparation for those who would cross the Channel, a stage at which family hopes and expectations for travel are particularly clear. These options also served as alternatives for those who would not see the Continent, however, young men who never appear in accounts of the Grand Tour but are just as important for establishing the position of travel in family strategy. In either case, domestic and foreign journeys overlapped in a wider culture of educational mobility. Chapter 2 recognises that trade-offs between means, ambitions and obligations similarly shaped encounters with the Continent. Travel gave the wealthiest families an opportunity to advertise their standing, but the expense of a lengthy voyage to Italy encouraged those of lesser means to seek out approximations, adaptations and alternatives. Here, the decision to follow generations of landed families rather than search for Grand Tours reveals more diverse forms of educational travel, by younger sons as well as heirs. Parents bought into elite practice as far as their finances and connections would allow, spreading continental experience deep into the gentry.

Chapters 3 and 4 argue that educational travel was not simply about conspicuous consumption, however, as sons went abroad to gather skills and acquaintances that would contribute towards family standing at home. To begin with, Chapter 3 prioritises settled learning over the mobile 'tours' that have dominated most histories of travel. It finds that periods at continental universities and academies, learning French, physical 'exercises' and other skills, were central to the experiences of those who subsequently made tours and those who came directly home. Whether families combined settled learning with optional periods of heightened mobility depended, again, on their means, ambitions and obligations. This conclusion shifts the scholarly focus northwards from Italy, casting educational travel between 1650 and 1750 as primarily an immersion in French culture. In turn, Chapter 4 recognises the parallel importance of fellow travellers. Other English-speakers get in the way of traditional accounts of the Grand Tour, offering

[62] Families who divided their lives and interests between England and Ireland were 'Anglo-Irish in a precise and strong sense': T. C. Barnard, 'Crises of Identity among Irish Protestants 1641–1685', *Past & Present*, 127 (1990), 39–83, at 47.

comic asides to the appreciation of classical and Renaissance culture, but networking among the British elite lay at the heart of educational travel. Here was no strict rite of passage, as young men remained within domestic webs of obligation and rubbed shoulders with family and friends abroad on shorter trips for health or leisure. Indeed, those of lesser standing hoped to secure patronage by travelling with social superiors, knowing that shared encounters with the foreign, and shared excesses abroad, promised lasting connections at home. By contrast, relationships with hosts and foreign travellers were more fleeting, based on the desire to observe and learn from continental cultures rather than form enduring personal links. Families invested in travel because of what its outcomes would mean in Britain and Ireland.

In Chapter 5, the work of travel continues at home. Attention to family strategy involves contextualising time abroad within individual lives, as the years after travel decided whether the investment paid off. It was then that travellers and their families directed their experiences towards social and political advancement. The first half of the chapter gauges the importance of travel at three milestones: entry into public life, marriage and attempts by Irish landowners to move to England. Where time abroad offered preparation for high office and a route into the English elite, its place in marriage negotiations was more complex, as an opportunity for improvement but also a test of character. The second half of the chapter recognises that the power that travel held at these points depended upon its presentation at home, through discussion, speech, deportment, consumption, souvenirs and the distribution of gifts. Meanwhile, travellers and their relations created transgenerational archives that sustained the identities of families who travelled. This section pursues the idea that travel formed English gentlemen by thinking about how it shaped later attitudes, conduct and self-perceptions.

Continental travel was widespread and habitual among the gentry of the late seventeenth and early eighteenth centuries, much of it taking place within a broader culture of educational mobility. Landed families knew far more of France and the Low Countries, and to a lesser extent Germany and Italy, than has been recognised, and this book normalises their foreign experiences, as a part of lives, identities and social history. Travel contributed to stability or upward mobility in individual families and, at the same time, to the coalescence of a shared landed identity. It helped to build and manifest the elite. Young men from Britain and Ireland brought home the 'parts' that made the 'compleat' English gentleman, as well as the connections and shared experiences that lodged them among the ranks of the privileged. Many even participated in virtuous circles, in which sons acquired skills and networks abroad that secured or advanced their positions at home, thereby encouraging travel by their descendants. Over generations, they used foreign experience to buy into the English elite. Grand Tour portraits may isolate young men abroad, but educational travel was a matter of paramount importance for the whole family.

1

Before Travel

Most work on early modern European travel—which has until very recently been to say work on the Grand Tour—has isolated time spent on the Continent from what came beforehand and afterwards. Books begin when their subjects arrive on the French or Dutch coast and end when they recross the Channel. Chapter 5 will follow travellers home, exploring the ways in which they made use of their foreign experiences, but this chapter reconnects young men to the broader family and educational contexts within which they ventured abroad. Movement around Britain and Ireland, whether for institutional learning or less formal (but no less formative) journeys, offered preparations for foreign travel or alternatives to European experience, depending on the means, ambitions and obligations of individual families. Domestic and foreign travel are usually studied separately by different historians, but this chapter argues that both were part of a culture of educational mobility that must be understood in its entirety.

Scholars have long written of 'travel as education', most incisively Michèle Cohen, but few have placed it within longer courses of upbringing. Most studies consider the preparations and preconceptions with which young men left home all too briefly, referring only to classical learning as a lens for viewing Italy.[1] Several historians, usually of the social elite rather than of travel, have nevertheless recognised that time abroad stood alongside other aspects of education, including discipline, rhetoric and the classics,

[1] Cohen, 'Grand Tour: Constructing the English Gentleman'; George B. Parks, 'Travel as Education', in Richard Foster Jones, ed., *The Seventeenth Century: Studies in the History of English Thought and Literature from Bacon to Pope* (Stanford, CA: Stanford University Press, 1965), pp. 264–90. G. C. Brauer places the Grand Tour among several stages of upbringing but bases his 'study in an ideal' on conduct literature and educational theory: *The Education of a Gentleman: Theories of Gentlemanly Education in England, 1660–1775* (New York: Bookman Associates, 1959), pp. 7, 156–94. Keith Dent focuses on domestic travel: 'Travel as Education: The English Landed Classes in the Eighteenth Century', *Educational Studies*, 1:3 (1975), 171–80.

some of them exploring practice as well as prescription.[2] Henry French and Mark Rothery have gone furthest, placing travel among other discrete stages—schools and academies, universities and apprenticeships, marriage and fatherhood—in the achievement of adult independence by gentry sons.[3] This chapter develops their argument, finding that schools, universities, academies and domestic journeys were integrated stages in a longer process of educational mobility, which often culminated in foreign travel but offered alternatives for those with different intentions.

Mobility was a central aspect of education throughout early modern Europe.[4] In late seventeenth-century Britain and Ireland, and indeed long afterwards, boys from landowning families were mobile from an early age, once they had passed a milestone that tended to come between the ages of 6 and 9.[5] Sir Alan Brodrick teased his son and namesake that 'You are now seven years old, and it is time to bethink yourself of applying to some trade or way of getting your livelyhood', jokingly weighing the merits of chimney-sweep, coachman, huntsman or scholar.[6] This was indeed a turning point in elite male upbringing, however, and the moment at which many boys left the family home for the first time; from then on, they spent little time there. The sons of landowners pursued institutional or residential education, rehearsed the practicalities of foreign travel and built a sense of 'home' against which some of them would judge the Continent.

The course of elite male education could take several routes. The expansion and diversification of pedagogical provision from the second half of the seventeenth century offered parents a 'mixed economy' from which they could assemble a series of appropriate stages.[7] Oxford, Cambridge and the Inns of Court continued to feature prominently, as did well-known public schools, and apprenticeships remained a popular option for younger sons or gentry families of lesser means.[8] Increasingly, these institutions were joined by a range of private schools and individual tutors that emerged in

[2] For example, Rosenheim, *Emergence of a Ruling Order*, pp. 33–45.

[3] French and Rothery, *Man's Estate*.

[4] Hilda de Ridder-Symoens, 'Mobility', in De Ridder-Symoens, ed., *A History of the University in Europe, Volume II: Universities in Early Modern Europe* (Cambridge: Cambridge University Press, 1996), pp. 416–48; Alan S. Ross, 'Pupils' Choices and Social Mobility after the Thirty Years War: A Quantitative Study', *Historical Journal*, 57:2 (2014), 311–41, at 322, 325.

[5] Plumb, 'New World of Children', p. 288; Houlbrooke, *English Family*, pp. 150–3; French and Rothery, *Man's Estate*, p. 51.

[6] SHC, 1248/2/352: Sir Alan Brodrick to Alan Brodrick II, 18 Jan 1708/9.

[7] French and Rothery, *Man's Estate*, p. 40.

[8] Patrick Wallis and Cliff Webb, 'The Education and Training of Gentry Sons in Early Modern England', *Social History*, 36:1 (2011), 36–53.

seventeenth- and eighteenth-century London.[9] At the same time, the growing importance of urban sociability to elite life, and the consequent desire for children to observe and practise successful participation, encouraged parents to choose options with easy access to the capital.[10] Many institutions and entrepreneurs advertised their services to the gentry and nobility, at least in theory, and the desire of provincial landowning families to take advantage of opportunities in south-eastern England accounts for much educational mobility in this period.[11] They had a full range of options with which to plan the upbringing of their sons, in a society in which educational opportunity was a function of social standing.[12]

Expanding possibilities have encouraged accounts of early modern education to emphasise change, particularly between 1650 and 1750. Small schools and academies are usually overlooked, but most studies chart the decline of large institutions and the rise of domestic tuition, an option favoured by famous contemporaries such as Locke and the Earl of Chesterfield. Illustrating this change, one recent survey estimates that 'by the eighteenth century perhaps a quarter of the peerage, a third of the gentry, a quarter of the clergy and a fifth of the professions were being educated at home' by domestic tutors.[13] These figures, while striking, actually show that the majority in each category still left home as part of their upbringing, and some historians emphasise continuity as well as change in elite education.[14] Between the seventeenth and twentieth centuries, French and Rothery find that families used incremental removal from the home to develop masculine independence, the ultimate objective of gentry education. Scholars who distinguish clearly between 'public' institutions and 'private' tuition pay too much attention to printed debate, as both options coexisted within a mixed economy of education.[15]

Families made similar decisions for different reasons, hoping that a younger son or an heir to a lesser fortune would begin a legal, clerical or military career at the same university, academy or Inn of Court that an

[9] Gallagher, *Learning Languages*, pp. 14–54; Margaret E. Bryant, *The London Experience of Secondary Education* (London: Athlone Press, 1986), pp. 66–70.
[10] Borsay, 'Children, Adolescents and Fashionable Urban Society'; Rosenheim, *Emergence of a Ruling Order*, pp. 221–2.
[11] For educators targeting the elite, see Gallagher, *Learning Languages*, pp. 26–33, 40, 42–3.
[12] David Cressy, 'Educational Opportunity in Tudor and Stuart England', *History of Education Quarterly*, 16:3 (1976), 301–20.
[13] Adam Fox, 'Words, Words, Words: Education, Literacy and Print', in Wrightson, ed., *Social History of England*, pp. 129–51, at p. 133.
[14] Fletcher, *Growing up in England*; Rosenheim, *Emergence of a Ruling Order*, pp. 33–45.
[15] French and Rothery, *Man's Estate*, pp. 40, 81–3.

elder brother or wealthier neighbour had attended to acquire skills and connections before foreign travel. Options were not interchangeable, however, especially once boys reached their mid-teens, as certain choices opened or closed other opportunities. A recent study cannot find a single gentry son who moved from an apprenticeship to Oxford, Cambridge or the Inns of Court, but traffic between the universities and the inns was common, with a third of university students progressing to the London institutions in the seventeenth century.[16] It is unclear whether most gentry sons went to Oxford and Cambridge between 1650 and 1750, when attendance levels recovered quickly from the mid-seventeenth-century low point established by older studies.[17] Most who travelled abroad nevertheless seem also to have gone to university, as Oxford and Cambridge were twice as common among the upper gentry than the lower and much more likely for elder than younger brothers.[18] In other words, university more often featured in the educational strategies of families who were better able to afford travel and in the individual courses of sons for whom they were more likely to make the investment.

Families could plan for the foreign travels of their sons well in advance, as part of an integrated course of mobility, because continental experience was increasingly expected at higher social levels. If many gentlemen had crossed the Channel during the Interregnum, Restoration 'Gallomania' encouraged further travel.[19] Educational entrepreneurs had sought to make time abroad unnecessary since the sixteenth century, but, from the 1660s onwards, they instead advertised domestic preparations for continental experience. Travel had become something to be improved rather than denied.[20] As a result, families could plan far ahead, even if they might often work out practicalities on the hoof. John Brydges was only 5 when friends recommended a potential tutor, ready for when the boy might be 'fitt for Travell'.[21] At 9, Edward Southwell was similarly too young to have an experienced traveller as his domestic tutor, but it was felt that he might make the upgrade around the age

16 Wallis and Webb, 'Education and Training', 46.

17 Stone and Stone, *Open Elite?*, p. 264; Wallis and Webb, 'Education and Training', 42–3.

18 Wallis and Webb, 'Education and Training', 48, 50. Towner, 'Grand Tour', 307, for travellers who had attended university. His figure is above 50 per cent for 1661–1700 and 1714–40, though it unites travellers of diverse age and background.

19 Brennan, ed., *Origins of the Grand Tour*, pp. 30–6; Gesa Stedman, *Cultural Exchange in Seventeenth-Century France and England* (Farnham: Ashgate, 2013), p. 63.

20 Richard Ansell, 'Foubert's Academy: British and Irish Elite Formation in Seventeenth- and Eighteenth-Century Paris and London', in Sweet, Verhoeven and Goldsmith, eds, *Beyond the Grand Tour*, pp. 46–64, at p. 50.

21 HL, mssST58, vol. 3, fo. 125: Charles Davenant to James Brydges, 19 Oct 1708.

of 12, years before any foreign experience.[22] Fathers also provided for travel in their wills, as an essential part of education. Sir Bourchier Wrey planned for his 11-year-old heir to attend Eton and Oxford before going 'beyond sea for the space of ffour yeares in those Countries that shall be in League of peace with the Kingdome of England'. Destinations were flexible but the fact of travel was not, meaning that, after his father's death, Wrey's son could take on a tutor intended for his travels before he had even started at Eton.[23] Travel had become an educational essential for those with aspirations to enter or remain among the elite.

The length of prior education dictated when a young man would leave for the Continent, so decisions over domestic options were also decisions over when to travel. The question was an enduring subject of printed debate, in which commentators criticised parents for sending their sons abroad either too old to learn or too young to understand. Published recommendations ranged from the early teens, or even the age of 7 in the case of Locke, to the mid-twenties. Varying opinions were to some extent the product of different priorities, with those who advocated youthful travel emphasising the language-learning and behavioural change more easily achieved at an earlier age and those favouring travel in adulthood calling for informed observation of neighbouring countries, at an age at which travellers would be less vulnerable to their vices. It must be said, though, that many who made part of their income as travelling tutors—including Richard Lassels, Jean Gailhard and indeed Locke—recommended travel in the mid-teens or earlier, a time of life when their services would be most necessary.[24] Practice ranged equally widely, with the Duke of Ormond sending his grandsons abroad separately at 12 and 15 but his client, Robert Southwell, only travelling at 23.[25] Differences related to wealth and social standing as much as pedagogical theories. Early travel allowed privileged young men to imbibe continental influences fully before returning for strategic marriage or speedy entry into public life, but it was much more expensive. At the very least,

[22] Kent Archives, U1713/C2/43: Sir Edward Dering to Sir Robert Southwell, 14 May 1681.

[23] HL, mssEL 9068: Florence, Lady Wrey to Earl of Bridgewater, 16 Dec 1697; 9084: Sir Bourchier Wrey's will, 16 Aug 1694; Wasey Sterry, *The Eton College Register, 1441–1698* (Eton: Spottiswoode, Ballantyne & Co., 1943), p. 380.

[24] Brauer, *Education of a Gentleman*, pp. 171–6; Cohen, 'Grand Tour: Constructing the English Gentleman', 245–6.

[25] Stuart Handley, 'Butler, James, second Duke of Ormond (1665–1745)', *Oxford Dictionary of National Biography*, Oxford University Press, 2004; online edn, 2006 (https://doi-org/10.1093/ref:odnb/4193); Éamonn Ó Ciardha, 'Butler, Charles (1671–1758), 2nd Earl of Arran, 3rd titular Duke of Ormond', *Dictionary of Irish Biography*, Royal Irish Academy, 2009–18; online edn, 2021 (https://www.dib.ie/biography/butler-charles-a1230).

it required a tutor, the best of whom could claim £100 a year.[26] Younger travellers might also need larger households, private transport, specialised teachers or equipment for fencing or riding. The later gentlemen's sons left for the Continent, by contrast, the more easily they could dispense with tutors, masters and other expenses, travelling with a single servant or among a group of peers. They also had greater need of the skills and connections on offer at Oxford and Cambridge, which they usually entered at 16 or 17, and so tended to travel only after university.[27] Broadly, the sons of gentlemen travelled in their late teens or early twenties. In what follows, several men of higher birth take a similar course; if a traveller ventures abroad before that point, though, he is probably the son of a nobleman.[28]

If a young man travelled as part of his education, then, it was likely that he had already attended school, university and perhaps an Inn of Court or another London establishment. This chapter explores how these options offered preparations for those who went abroad and alternatives for those who stayed at home. Tracking attendance at small schools and academies, which rarely outlived their founders and generally lack institutional records, involves significant archival research. This is probably why such options are underplayed in accounts of elite education that emphasise public schools and universities, whose ample records were usually in print by the early twentieth century. Attention to family strategy, however, affords equal weight to all options that featured in contemporary education. Parents rationalised their decisions, which demanded lengthy separation from the family, as tending towards the improvement of their children. With his sons in France, Sir George Rawdon was 'well con[tent to] deny myself for their good and advantage of educa[tion]', a sentiment that Lady Wrey, Sir Bourchier's widow, echoed with regard to her son's progression from Eton to Oxford: she had lost 'the satisfaction of my chilldreens company, that they may have

26 The figure was consistent across the period: NLI, MS 7863, fo. 34: Sir Robert Southwell to Jean Gailhard, 5 Jun 1676; BL, Add MS 46954B, fo. 3v: Southwell to Alexandre de Rasigade, 9 Jul 1677; NAL, Forster MS 47.A.40, fo. 7v: Duke of Ormond to Southwell, 14 Aug 1678; WYAS, WYL100/C/9A: George Townshend to John Roades, 4 Jul 1702; Richard Thornton to same, 26 Feb 1704[/5]; Jennifer Mori, *The Culture of Diplomacy: Britain in Europe, c. 1750–1830* (Manchester: Manchester University Press, 2010), p. 152.

27 Wallis and Webb, 'Education and Training', 48.

28 Of 20 educational travellers of gentry status mentioned regularly here, defined as a first journey abroad without parents, before or during their mid-twenties, so excluding Robert Molesworth (28) but including Samuel Waring (27) and the outlier Arthur Rawdon (8), the average age is 20.1. For those 17 of noble status, including the sons of noblemen, it is 16.4.

good education'.[29] Parents reached familiarly for such consolation because the major constant, in a century of pedagogical change, was that boys tended to move for their education.

A Mobile Education

Education among the early modern gentry was highly mobile. Boys often began with a few other pupils at the house of a private tutor, usually within a day's travel, or received a tutor in their own home. Both options preparing them for a variable but relatively short period at a larger public school.[30] Edward, the 9-year-old future fourth Viscount Irwin, lived from 1695 with the vicar of Normanton, 10 miles from his Temple Newsam home, and was joined by his brothers, Rich and Arthur, two years later.[31] Edward and Rich may have moved another 25 miles to St Peter's School in York, where six younger brothers would eventually board, but the eldest two certainly ventured to Eton in 1699. They lodged in Windsor, visited London and made early connections with other young noblemen. After three years, Edward and Rich left together for Cambridge.[32] John and Henry Brydges similarly attended a small school at Isleworth, west of London, before moving to Westminster, where John spent two years and Henry six before university. Their father, James Brydges, thought 'a London Publick School to be the best Place for Educating of Youth', an opinion probably dependent on his own four years at Westminster. He nevertheless recognised, in early 1724, that his younger son was 'as full of Joy as I was 34 years ago; for he hath this Day been told he is to go no more to School & is preparing for the University'.[33] Public schools provided targeted preparation for this next stage of education. Sir George Rawdon sent his 15- and 16-year-old

[29] *Calendar of State Papers, Domestic Series, of the Reign of Charles II*, 28 vols (London: Longman, Green, Longman & Roberts, 1860–1938), xviii, pp. 354–5: Sir George Rawdon to Conway, 4 Oct 1676; HL, mssEL 9068: Lady Wrey to Earl of Bridgewater, 16 Dec 1697.

[30] French and Rothery, *Man's Estate*, pp. 46–8, 51.

[31] WYAS, WYL100/EA/12/17/3: bills for boarding with William Herbert, 1695–7; WYL100/C/7: Herbert to John Roades, 20 Dec 1697.

[32] WYAS, WYL100/EA/12/17/3: bills for boarding with William Thomlinson, York, 1701–11; WYL100/EA/12/17/4: bills for Eton, 1699–1702; WYL100/C/8: Charles Drury to John Roades, 4 Jan 1699; same to same, 28 Jan 1699.

[33] HL, mssST57, vol. 14, fos 356–7: James Brydges to John Brydges, 11 Feb 1716/17; vol. 16, fo. 194: same to Edmund Chamberlayne, 16 Jun 1719; vol. 23, fo. 209: same to William Brydges of Tibberton, 13 Jan 1723/4; G. F. Russell Barker and Alan H. Stenning, *The Record of Old Westminsters*, 2 vols (London: Chiswick, 1928), i, p. 134.

sons to Kilkenny College, an institution popular with Ireland's Protestant elite, as a one-year preface to Cambridge.[34] Alan Brodrick similarly went to Eton so that he would be ready for a 'more manly' university education 'by Christmas next'.[35] Eton and Westminster rose to dominance over the late seventeenth century, aided by the new importance that parents ascribed to social exclusivity in education, and attendance had become lengthier among the Brodrick, Brydges, Ingram and Southwell families by the 1740s.[36] Throughout the period, such schools offered classical knowledge, which distinguished young gentlemen from sisters and social inferiors, and an opportunity to form enduring acquaintances.[37]

Families regularly blended periods at school with domestic tuition, rendering the debate between 'public' and 'private' options less clear in practice.[38] Sir Robert Southwell used elements of both to raise his only surviving son, Edward, born in 1671. Edward split his early years between relations in London, Kent and Ireland, with Sir Robert then home schooling his son during his Whitehall career, his political retirement in Gloucestershire and his subsequent return to London.[39] Southwell employed a tutor and a rotating cast of live-in masters for French, mathematics and music, often sourced from the Huguenot community, while relations donated the 'old musty' books and prints with which they had been taught.[40] Sir Robert turned to institutions only when this approach became impracticable. Edward briefly studied at a school in Kensington while his father was envoy to the Elector of Brandenburg in 1680 and attended another London institution, run by a French convert to Anglicanism, Louis-Compiègne de Veil, while Sir Robert sat

[34] *CSPD, Charles II*, xi, p. 74: Sir George Rawdon to Conway, 8 Feb 1671.

[35] SHC, 1248/4/19: Sir Alan Brodrick to Alan Brodrick II, 7 May 1717.

[36] Barker and Stenning, *Record of Old Westminsters*, i, pp. 134; ii, pp. 870–1; R. A. Austen-Leigh, *The Eton College Register, 1698–1752* (Eton: Spottiswoode, Ballantyne & Co., 1927), pp. 44, 189; Geoffrey Holmes, *Augustan England: Professions, State and Society, 1680–1730* (London: Allen & Unwin, 1982), pp. 46–7.

[37] Anthony Fletcher, *Gender, Sex and Subordination in England, 1500–1800* (New Haven, CT, and London: Yale University Press, 1995), pp. 300–3; French and Rothery, *Man's Estate*, pp. 80–1.

[38] French and Rothery, *Man's Estate*, pp. 52, 55, 81–2.

[39] Kent Archives, U1713/C1/21: Sir Edward Dering to Sir Robert Southwell, 21 Jul 1673; BL, Add MS 46953, fo. 166: Sir Philip Perceval to same, 9 Sep 1676.

[40] BL, Add MS 46956C, fo. 148: John Perceval to Sir Robert Southwell, 13 Aug 1680; 46958A, fos. 73–4: Southwell to Perceval, 19 Mar 1680/1; 46959B, fo. 31: Perceval to Southwell, 26 Aug 1682; 46960A, fo. 127: same to same, 19 May 1683; 46961, fo. 113v: Southwell to Perceval, 26 Sep 1684; Blackburn CL, Southwell Papers, vol. 1: Sir Robert Southwell to Edward Southwell, 28 Sep 1685; same to same, 29 Nov 1686. Edward shared French, music and dancing masters with his three sisters.

in the 1685 parliament.[41] The most detailed evidence of Edward's education dates from this second period, including French and Latin translations and notes on logic and rhetoric.[42] He left de Veil after a year but stayed in London for another six months, pursuing mathematics and 'other good preparations' to 'Compleat' him for Oxford. These included domestic tuition in French, attendance at the Old Bailey and discussions with the natural philosopher Robert Boyle, 'from whom I have gott my experiments'.[43] Sir Robert intended that his son should travel after university, an eventuality for which he had been planning since Edward's time with de Veil. In the end, when he withdrew him from Oxford on account of James II's interference in university government, the 1685 Revocation of the Edict of Nantes had made a crossing to France inadvisable, having formalised the persecution of its Protestant community.[44] Edward would travel to the Low Countries only in 1691 and 1696.

While Edward Southwell's mixture of domestic and institutional options kept him mostly at home, other families sent sons to live with relations. To an extent, they continued the medieval system of fosterage, in which parents placed children in the households of friends, patrons or kin for training and networking, even if boarding schools are usually said to have assumed these functions over the late sixteenth and early seventeenth centuries.[45] Naomi Tadmor has called for more research on how early modern educational change interacted with established patterns of kin support, questioning assumptions of disruption.[46] The continuation of fosterage into the seventeenth and eighteenth centuries—above all, with uncles and aunts based in London—suggests that kin networks in fact allowed families to meet the rising demands of elite education. An old practice served new ends, as fosterage introduced children to the capital and gave them access to its multiplying pedagogical options.

[41] BL, Add MS 46956C, fo. 62: John Perceval to Sir Robert Southwell, 23 Mar 1679/80; 46961, fo. 247v: Southwell to Perceval, 24 Mar 1684/5; Blackburn CL, Southwell Papers, vol. 1: Sir Robert Southwell to Edward Southwell, 17 Aug 1680; Edward Southwell to Sir Robert Southwell, 23 Mar 1685/6; Fernand de Schickler, *Les églises du refuge en Angleterre*, 3 vols (Paris: Fischbacher, 1892), ii, p. 335.

[42] Blackburn CL, Southwell Papers, vol. 1: Edward Southwell, schoolwork, 1685–7.

[43] BL, Add MS 38015, fo. 387: 'Sr Roberts advice to his son', [*c.* Mar 1686]; 46962, fo. 263: Sir Robert Southwell to Sir John Perceval, 31 Mar 1686; Blackburn CL, Southwell Papers, vol. 1: Edward Southwell to Sir Robert Southwell, 22 May 1686; same to same, 30 Jul 1686.

[44] BL, Add MS 46962, fo. 26: Sir Robert Southwell to Sir John Perceval, 9 May 1685; fo. 196v: Perceval to Southwell, 2 Feb 1685/6; D. W. Hayton, 'Southwell, Edward (1671–1730)', *ODNB*, online edn, 2004 (https://doi-org/10.1093/ref:odnb/26060).

[45] Houlbrooke, *English Family*, pp. 150–3, 176–7.

[46] Naomi Tadmor, 'Early Modern English Kinship in the Long Run: Reflections on Continuity and Change', *Continuity and Change*, 25:1 (2010), 5–48, at 29.

Sir Robert Southwell also offered a base for two generations of boys in his extended family, developing his 'powerfull knack of Education'.[47] Southwell had lacked such resources in his own youth, when his father had turned to acquaintances in the absence of useful kin. The elder Robert, rooted in Kinsale and Dublin, initially lodged his son with a merchant in Cork to attend a school there in 1648, as Kinsale lacked a settled institution during 'those times of distraction', but plague and political volatility pushed the boy to an Irish Protestant contact in London two years later. The younger Robert spent a year and a half at a boarding school on the Hampshire coast before moving to Middlesex, where he lived for another year with a local minister. There, he and a handful of other pupils prepared for Oxford.[48] In later generations, Sir Robert Southwell built on his connections and his divided residence in London and Gloucestershire, resources that his English education and foreign travels had helped to furnish. He provided young kinsmen with domestic tuition and oversaw periods at school and university.

Katherine, Lady Perceval sent her three sons to Southwell, her brother, upon the death of her husband in 1669. Sir Philip (13), Robert (12) and John (9) had attended a grammar school in Kinsale before coming to England.[49] Their uncle then supervised their education from Whitehall, sending them to establishments in Chelsea and Islington run by masters formerly of Westminster and Eton. The boys attended the Islington school with their own French tutor and paid for masters in fencing, dancing, writing, music and handling a pike.[50] They stayed for three years, during which they sat for matching portraits (Figures 1, 2 and 3), probably recognising that their courses would soon diverge. Afterwards, the brothers used periods at Westminster to stagger their attendance at university, despite Lady Perceval's worries that the school was 'loose and visious'. Sir Philip went immediately to Oxford and Robert soon left for Cambridge, but John spent two years at

[47] William Byrd II, 'Cavaliero Sapiente. Southwell', in *Another Secret Diary of William Byrd of Westover, 1739–1741. With Letters and Literary Exercises, 1696–1726*, ed. Maude H. Woodfin, trans. Marion Tinling (Richmond, VA: Dietz Press, 1942), pp. 206–8.

[48] BL, Egerton MS 1634: Helena Le Grand, 'Life of Sir Robert Southwell', [before 1731], fos 4v–6; Add MS 46934, fo. 27: Southwell to Francis Jacob, 23 Apr 1653.

[49] BL, Add MS 46951A, fo. 51v: Lady Perceval to Sir Robert Southwell, 20 Mar 1673[/4].

[50] For Adam Littleton in Chelsea, BL, Add MS 38015: Sir Robert Southwell to Littleton, 3 Aug 1670; Ambrose Heal, *The English Writing-Masters and their Copy-Books, 1570–1800* (Hildesheim: Georg Olms Verlagsbuchhandlung, 1962), p. 71. For Thomas Singleton in Islington, BL, Add MS 47947B, fo. 9: Singleton to Sir Robert Southwell, [*c.* 3 Aug 1670]; 46948, fo. 127: same to same, 22 Jun 1671; Add MS 46950, fos 172–3v: Accounts for the Perceval brothers, 1672–3. Singleton soon founded a school at Clerkenwell: Bryant, *London Experience*, p. 67.

Figure 1 Portrait of Sir Philip Perceval, second Baronet (1656–80) by Thomas Pooley. © National Gallery of Ireland.

Figure 2 Portrait of Robert Perceval (1657–77) by Thomas Pooley. Private collection.

Figure 3 Portrait of Sir John Perceval, third Baronet (1660–86) by Thomas Pooley. © National Gallery of Ireland.

Westminster before following his eldest brother.[51] George Perceval, their less active uncle, observed that Southwell had done 'very advisedly in seperating the three brothers, for as there estates are different, so I thinke there education ought to bee'.[52] Their trajectories would truly part after university and separate periods at Lincoln's Inn, when only Sir Philip would travel abroad.[53]

Other relations recognised Southwell's educational expertise and 'fatherly care', as well as his convenient English residence and connections.[54] When Richard Southwell, of the family's Limerick branch, died in 1680, his trustees passed responsibility for the education of his two sons to Sir Robert, who sent the 15-year-old Sir Thomas to Christ Church as quickly as possible.[55]

[51] BL, Add MS 46950, fo. 227v: Lady Perceval to Sir Robert Southwell, 13 Nov [1673]; fo. 244: Sir Philip Perceval to same, 20 Dec 1673; Add MS 46951A, fo. 64: Robert Perceval to same, 24 Mar 1673/4; Add MS 46953, fo. 65: John Perceval to same, 14 May 1676.
[52] BL, Add MS 46951B, fo. 52: George Perceval to Sir Robert Southwell, 3 Nov 1674.
[53] BL, Add MS 46952, fo. 41: Sir Philip Perceval to elder Robert Southwell, 9 Feb 1674/5; 46954A, fo. 68: Robert Perceval to Sir Robert Southwell, 21 Mar 1677; 46956B, fo. 129: John Perceval to same, 20 Dec 1679. Sir Philip initially split his time between Oxford and London.
[54] BL, Add MS 46950, fo. 227v: Lady Perceval to Sir Robert Southwell, 13 Nov [1673].
[55] C. Ivar McGrath, 'Southwell, Sir Thomas (*c.* 1665–1720)', *DIB* (https://www.dib.ie/biography/southwell-sir-thomas-a8200).

Southwell had built up contacts at the college during the 1670s, when he had sent two of his Perceval nephews there and overseen the Oxford education of Ormond's Catholic ward, the 12-year-old Baron Kingsale.[56] He may also have used existing links to Westminster, where he discussed sending Sir Thomas's brother, John, that he might be 'in a little while fit' for Christ Church.[57] Sir Robert corresponded with the boys and their tutors, took them into his homes during vacations and consulted a Huguenot academy master, Solomon Foubert, on their behaviour.[58] Sir Robert's expertise, connections and location meant that he was, Sir Thomas's Irish guardians recognised, 'better able to prescribe & direct the way & manner of his Education than wee are'.[59]

Southwell had honed his approach by the time that the last of his Perceval nephews died in 1686, trusting him with the upbringing of his two sons, Sir John and Philip.[60] The eldest would remember Sir Robert's 'peculiar art in directing the education of youth', recalling a resident tutor and a live-in French master alongside his great-uncle's tests on French and Latin. Southwell also emphasised the body in a strikingly functional manner, employing a famous actor, Thomas Betterton, 'to form their accents' and model 'good Action', alongside a soldier who taught 'military exercise'. When there was 'little left to be learnd from Tutors at home', Sir Robert sent his great-nephews to Monsieur Meure, a Frenchman who ran a London boarding school 'then in great vogue', where they learned French, Latin and Greek and may have practised fencing and dancing. The brothers moved to Westminster after a year and a half, lodging with a Huguenot master in order to keep their French and avoid 'bigger boys, who in all great schooles are vicious'.[61] Their childhood tutor, Henry Roby, remained in touch during this period, immersing them in travel books and geography alongside their schoolboy classics: Roby's letters plotted voyages around the world for his

[56] Christ Church, MS 426: Benson of Oxford's letters relating to Lord Kinsale [*sic*], 1676–81.

[57] Christ Church, MS 427, letter 36: Sir Thomas Southwell to Sir Robert Southwell, 10 Sep 1682.

[58] Christ Church, MS 427: Letters relating to the education of Sir Thomas Southwell at Oxford, 1681–4; NLI, MS 664: Correspondence of Sir William King with Sir Robert Southwell, 1681–4.

[59] Christ Church, MS 427, letter 27: Sir William King, John Pigott and John Bourke to Sir Robert Southwell, 30 Sep 1681.

[60] BL, Add MS 46963, fos 7–8: Lady Perceval to Sir Robert Southwell, 5 May 1686.

[61] BL, Add MS 47025, fo. 9: Sir John Perceval to Sir Robert Southwell, 10 Feb 1697/8; fo. 10v: same to same, 3 May 1698; fo. 21: Louis Payzant to Perceval, 13 Feb 1699/1700; Add MS 47072: Sir John Perceval, Autobiography and notes, [*c.* 1729], pp. 6–7; David Roberts, *Thomas Betterton: The Greatest Actor of the Restoration Stage* (Cambridge: Cambridge University Press, 2010), pp. 164–5. For Meure, see Gallagher, *Learning Languages*, pp. 27–8.

former pupils, recalling journeys that they had imagined together, referring to 'a Map of the World hanging by their bedside in Town' and reminding them of familiar travel writers such as Le Comte, Dampier, Rycaut and the Earl of Carlisle.[62] Sir John left Westminster for Oxford after a year, but Philip only followed after two more years at school and a lengthy period of home tuition by Southwell himself and visiting masters for fencing and writing.[63] The households of wealthy early modern families could be 'sophisticated educational hubs'.[64]

A succession of untimely deaths handed Sir Robert Southwell a prominent role in the upbringing of several kinsmen, a common position given contemporary mortality, but rich and well-situated relations encouraged educational mobility in various circumstances. Early modern kinship provided 'a store of wealth, like a reserve account to be drawn upon as need arose'.[65] From the ages of 7 and 6, William and James Leigh spent five years with their grandparents, Lord and Lady Chandos, in order to attend a school in Hereford and receive private tuition in writing and casting accounts. Embarrassed by the fact that his parents-in-law had assumed their 'whole charge & care', Theophilus Leigh eventually moved his sons to board with the rector of Kiddington in Oxfordshire, where they stayed for another two years before university and the army respectively.[66] As with the Southwells and Percevals, a single educational course could incorporate domestic instruction at home or in other households alongside attendance at school while living with parents or relations. Families varied the balance between these elements, such that boarding schools and fosterage overlapped in educational strategies, offering education, independence and social connections.[67]

Kinship networks allowed provincial families to access educational opportunities in London. Sir Alan Brodrick's eldest son, St John, had spent a year at Eton and another at Cambridge under the supervision of

62 BL, Add MS 47025, fos 2–3: Henry Roby to Sir John Perceval, 23 May 1697; fos 3–4v: 'A paper of Geography'; fos 5–9: Roby to Perceval, 14 Jan 1697/8; fo. 16v.

63 BL, Add MS 47025, fo. 27v: Philip Perceval to Sir John Perceval, 4 Jun 1700; Barker and Stenning, *Record of Old Westminsters*, ii, p. 733.

64 Gallagher, *Learning Languages*, p. 18.

65 David Cressy, 'Kinship and Kin Interaction in Early Modern England', *Past & Present*, 113:1 (1986), 38–69, at 69.

66 SBT, DR18/20/18/26: Lord Chandos to Theophilus Leigh, 22 May 1699; 28: William Leigh to same, 1 Dec [1700]; 39: Lady Chandos to same, 2 Jul 1703; 41: Theophilus Leigh to Lady Chandos, 23 Jun 1703; 43: Lady Chandos to Theophilus Leigh, 26 Feb 1703/4; 45: same to William Leigh, 8 Feb 1706/7.

67 Houlbrooke, *English Family*, p. 152.

his grandfather in the capital, but a family dispute led Sir Alan to worry from Dublin that 'My child is got into a vicious world and under no freinds guidance'. He asked his brothers in London to 'have an eye after' St John as he began at Middle Temple, concerned that his son had 'cost me about 200£ in one year at the Universitye'. Sir Alan hoped, during St John's several years in the capital, that the boy's uncles would 'find by discoursing him whether he follows his study or not'.[68] A similar arrangement held for his second son, the younger Alan, who had lived with his great-aunt in London for several years by the age of 7 and followed the same course through Eton, Cambridge and Middle Temple. When Alan, unlike St John, came to travel abroad, it would be his uncle and great-aunt in the capital who oversaw many of the practicalities, including his finances.[69] For landowning families such as the Southwells, Percevals, Brodricks and Leighs, close relations facilitated the childhood mobility required to access desirable forms of education.

Preparing for Travel at University

After a patchwork of schooling and tuition, many young men went to Oxford or Cambridge. Although university often prefaced travel, surveys of education or the social elite tend to present it in opposition to foreign experience, especially in work that uses the Grand Tour to explain declining university attendance among the gentry. Many studies of educational travel, meanwhile, trace its popularity to falling standards at Oxford and Cambridge.[70] The families examined here demonstrate, however, that university and foreign travel stood together in educational strategies. Time at university, like public school, was often short for those who planned to see the Continent, with James Brydges recognising that it was 'very seldom that any Young Gentleman stays above two Years in the University'.[71] Like schooling, it could be a targeted preparation for the next stage of education. Oxford or Cambridge sometimes featured as an alternative to travel, as in

[68] SHC, 1248/2/10: Sir Alan Brodrick to Thomas Brodrick, 8 Oct 1700; fo. 27: same to elder St John Brodrick, 28 Sep 1701; fo. 191: same to same, 8 Apr 1705; Austen-Leigh, *Eton College Register, 1698–1752*, p. 44.

[69] SHC, 1248/2/250: Sir Alan Brodrick to Alan Brodrick II, 16 Oct 1706; fo. 352: same to same, 18 Jan 1708/9; 1248/7/13: same to Martha Courthope, 2 May 1726; D. W. Hayton, *The Anglo-Irish Experience, 1680–1730: Religion, Identity, and Patriotism* (Woodbridge: Boydell, 2012), p. 96.

[70] For example, Rosenheim, *Emergence of a Ruling Order*, p. 195; Redford, *Venice and the Grand Tour*, p. 15.

[71] HL, mssST57, vol. 24, fos 188–9: James Brydges to Robert Jenkin, 4 Aug 1724.

the deliberations of relatives and trustees over whether to send the 16-year-old Viscount Irwin 'upon Travillinge or the university' after the death of his father, but they more often worked as complementary stages in a longer course. In the end, Irwin's friends decided that continental travel should follow two years at Cambridge.[72]

University students and their families discussed a European voyage as the next logical step. Sir Thomas Southwell would 'presently goe to travell' after Oxford in the mid-1680s, Sir John Perceval 'quitted the University' in 1701 'with design to travell' and, two years later, Irwin 'had Thougts of leaving Cambridge shortly to Travell', planning his time in Utrecht from East Anglia.[73] The closeness of university and continental experience meant that families might set up a household fit for travel while sons remained at their colleges. Irwin's trustees chose the German Protestant John Haccius as his live-in Cambridge tutor because he would serve well during the travels to come.[74] James Brydges similarly arranged for John to acquaint himself with his travelling tutor over a few months at Oxford and sent them a servant who had 'travelled with several Gentlemen abroad'. Brydges recognised that John's 'stay in the University is not to be much longer & that your advancement in years, makes me judge it necessary of sending you soon to travell for your further Improvement'.[75] Young men would regularly go abroad with university contemporaries, often leaving Oxford or Cambridge directly for London and the Continent.

The universities were aware of their close relationship with travel, as sites where young men deepened their immersion in travel culture. Students from Magdalen and Merton even debated 'whither travelling be good for English Gentlemen' at the Oxford celebration of Encaenia in 1677. Their Latin texts appeared in print later that year, but the question had already been settled within and beyond the university.[76] Travel was absent from formal curricula at Oxford and Cambridge, despite the condensed survey of the arts and sciences offered for those gentlemen who intended to leave for London

[72] WYAS, WYL100/C/9A: George Townshend to John Roades, 2 Jul 1702; same to same, 21 Jul 1702.

[73] NLI, MS 664, letter 7: Sir Robert Southwell to Sir William King, 5 Oct 1683; BL, Add MS 47072: Perceval, Autobiography, p. 8; NYCRO, ZON13/1/259: Thomas Worsley to Irwin, 27 Nov 1703.

[74] WYAS, WYL100/C/9A: William Thompson to [John Roades], 18 Jul [1702].

[75] HL, mssST57, vol. 17, fos 203–4: James Brydges to John Brydges, 11 Oct 1720; fo. 274: same to same, 25 Nov 1720; fo. 337: same to Dr Charles Stuart, 30 Jan 1720/1.

[76] Northants RO, IC 1067: Richard Richardson to Sir Thomas Isham, [undated]; IC 1073: Gilbert Clerk to same, 15 Dec 1677; *Theatri Oxoniensis Encaenia* (Oxford: 1677).

or the Continent after a couple of years.[77] Students of high social standing nevertheless read about and discussed their next educational stage. Robert Southwell kept two commonplace books during his time at Oxford that reflect his studies, containing philosophical notes and his own Latin poetry, but a third document, representing reading at university and Lincoln's Inn, reveals parallel engagement with the literature of educational travel.[78] Southwell compiled unattributed extracts from Francis Bacon's essay 'Of Travel' and Francis Osborne's *Advice to a Son* alongside quotations from Juvenal and Cicero, assembling an introduction to the theory and practice of travel.[79] His accompanying notes on French verse and conversation, as well as a copy of a letter of thanks to an Italian host, show that he kept the manuscript close by him on the Continent.

Colleges such as Christ Church at Oxford and Christ's at Cambridge employed academics with their own experience of—or interest in—foreign travel, and they may have imparted knowledge in tutorials. It remains difficult to reconstruct what was shared in such settings, but the globes and maps requested of parents suggest geographical study.[80] John Covel was a noted traveller in the Ottoman Empire while chaplain to the ambassador at Istanbul and returned via Italy and France, where Sir Philip Perceval found that he could give 'a very good account' of Turkey.[81] Covel was Master of Christ's for much of the rest of his life, sharing his expertise with those who were planning travel. He weighed the merits of France and the Netherlands in the recovery of Thomas Worsley's fragile health, suggesting that, having found a 'discreet and carefull' tutor, he might 'try one summer first in seing England'. Covel also had Viscount Irwin, Worsley's eventual companion abroad, 'every day with me or I with him, so as he shall not want all the good

[77] Mordechai Feingold, 'The Humanities', in Nicholas Tyacke, ed., *The History of the University of Oxford: Volume IV, Seventeenth-Century Oxford* (Oxford: Clarendon, 1997), pp. 211–358, at p. 217.

[78] Bodleian, MS Eng. Poet. f 6: Robert Southwell, Poetic commonplace book, [1650s]; Beinecke, Osborn MS b112: Robert Southwell, Commonplace book, [1650s].

[79] King's College, Cambridge, Keynes MS 152: 'An Abridgement of a Manuscript of Sr Robert Southwell's concerning travelling. Writt. 1658. Feb. 20', [1658–61]; Francis Bacon, 'Of Travel', in *The Essays*, ed. John Pitcher (London: Penguin, 1985), pp. 113–14; Francis Osborne, *Advice to a Son* (Oxford: 1656). The manuscript is in Southwell's hand and is evidently the original that he took to the Continent, despite its title.

[80] BL, Add MS 46951A, fo. 1: Benjamin Woodroffe to Sir Robert Southwell, 6 Jan 1673/4; 46951B, fo. 34: Robert Perceval to same, 22 Oct 1674; 46952, fo. 87: same to same, 6 May 1675; Christ Church, MS 427, letter 36: Sir Thomas Southwell to same, 10 Sep 1682.

[81] Elisabeth Leedham-Green, 'Covel [Colvill], John (1638–1722)', *ODNB*, online edn, 2008 (https://doi-org/10.1093/ref:odnb/6471); BL, Add MS 46956A, fo. 9: Sir Philip Perceval to Sir Robert Southwell, 18 Jan 1678/9.

advice I can give him'.[82] At Christ Church, too, academics probably shared knowledge of the Continent, whether or not they had travelled themselves. Henry Aldrich, who acted as tutor to many noble—and especially Tory—students, visited Cologne while at the college and probably spent time in France and Italy.[83] Francis Atterbury had not ventured abroad by his time as tutor to the well-born at Christ Church, despite his later banishment to France for Jacobitism.[84] He nevertheless compiled a commonplace book of his travel reading, rearranging passages by Lassels on Italy and Brown on the Netherlands, Germany, central Europe and Greece.[85] Perhaps he shared the notes with pupils. Sir John Perceval certainly answered his old French master's hope that he would continue reading travels at Magdalen in Oxford, where he 'composed an Essay upon travelling' for his tutor, Richard Smalbrook. Smalbrook would himself write of another student that 'travelling is one of the best methods he can use for his education'. More practically, he taught Perceval the principles of fortification so that he might make the most of a visit to Portsmouth on a coming domestic tour.[86] The discussion of travel added to the aristocratic feel developed at these colleges, where tutors surely recognised an informal duty to prepare wealthier students for the next stage of their education.[87] Parents placed great importance on the capability of college tutors and the prominence in this role of men who had travelled, or at least took an interest in travel, was probably no coincidence.[88]

The position of the French language at university also reveals that families, students and tutors kept an eye on travel. It would be misleading

[82] NYCRO, ZON13/1/260: John Covel to Thomas Worsley, 25 Dec 1703; Christ's College, Cambridge, Fellows' Papers, Box 24, item i: same to John Roades, 3 Nov 1702.

[83] W. G. Hiscock, *Henry Aldrich of Christ Church, 1648–1710* (Oxford: Holywell Press, 1960), p. 13.

[84] D. W. Hayton, 'Atterbury, Francis (1663–1732)', *ODNB*, online edn, 2004 (https://doi-org/10.1093/ref:odnb/871).

[85] Christ Church, MS 475: Francis Atterbury, 'Regiones, Urbes, Montes, Flumina, &c.', [after 1685]. Nothing supports the attached claim that it was made 'during a foreign tour'. Its descriptions are wholly lifted from print, mainly Richard Lassels, *The Voyage of Italy, or A Compleat Journey through Italy* (Paris: 1670) and Edward Brown, *A Brief Account of Some Travels in Divers Parts of Europe* (London: 1685).

[86] BL, Add MS 47025, fo. 21: Louis Payzant to Sir John Perceval, 13 Feb 1699/1700; fo. 24: Richard Smalbrook to Sir Robert Southwell, 5 Apr 1700; fo. 40v: same to same, 11 Mar 1700/1; Borsay, 'Children, Adolescents and Fashionable Urban Society', p. 57.

[87] For an 'informal', 'aristocratic' curriculum requiring individual tutors, see Victor Morgan, *A History of the University of Cambridge: Volume 2, 1546–1750* (Cambridge: Cambridge University Press, 2004), pp. 133–8, 328. On the deliberately aristocratic character of Christ Church, see Stephen Porter, 'University and Society', in Tyacke, ed., *History of the University of Oxford: Volume IV*, pp. 25–104, at pp. 62–3.

[88] French and Rothery, *Man's Estate*, p. 93.

to treat any language-learning as evidence of preparation for the Continent, though, since contemporaries valued French, in particular, for its social cachet at home.[89] Many boys and girls who learned it from an early age had no intention of travelling. Sir Alan Brodrick urged his young namesake, still at Eton, to make himself 'master of French' in case he should find himself 'near Court or in any thing which relates to publick affairs' in London, in which circles he knew 'not any one thing you can be lesse without'. Brodrick soon begged his brother to make sure that his son did not lose any French during time in the capital between school and Cambridge.[90] Children as young as 5 learned the language and some, such as the sons of Robert Molesworth or Brodrick's cousin, Laurence Clayton, attended establishments with French masters. Others employed additional language tutors during schooling or recruited live-in 'Frenchmen', as did the Southwells and Percevals.[91] Sir Robert Southwell's eldest son, Rupert, had been 'an absolute Frenchman' before his death at the age of 8, speaking the language 'with the same ease almost as English'.[92] These lessons were for use at home, perhaps even more than abroad.

The acquisition of French nevertheless stepped up at university in preparation for impending travel. Since modern languages would not feature on Oxford or Cambridge curricula until the mid-1720s, and even then as a failed experiment unrepeated for another century, students made progress with private tutors. Katherine Southwell advised her son to keep up his linguistic skills by reading Latin and Greek books in French translation, since the latter tongue would be 'of more use'. As late as 1755, though, she knew that her recommendation 'woud be thought Heresy at Cambridge'.[93] Throughout the seventeenth and eighteenth centuries, both towns nevertheless hosted communities of linguists, often including more Huguenot immigrants, who catered less officially to the requirements of

[89] Kathleen Lambley, *The Teaching and Cultivation of the French Language in England During Tudor and Stuart Times* (Manchester: Manchester University Press, 1920), pp. 361–400.

[90] SHC, 1248/4/19: Sir Alan Brodrick to Alan Brodrick II, 7 May 1717; fo. 93v: same to Thomas Brodrick, 11 Nov 1717.

[91] NLI, microfilm P.3752: Robert Molesworth to Laetitia Molesworth, 17 May 1700; SHC, 1248/3/35: Laurence Clayton to Thomas Brodrick, 16 Jan 1710; BL, Add MS 46954A, fo. 135: Sir Robert Southwell to Sir Philip Perceval, 30 Apr 1677; 47072: Perceval, Autobiography, p. 6; Máire Kennedy, *French Books in Eighteenth-Century Ireland* (Oxford: Voltaire Foundation, 2001), pp. 22–48; Gallagher, *Learning Languages*, pp. 16–19; Cohen, *Fashioning Masculinity*, pp. 83–5.

[92] BL, Add MS 46954A, fo. 118v: Sir Philip Perceval to Sir Robert Southwell, 1 May 1677; 38015, fo. 204: Sir Robert Southwell to Helena Southwell, 31 May 1678.

[93] Bristol Archives, 45317/2/5/1/24: Katherine Southwell to Edward Southwell III, 29 Nov 1755.

genteel education.[94] Sir Philip and John Perceval used 'a little french man who dos teach att several Colledges' in Oxford, Hippolytus du Chastlet de Luzancy, who visited 'once or twice a week after supper time to read or talke'. He also helped with letter-writing.[95] Decades later, John's son availed himself of 'two or three Gentlemen that study french', while the Leigh family similarly turned to one Peter Dubois.[96] Robert Perceval had a harder time finding an equivalent at Cambridge, a regular complaint, and almost lost his French 'for want of discoursing with some body'.[97] Students also read French histories and subscribed to French gazettes, which Irwin and Worsley shared at Cambridge. Depending on their abilities, young men worked through them alone or with tutors.[98]

Families balanced the need for French with other priorities, worrying that it might interfere with school and university curricula.[99] They were particularly concerned for younger sons, who would have to support themselves through work and often travelled less ambitiously, if at all. Financial accounts for the education of Irwin's younger brothers show that the family only started paying for a French master relatively late in periods at Oxford.[100] For those destined to travel, however, the language became urgent. Sir Robert Southwell worried that his kinsman, Sir Thomas, was ill prepared for the voyage that his rank demanded, urging his Oxford tutor to find 'any way for him to gett a little ffrench without the interrupting of other things'. He would then be able to 'travell when he goes into France, whereas

[94] C. H. Firth, *Modern Languages at Oxford, 1724–1929* (London: Oxford University Press, 1929), pp. 1–19; Morgan, *History of the University of Cambridge, Volume 2*, p. 549; Lambley, *Teaching and Cultivation*, pp. 198–210; Gallagher, *Learning Languages*, pp. 19–23.

[95] BL, Add MS 46951A, fo. 8: Sir Philip Perceval to Sir Robert Southwell, 21 Jan 1673/4; 46954A, fo. 51: Southwell to John Perceval, 3 Mar 1676/7; 46954B, fo. 145: same to same, 9 Oct 1677; Feingold, 'Humanities', p. 272. Luzancy was a Catholic convert to Anglicanism.

[96] BL, Add MS 47025, fo. 20v: Sir John Perceval to Sir Robert Southwell, 10 Feb 1699[/1700]; HL, mssHM 81620: Edward Leigh to Theophilus Leigh, 12 Jun 1736.

[97] BL, Add MS 46951B, fo. 142: Robert Perceval to Sir Robert Southwell, 16 Dec 1674; Lambley, *Teaching and Cultivation*, pp. 199, 205, 207.

[98] BL, Add MS 46951A, fo. 1: Benjamin Woodroffe to Sir Robert Southwell, 6 Jan 1673/4; fo. 8: Sir Philip Perceval to same, 21 Jan 1673/4; 47025, fo. 17: Sir John Perceval to same, 20 Nov 1699; Blackburn CL, Southwell Papers, vol. 1: Edward Southwell to same, 30 Jul 1686; WYAS, WYL100/C/9A: John Haccius to John Roades, 26 Jan 1703; Irwin to same, 6 Mar 1703.

[99] BL, Add MS 46951A, fo. 8: Sir Philip Perceval to Sir Robert Southwell, 21 Jan 1673/4; WYAS, WYL100/C/9B: Ralph Bragg to John Roades, 5 Dec 1702.

[100] WYAS, WYL100/C/10: Richard Ibbetson to John Roades, 11 Aug 1707; WYL100/EA/12/17/6: Accounts for Henry Ingram at Oxford, 25 Mar 1709. Henry funded his own travels, outside family strategy.

else he must for a yeare be there only at Schoole'.[101] Students practised French at university in order to save time and money during their travels, but the social importance of accent and gesture, as well as verbal fluency, still encouraged linguistic learning abroad.

Preparations rarely included Italian. Travel was supposed to impart skills that would be useful at home, a value that parents ascribed, above all, to French. Besides its cachet, it was the tongue of a closer neighbour and a more important European power. Ormond prioritised 'the French language & mien' in the education of his grandson 'because they are undeniably the most gracefull, & because it may be an advantage against an enemy to be perfect in their dialect'.[102] Alongside such demand, Huguenot exiles offered an improving supply of tuition to late seventeenth-century Britain and Ireland. Equally, French was more useful than Italian on the Continent, ubiquitous among courtly and learned circles by the same decades. If preparations aimed at saving time abroad, moreover, they could target the standard period of settlement in France or the Netherlands, which Chapter 3 will examine primarily as a stage for learning French. At the most basic level, though, choices at university reflected the fact that France was a more popular destination than Italy among the gentry.

London Inns of Court and Academies

Family correspondence filled with thoughts of impending travel as young men prepared to leave Oxford or Cambridge. Debates over the correct age for foreign experience had long simmered in print and in families, but this was a point at which they became particularly urgent: should sons travel immediately after university? Many, such as Irwin and Worsley, went straight to the Continent, but others passed through an additional stage. The time between university and travel allowed sons in their late teens to deepen their acquaintance with London, though at an age when many began to resent parental oversight. Independence was a major aim of elite male upbringing, but the capital represented, as one educational entrepreneur put it, 'the most dangerous step of youth'.[103] Families used institutions for a managed introduction, particularly if they lacked metropolitan kin networks. This

[101] Christ Church, MS 427, letter 21: Sir Robert Southwell to William Breach, 5 Oct 1683.
[102] Bodleian, MS Carte 70, fos 480–1: Ormond to Sir Robert Southwell, 21 Jan 1678/9.
[103] TNA, SP29/411, fos 417–20: 'The Humble Proposalles of Solomon de Fobert Esq', 14 May 1679.

stage could be the last before adulthood for those who never crossed the Channel, just as university had been for others; for many of those who did travel, however, it was a crucial prior step.

Across the sixteenth and seventeenth centuries, families turned to the Inns of Court for a semi-structured period in London. They combined ‘a little sound of the Law’, so useful to landowners, with the chance to meet counterparts from across Britain and Ireland.[104] The inns were on the doorstep of the capital’s social life, with oversight that was appropriately laxer than at Oxford and Cambridge. They were also surrounded by dancing and fencing masters, offering a kind of gentlemanly training that was harder to come by at the universities.[105] When Edward Rawdon turned 18, Sir George considered sending him from Cambridge to the Inns of Court with his younger brother, John, who would ‘only accompany him half a year to see the town’. Sir George’s brother-in-law, the Earl of Conway, thought that both boys should ‘come away to London and stay there at the Inns of Court one yeare, before they travell’.[106] In the end, Edward returned to Christ’s for the rest of the year before moving to Middle Temple, where he regularly visited his uncle Conway, but John went straight from Cambridge to a Parisian academy. The brothers reunited in France just over a year later.[107] Families might use the inns differently for each of their sons: Sir Philip Perceval spent a year at Lincoln’s Inn before travel, ‘more addicted to the exercises of the body’ than ‘fixt upon the Book’; Robert followed a few years later, a last effort to reform his wayward conduct, but his descent into ‘Drink and Gaming’ ended within months in his murder on the Strand; John, whose ‘parts’ were ‘naturally adapted for the study of the law’, prepared at Lincoln’s Inn for a legal career long ‘forecasted & prepared’.[108] Historians argue over precisely when the inns lost their role as ‘a finishing school for wealthy amateurs’, with some charting slow decline from the early seventeenth century, others

[104] NLI, MS 664, letter 7: Sir Robert Southwell to William King, 5 Oct 1683.

[105] Wilfrid R. Prest, *The Inns of Court under Elizabeth I and the Early Stuarts, 1590–1640* (London: Longman, 1972), pp. 21, 23–4, 39–40, 138, 154, 166; David Lemmings, *Gentlemen and Barristers: The Inns of Court and the English Bar, 1680–1730* (Oxford: Clarendon, 1990), pp. 20, 23.

[106] *CSPD, Charles II*, xvi, pp. 212–13: Sir George Rawdon to Conway, 28 Mar 1674; HL, mssHA 14525: Conway to Rawdon, 31 Mar 1674.

[107] HL, mssHA 14527: Conway to Sir George Rawdon, 6 Jun 1674; 14536: same to same, 5 Dec 1674; 14539: same to same, 15 May 1675; 14546: same to same, 14 Mar 1675/6; H. A. C. Sturgess, *Register of Admissions to the Honourable Society of the Middle Temple: From the Fifteenth Century to the Year 1944*, 3 vols (London: Butterworth, 1949), i, p. 191.

[108] BL, Add MS 46953, fo. 68: Sir Robert Southwell to Lady Perceval, 20 May 1676; 46954A, fo. 102: Roger Altham to Southwell, 9 Apr 1677; fo. 239: Southwell to Sir Philip Perceval, 7 Jun 1677; 46954B, fo. 242: John Perceval to Southwell, 28 Nov 1677.

pronouncing it terminal after 1660 and still more finding signs of life in the mid-eighteenth century.[109] The rising costs of a gentlemanly education may have meant that some families had to choose between university, an Inn of Court and foreign travel, or to decide which of their sons could attend these institutions. The evidence of family practice—from as late as the 1670s—suggests that these options were not always mutually exclusive, however.

Another explanation for the decline of 'amateur' attendance at the Inns of Court is that newer institutions began to offer better access to London life and gentlemanly accomplishments, a role in which the inns had never been entirely comfortable. Sir Robert Southwell had sent his Perceval nephews to Lincoln's Inn, where he himself had gone before travel. He found another possibility a few years later, though, where 'there was formerly the Inns of Court alone'. In 1683, he informed Sir Thomas's guardians that 'of late ther is erected a very famous Academy in London Governed by a french Gentleman Monsr. Faubert, where riding, ffrench, Mathematicks, & all exercises are tought which are usually learnt in Travell'. Solomon Foubert, who had hosted the Rawdons in Paris, had moved to England four years previously, fleeing persecution of the Protestant community in France. Sir Thomas might 'passe 6 Monthes' with him, 'as many Englishmen doe, the better to prepare them for the improovement of Travell'.[110] English entrepreneurs had tried to found similar institutions since the late sixteenth century, tending to fail because of inadequate official backing, the opposition of the universities and the growing cachet of educational travel, but Restoration Francophilia and the influx of French Protestant refugees in the decades either side of 1685 gave renewed impetus to such schemes.[111] Foubert and his fellow Huguenots enriched London's French-speaking educational economy, particularly in the fashionable West End, and featured variously in family strategies.

Foubert has left more traces than most, not least because he periodically sought royal support. Extensive research in family archives reveals that many young men looked to him for an introduction to London life and gentlemanly accomplishments, seeking direct preparation for foreign experience. Foubert had indeed advertised his services to the nobility and

[109] Stone and Stone, *Open Elite?*, p. 265; Prest, *Inns of Court*, pp. 44–5; Lemmings, *Gentlemen and Barristers*, pp. 9, 258; Paul Lucas, 'A Collective Biography of Students and Barristers of Lincoln's Inn, 1680–1804: A Study in the "Aristocratic Resurgence" of the Eighteenth Century', *Journal of Modern History*, 46:2 (1974), 227–61.

[110] NLI, MS 664, letter 7: Sir Robert Southwell to William King, 5 Oct 1683; Ansell, 'Foubert's Academy'.

[111] Ansell, 'Foubert's Academy', pp. 47–50.

gentry as an immediate preface to continental travel. Although questions remain about the size, composition and activities of the student body, his academy was small, with a high turnover of perhaps 20 to 40 students at any one time, and it echoed other Huguenot establishments by immersing its pupils in a Francophone environment.[112] Little survives for other institutions, which similarly lack official records. Even where young men attended a French-run school or academy in London, parents often refer obliquely to its identity, while relatively brief attendance works against large caches of student letters. James Brydges's nephew, John Chamberlayne, moved from Oxford to a year of French and mathematics at an unknown establishment in the capital, intending to prepare for the German academy of Wolfenbüttel. Later, it seemed that his time there would allow him to cut a planned trip to France down to three or four weeks, since he could already read the language.[113] It can be difficult to decide whether French-run establishments in London should be classed as academies, as it is not often clear what differentiated them from the small schools attended at an earlier age by the likes of the Percevals. At any rate, families commonly used them as a preface to imminent travel.

It makes more sense to assess establishments from the viewpoint of their patrons, examining how they were used rather than how they were run. Robert Molesworth initially divided his time between Dublin and London, aside from a period as ambassador to Denmark between 1689 and 1692, and he bought an estate in Yorkshire a few years later. He prepared several sons for travel by boarding them in the capital with Monsieur La Place of Queen Square. Molesworth's eldest sons, John and Richard, moved there after home schooling in the late 1690s, travelling abroad immediately afterwards.[114] Their sixth brother, Coote, followed them to La Place from a school near Sheffield over a decade later. John, newly appointed envoy to Florence, paid £30 for Coote to board at the 'famous french school' in order 'to file & finish him that he may be fit to be sent to him about a year hence'. La Place taught him 'Latin, but Chiefly French, to qualify him for going abroad, in Publick Business'. Changed plans took Coote to Westminster School two years later, however, before extensive continental travel, some

[112] Ansell, 'Foubert's Academy'; Gallagher, *Learning Languages*, pp. 27–8, 30–5.

[113] SBT, DR18/20/18/47: Theophilus Leigh to James Leigh, 4 Jul 1709; DR18/20/18b: same to same, 16 Apr 1711; same to same, 27 Sep 1712.

[114] NLI, microfilm P.3752: Robert Molesworth to Laetitia Molesworth, 4 Apr 1690; same to same, 8 Sep 1692; same to same, 23 Feb 1697; same to same, 17 May 1700; Bysse Molesworth to John Molesworth, 25 Jul 1719.

of it in his brother's retinue.[115] Robert Molesworth had intervened in the debate over educational travel in the preface to his 1694 *Account of Denmark*, arguing that too many parents sent sons abroad as 'Children, and bring them home great Boys', failing to justify 'the Expenses laid out'.[116] He allowed his own sons to travel only in their late teens and early twenties. Despite their age, however, none of Molesworth's seven boys studied at Oxford or Cambridge. John, Richard and Coote would only enter Middle Temple after travel, having prepared for the Continent with the Huguenots of London.[117]

French-run institutions also provided alternatives for those unwilling or unable to send sons abroad, fulfilling several roles in elite travel culture. Foubert arrived in an England shaped by the 'Popish Plot' and immediately offered a safer option to those who feared the Catholicising influence of a journey to France. He protested the staunch Protestantism of his entire household.[118] In the longer term, he also catered to those who could not afford to send young men to the Continent. Sir Robert Southwell introduced the academy as a preparation for Sir Thomas's time abroad, but it reappeared as an alternative when the latter's estate failed to furnish adequate funds for him to travel with a tutor. Sir Thomas preferred the Inns of Court, which Sir Robert and his guardians suspected would allow him to 'shake off Government and live without Controule' among the 'seeminge delights' and 'vanity' of London. Sir Robert, who knew Sir Thomas's misbehaviour first hand from his university vacations, preferred Foubert's academy as 'a place of some Discipline, & Government'. The young man rejected it, however, as 'a place of Pedantrie'. It is unclear who prevailed, as letters cease with Sir Thomas's move to London and he does not appear in the records of any Inn of Court.[119] Foubert certainly offered a base for the sons of other Irish landowners, who learned gentlemanly exercises, toured the capital, made lasting connections and returned home without seeing the Continent. For

[115] NLI, microfilm P.3752: Robert Molesworth to Laetitia Molesworth, 6 Nov 1710; Coote Molesworth, Family history and pedigree, 1765, in Hugh Mayo, 'Robert Molesworth's *Account of Denmark*: Its Roots and Its Impact', Ph.D. thesis (University of Southern Denmark, 2000), pp. 219–20.

[116] Robert Molesworth, *An Account of Denmark, as it was in the Year 1692* (London: 1694), sig. A4v.

[117] Sturgess, *Register of Middle Temple*, i, pp. 248, 286; Coote Molesworth, Family history, in Mayo, 'Molesworth's *Account of Denmark*', pp. 219–20.

[118] Ansell, 'Foubert's Academy', pp. 54–5, including Ormond's ward, Baron Kingsale, and Robert Harley, the future Earl of Oxford.

[119] Christ Church, MS 427, letter 3: Sir Robert Southwell to Henry Aldrich, 22 Jan 1682/3; BL, Add MS 46961, fo. 121: same to John Pigott, 18 Oct 1684; NLI, MS 664, letter 12: William King to Southwell, 24 Oct 1684.

those English gentry families who could not afford foreign travel for any of their sons, such as the Johnsons of Spalding, Foubert similarly offered accomplishments that had previously lain beyond their reach.[120]

The academy also hosted the younger brothers of those who travelled. It often combined an introduction to French manners and metropolitan society with preparation for the army, especially as pretensions to a fuller academy formation faded under Foubert's son, Henri.[121] Where William Leigh had travelled to Italy, his 15-year-old younger brother, Henry, attended the academy for one quarter in order to see the town and prepare for the army. Their father overcame his reservations that 'Riding the great Horse & Dancing' would not allow Henry to make 'his fortune in the world', for all that they would make him 'a Beau'. In the end, Theophilus Leigh was satisfied that his son was 'well improved for so short a time, at least in the outward man'.[122] A similar course shaped John Ingram, whose elder brothers had travelled abroad. His mother moved him from St Peter's in York to 'a good french house' in London, where she had taken a home on Leicester Street with her eldest son, recently returned from Italy. It seems likely, given John's bills for tuition in French and fencing, as well as his imminent military career, that he had moved to Foubert's nearby establishment.[123] The Huguenot schools and academies of London deserve prominence in a full account of educational mobility, as families used them for early immersion in French culture, an immediate preface to travel or an alternative to time abroad.

Domestic Travel

Sons also prepared for continental travel by touring parts of Britain. Domestic and foreign journeys were intimately linked in family educational strategies, but they are usually studied separately. On the one hand, historians of continental travel rarely concern themselves with the decisions and processes that sent young men abroad. On the other, work on domestic travel is split between older books on its practicalities and more recent analytical

[120] Ansell, 'Foubert's Academy', pp. 55–9.
[121] Ansell, 'Foubert's Academy', pp. 47, 52.
[122] SBT, DR18/20/18b: Theophilus Leigh to James Leigh, 27 Apr 1713; same to same, 6 Jul 1713; same to same, 24 Aug 1713.
[123] WYAS, WYL100/C/10: Dowager Viscountess Irwin to John Roades, 8 Oct 1709; same to same, 4 Feb 1709[/10]; WYL100/EA/12/17/7: Bills relating to the education of Cornet John Ingram.

treatments. The latter focus on two important developments that they date to the mid- to late eighteenth century: the rise of country-house tourism and the pursuit of the 'picturesque'. They tend to see domestic journeys as a direct alternative to the Grand Tour, then suffering from warfare and waning exclusivity.[124] Periodic calls to treat domestic and foreign travel together go unheeded, with studies of the 'home tour' actively distancing themselves from those of continental journeys in order to carve out a separate niche.[125] The few studies that take up the invitation have dwelt on parallels, as in the educational and leisure roles shared by domestic and foreign travel, rather than on links between the two forms of mobility.[126] Attention to family strategy, instead of preformed notions of 'home tour' and 'Grand Tour', shows, by contrast, that young men often undertook British travel with later continental experience firmly in mind.

Gentry and noble families sent sons on domestic journeys from an early age. Young Southwells used short trips, often in their early teens with parents or servants, to acquaint themselves with the bad weather, limited food and early hours endured by travellers. They wrote descriptions of Cirencester, Monmouthshire and London.[127] Sir William Petty prosecuted a more ambitious tour through Britain and Ireland with his son, inoculating Charles against similar discomforts and encouraging him, 'before hee bee 8 yeares old', to 'give a good account of all that is remarkable in the 3 kingdomes'.[128] Older students fitted journeys around university terms.

[124] Esther Moir, *The Discovery of Britain: The English Tourists, 1540–1840* (London: Routledge & Kegan Paul, 1964); Malcolm Andrews, *The Search for the Picturesque: Landscape Aesthetics and Tourism in Britain, 1760–1800* (Aldershot: Scolar, 1989); Ian Ousby, *The Englishman's England: Taste, Travel and the Rise of Tourism* (Cambridge: Cambridge University Press, 1990). For alternative framings, see Andrew McRae, *Literature and Domestic Travel in Early Modern England* (Cambridge: Cambridge University Press, 2009); John Cramsie, *British Travellers and the Encounter with Britain, 1450–1700* (Woodbridge: Boydell, 2015).

[125] For such calls, see Ilaria Bignamini, 'The Grand Tour: Open Issues', in Wilton and Bignamini, eds, *Grand Tour*, pp. 31–5, at p. 32; Rosemary Sweet, Gerrit Verhoeven and Sarah Goldsmith, 'Introduction', in Sweet, Verhoeven and Goldsmith, eds, *Beyond the Grand Tour*, p. 10. For distancing, see Zoë Kinsley, *Women Writing the Home Tour, 1682–1812* (Aldershot: Ashgate, 2008), pp. 1–2; Jocelyn Anderson, *Touring and Publicizing England's Country Houses in the Long Eighteenth Century* (New York: Bloomsbury Academic, 2018), pp. 25–6.

[126] Conway, *Britain, Ireland and Continental Europe*, pp. 192–3; French and Rothery, *Man's Estate*, pp. 155–8; Dent, 'Travel as Education', 173–4.

[127] Christ Church, MS 427, letter 2: Sir Thomas Southwell to Sir Robert Southwell, 17 Jan 1682/3; Blackburn CL, Southwell Papers, vol. 1: Sir Robert Southwell to Edward Southwell, 28 Sep 1685; Beinecke, Osborn MS fb90: Edward Southwell, Accounts of travels, 1684–1715.

[128] *The Petty–Southwell Correspondence 1676–1687*, ed. Henry William Edmund Petty-FitzMaurice, Marquis of Lansdowne (New York: Augustus M. Kelley, 1967), pp. 79–80: Sir William Petty to Sir Robert Southwell, 18 Oct 1679.

Edward and John Rawdon, for instance, went from Ireland to Cambridge 'by way of Scotland, with a good guide', their father determined that they should profit by the journey.[129] Once matriculated, brothers or friends visited each other at Oxford and Cambridge, which offered opportunities for sightseeing, sociability and epistolary description, with correspondents expecting a comparison of the two universities.[130] John Covel took an alternative summer trip to Suffolk with his most prestigious Christ's student, Viscount Irwin, and Thomas Worsley planned a tour of Yorkshire later that year with 'a couple of Gentlemen' from the college.[131] Journeys were an inevitable feature of educational courses that took young men increasingly far from home, but families tried to make the most of them.

The Easter and summer vacations permitted longer itineraries, such as the increasingly ambitious tours undertaken by John and Henry Brydges in 1718, 1719 and 1720. John (15) requested the first trip himself, though James Brydges had been planning something similar for the 'improvement' of his eldest son. John Hudson, an Oxford scholar who had just dedicated a Latin edition of *Aesop's Fables* to John, was recruited to accompany the young nobleman and his three servants. The party saw everything 'remarkable' at Oxford, where Hudson brought John into 'the most valuable Company', and they toured regional seats.[132] Brydges intended the journey to prepare his eldest son for university, evidently, but also for the travels that would follow. Little survives of a second journey that John and Henry took together to Bath the following summer, beyond their stepmother's comment that it was 'for their divertion &c', but the boys continued to fit travel around their studies.[133]

The most ambitious journey came over three months in the summer of 1720, when 17-year-old John had finished his first year at Oxford and 12-year-old Henry was at Westminster. The boys set off with two tutors, an army officer and a retinue of servants, travelling from the family home at Cannons to Oxford and the country houses of the Midlands and Yorkshire. They moved up the east coast to Edinburgh, where they spent a week and a

129 *CSPD, Charles II*, xii, p. 27: Sir George Rawdon to Conway, 12 Dec 1671.

130 BL, Add MS 46952, fo. 104: Sir Philip Perceval to Sir Robert Southwell, 7 Jul 1675; fo. 119: Robert Perceval to same, 20 Jul 1675; 46955B, fo. 51: Southwell to John Perceval, 30 Jul 1678; WYAS, WYL100/C/9B: John Haccius to John Roades, 29 Jun 1703.

131 WYAS, WYL100/C/9A: John Haccius to John Roades, 26 Jan 1703; NYCRO, ZON13/1/259: Thomas Worsley II to Thomas Worsley, 19 Aug 1703.

132 HL, mssST57, vol. 15, fo. 153: James Brydges to John Brydges, 31 Mar 1718; fos 156–7: same to same, 2 Apr 1718; fos 183–4: same to same, 10 Apr 1718; John Hudson, *Fabularum Aesopicarum Collectio, Quotquot Graece Reperiuntur. Accedit Interpretatio Latina* (Oxford: 1718).

133 *Cassandra Brydges, Duchess of Chandos, 1670–1735: Life and Letters*, ed. Rosemary O'Day (Woodbridge: Boydell, 2007), p. 103: Cassandra Brydges to Martha Perrot, Jul 1719.

half before touring the noble seats of central Scotland and the Borders. John and Henry returned home via more country houses in Lancashire, Yorkshire and the Midlands, completing an initiation to the British ruling classes, to whom their father wrote regular introductions.[134] Their itinerary is laid out in Henry's detailed diary, its cover improvised from a printed souvenir of the Sheldonian Theatre in Oxford (Figure 4).[135] The three-month tour of England and Scotland offered him early practice in observation and journal-writing, essential skills for much later continental experience. Preparations were more immediate for John, who would soon leave the university, and he made a final month-long tour through 'the West' in 1721 before crossing the Channel. His party, which included his travelling tutor and an army officer with West Country expertise, visited Bath, Somerset, Devon, Winchester and Portsmouth, where they may have embarked for France.[136] English and Scottish journeys prepared the Brydges brothers for the Continent.

Figure 4 Cover of Henry Brydges's Journal, 1720. The Huntington Library, San Marino, California, mssHM 30301.

[134] HL, mssST57, vol. 18, fos 37–8: James Brydges to John Brydges, 7 May 1720.
[135] HL, mssHM 30301: Henry Brydges, Journal, 1720.
[136] HL, mssST57, vol. 19, fo. 29: James Brydges to Charles Stuart, 29 Mar 1721; fo. 30: same to John Brydges, 3 Apr 1721; fo. 47: same to Stuart, 22 Apr 1721; fos 54–5: same to same, 8 May 1721.

Contemporary writers similarly felt that domestic and foreign travel were closely related, with two kinds of text considering the link across the period. Topographers, antiquarians and domestic travellers argued that sights at home equalled anything abroad, while educational authors wanted travellers to have frames of reference for what they would see on the Continent. Edward Leigh's 1659 sentiment that 'our Englishmen ... do not sufficiently prize the Excellencies of their own native Soil' runs through James Brome's 1700 claim that 'England is not destitute of those many taking Things, which all Travellers so passionately admire Abroad' and William Stukeley's 1724 desire to show that 'tis not impossibl to make a classic journey on this side the straits of Dover'.[137] Meanwhile, tutors, foreign travellers and educational entrepreneurs from Richard Lassels (1670) to Thomas Sheridan (1769) built on earlier seventeenth-century predecessors to argue that domestic travel offered essential comparisons for what travellers would find abroad. Lassels advised that a tour of England would 'enure' a young man to the hardships of travel 'and sharpen his appetite for forrain curiosityes', preparing him, Gailhard hoped, for the probing questions of continental hosts. All the while, he would maintain what John Macky called 'the true Value of [his] own Kingdom' as protection against undue foreign influence.[138] The idea of 'domestic travel' may be anachronistic, since 'travel' tended to mean going abroad, but the absence of a shared term did not prevent contemporaries from connecting foreign and domestic mobility.[139]

Gentry and noble families regularly sent sons on domestic journeys in preparation for continental experience. British tours do not leave as much evidence as foreign travel, whether because young men did not feel the same pressure to keep diaries or because these shorter journeys did not require the long-term financial supply that generated much correspondence. Traces, however, include those of Sir Philip Perceval, who saw parts of western England, including Salisbury, Stonehenge and several noble seats, on his way to Bristol and a tour of Dublin and Irish family lands. He returned to London through Oxford, having travelled between June and September 1676, and left for France within weeks. His

137 Edward Leigh, *England Described* (London: 1659), sig. A2v; James Brome, *Travels over England, Scotland and Wales* (London: 1700), sig. A3r; William Stukeley, *Itinerarium Curiosum* (London: 1724), p. 3.

138 Lassels, *Voyage of Italy*, sig. E4r; Gailhard, *Compleat Gentleman*, ii, p. 23; John Macky, *A Journey through England. In Familiar Letters from a Gentleman Here, to his Friend Abroad* (London: 1714), p. i; Thomas Sheridan, *A Plan of Education for the Young Nobility and Gentry of Great Britain* (Dublin: 1769), p. 144; Brauer, *Education of a Gentleman*, pp. 170–1.

139 McRae, *Literature and Domestic Travel*, p. 14.

tutor, Gailhard, practised the domestic preparations that he would soon advocate in print, but the trip also tested, 'by way of Experiment, how they would agree together'.[140] William and John Blathwayt, whose father was a neighbour and colleague of Perceval's uncle, Sir Robert Southwell, chose a different itinerary in 1703 before their continental travels of 1705–8. An incomplete journal by their Huguenot tutor covers observations and gentry sociability in East Anglia, Lincolnshire, Yorkshire and the north as far as Newcastle and Carlisle. Along the way, William and John wrote to their father in French.[141]

Two routes emerged in these and other families in the decades either side of 1700. Both began in London, where sons boarded at schools or academies or lodged with parents or close relations during university vacations. First, the Great North Road offered access to the Midlands, Yorkshire and Scotland, with detours to Norfolk and the Peak District. From the late seventeenth century onwards, strip maps—effectively road atlases—emphasised the country houses that lined the way.[142] A second route took others to western England. Communications with London were well established, having long allowed invasions and risings in the West Country to threaten the capital, and the region offered a cluster of notable seats and other attractions. In particular, the rise of Bath as a health resort after the Restoration coincided with growing interest in Stonehenge, conveniently located near Salisbury and Wilton House.[143] Historians of domestic travel often delay such itineraries to the mid-eighteenth century, awaiting further improvements to the roads and a weight of published encouragement.[144] By that point, though, the convenience and interest of northern and western routes had long made them popular among those preparing for foreign travel.

[140] BL, Add MS 46953, fo. 104: Jean Gailhard to Sir Robert Southwell, 9 Jun 1676; fo. 106: same to same, 14 Jun 1676; fo. 141: Sir Philip Perceval to same, 25 Jul 1676; fo. 152: same to same, 18 Aug 1676; fo. 156: Gailhard to same, 30 Aug 1676; fo. 158: same to same, 7 Sep 1676; James Anderson, *A Genealogical History of the House of Yvery, in its Different Branches of Yvery, Luvel, Perceval and Gournay*, 2 vols (London: 1742), ii, p. 379.

[141] [P. de Blainville], *A Diary of the Journey Through the North of England made by William and John Blathwayt of Dyrham Park in 1703*, ed. Nora Hardwick (Bristol: N. Hardwick, 1977); [P. de Blainville], *The Grand Tour: Letters and Accounts Relating to the Travels Through Europe of the Brothers William and John Blathwayt of Dyrham Park, 1705–1708*, ed. Hardwick (Bristol: N. Hardwick, 1985).

[142] Ousby, *Englishman's England*, pp. 12–18. On the Peaks, pp. 131–7.

[143] Sylvia McIntyre, 'Bath: The Rise of a Resort Town, 1660–1800', in Peter Clark, ed., *Country Towns in Pre-Industrial England* (Leicester: Leicester University Press, 1981), pp. 198–249; Rosemary Hill, *Stonehenge* (London: Profile, 2008), pp. 26–37.

[144] For example, Anderson, *Touring and Publicizing*, p. 54.

A few young men have left detailed impressions and itineraries. In 1701, Sir John Perceval toured much of England with William Byrd II, a Virginian planter known to his great-uncle, Sir Robert Southwell, from the Royal Society. Southwell, who had directed Perceval's education, intended that the young man 'should first make a progress round England, that after seeing what was most remarkable at home in arts and natural curiosities, he might be better qualified to make observations abroad, and more welcome to the conversation of foreigners'.[145] Southwell's diplomatic protégé, Francis Parry, similarly recommended to Perceval 'some observations … relating to your own country which in no wise you ought to be a stranger to, when you launch out into foreign parts'. His 'Survey' of England would serve as 'a good foundation for the conduct of your whole life'.[146] More pressingly, the journey was supposed to preface continental travel, as only Southwell's death in 1702 prevented earlier departure for the Netherlands. Without his mentor, Perceval 'was a considerable time in suspence whether to comence his travells or stay at home till he had more knowledge of the World, whereby he might be better received abroad, and be more capable of making such observations as at his return might give him Credit'.[147] In the end, he would not travel until 1705.

Perceval and Byrd recorded their tour in letters and a journal.[148] Between July and October 1701, they moved from London, through East Anglia, up the east coast to York and Newcastle. After a short foray into Scotland, from which they retreated after a night of stereotypically bad accommodation, they travelled down the west coast through Chester and the Midlands to London.[149] Evidently, the pair cancelled a projected south-western leg.[150] Perceval used the journey to reconnect with family history, following instructions from the antiquary Peter Le Neve on the Norfolk roots of the Southwells. He and Byrd also enjoyed the hospitality of local clergymen and notables, though they were never received by landowners of a higher standing than Perceval's own baronetcy. When they did visit noble seats such as Chatsworth, it was as tourists.[151] Byrd prepared Perceval, nine years his

145 BL, Add MS 47072: Perceval, Autobiography, pp. 6–8.
146 BL, Add MS 47025, fo. 42v: Francis Parry to Sir John Perceval, 24 Jun 1701.
147 BL, Add MS 47072: Perceval, Autobiography, p. 14.
148 *The English Travels of Sir John Percival and William Byrd II: The Percival Diary of 1701*, ed. Mark R. Wenger (Columbia: University of Missouri Press, 1989).
149 BL, Add MS 47025, fos 54v–55v: Sir John Perceval to Digby Cotes, 18 Sep 1701.
150 BL, Add MS 47025, fos 42v–43: Francis Parry to Sir John Perceval, 24 Jun 1701.
151 *English Travels*, pp. 15, 151–7; BL, Add MS 47025, fos 45v–53: Peter Le Neve to Sir John Perceval, 24 Jun 1701.

junior, assuring Southwell that 'We are not negligent in our observations, but lay about us for matter to put into our journall.' The pair used print for their itinerary and descriptions, drawing on Thomas Smith's *De Republica Anglorum*, William Camden's *Britannia* and James Brome's recent *Travels over England, Scotland and Wales*. Smith offered 'general information' and Camden greater detail, but Perceval and Byrd preferred their own journals—'like to swell into a volume'—to the offerings of 'the florid Mr. Brome'. Their pursuit of the 'seven wonders' of the Peaks, including Chatsworth, also shows the influence of seventeenth-century poetry by Thomas Hobbes and Charles Cotton.[152] Domestic travel rehearsed the key skills of observation and description, then, but long hours on the road also offered the chance to practise French for the journey ahead.[153]

The younger Edward Southwell, Perceval's second cousin, has left fewer letters but a similarly detailed journal for his English travels of 1724.[154] The journey occupied a more sophisticated position in Southwell's education, falling between a voyage to the Low Countries, Germany, France and Lorraine undertaken with his father, the elder Edward, and a journey to Italy prosecuted with a Westminster schoolfellow. The elder Edward had returned his son to Oxford after their shared trip, intending another once he had 'digested' the first. He then sent him 'to see every thing between London & Yorke' in the summer of 1724, deeming it 'necessary to be acquainted with his own Country: before he takes another Voyage abroad'.[155] The 19-year-old set off from London with another second cousin, Wyndham Knatchbull, touring noble seats in the East Midlands, East Anglia, Lincolnshire and Yorkshire but mentioning few encounters with their owners, dining often with clergymen. He wrote a standard description of Cambridge.[156] The domestic journal ends abruptly in Yorkshire, but its author was well prepared, and his father reassured, when he set out for Italy the following spring. The younger Edward's only son, a third Edward, in turn made a similar journey through

[152] *The Correspondence of Three William Byrds of Westover, Virginia, 1684–1776*, ed. Marion Tinling, 2 vols (Charlottesville: University Press of Virginia, 1977), i, pp. 210–13: William Byrd II to Sir Robert Southwell, 17 Jul 1701; same to same, 26 Jul 1701; same to same, 5 Aug 1701; *English Travels*, pp. 18–21, 146–51; Thomas Smith, *De Republica Anglorum* (London: 1583); William Camden, *Britannia* (London: 1586); James Brome, *Travels over England, Scotland and Wales* (London: 1700).

[153] *Byrd Correspondence*, i, p. 210: William Byrd II to Sir Robert Southwell, 17 Jul 1701.

[154] Brotherton Library, YAS/MS328: Edward Southwell II, Account of my Journey Begun 6 Aug[us]t 1724.

[155] LLRRO, DG7, Bundle 28/75: Edward Southwell to [Earl of Nottingham], 26 Nov 1723; TCD, Lyons MSS 1995–2008, fo. 2114: same to William King, 3 Sep 1724.

[156] Brotherton Library, YAS/MS328: Southwell, Account of my Journey.

northern England and lowland Scotland over the summer of 1755, during his time at Cambridge and before his continental voyage of 1758–61. He travelled with his university tutor and made the most of family connections along the way.[157]

Itineraries followed in preparation for continental travel played an unexamined role in the rise of domestic tourism. They had combined natural curiosities, historical sites and landed estates since at least the 1670s, where most historians find such routes only in the eighteenth century. Common examples of early country-house visiting or a nascent 'home tour' are drawn from the same class, even the same families, as those who travelled through Britain with the Continent in mind. They relied on the same trends, such as a creeping willingness among seventeenth- as well as eighteenth-century landowners to open their homes to genteel visitors, and on the same antiquarian and literary texts.[158] Often only the purposes with which Sir John Perceval, Edward Southwell or the Brydges brothers travelled through Britain set them apart, as they tested tutor–pupil relationships, practised observation and description, gathered conversational material and acquired a frame of reference for the Continent. Not everyone toured Britain before travelling abroad, and it was certainly cheaper to head straight for France or the Netherlands, but domestic travel joined the range of options from which families assembled educational courses. The determination of many to rehearse continental travel in Britain, not least by keeping a record of their experiences, created some of the best examples of the early 'home tour'. It was useful for contemporary authors of guides to Britain to set themselves up against continental voyages, just as it is for their historians, but domestic tourism developed as much in collaboration with foreign travel as in opposition to it.

Conclusion

Gentlemanly education moved through several stages in the seventeenth and eighteenth centuries. Boys were highly mobile from a young age, between childhood homes, boarding schools and the houses of close relations, and

[157] Bristol Archives, 45317/2/5/1/5–16: Katherine Southwell to Edward Southwell III, 3 Jun–19 Aug 1755.

[158] Felicity Heal, *Hospitality in Early Modern England* (Oxford: Clarendon, 1990), pp. 208–9; Anderson, *Touring and Publicizing*, pp. 64–6. Cassandra Willoughby, stepmother to John and Henry Brydges, often features as an example of domestic tourism: p. 67.

they accustomed themselves early on to the rhythms of travel. Families assembled educational courses from a variety of options, but those who travelled abroad tended to get there via some or all of university, a semi-structured period in London and a journey through parts of Britain. Each of these stages could serve as the final step towards adulthood, but they shaped the foreign experiences of those who continued their education across the Channel. Students discussed the theory and practice of travel at Oxford and Cambridge, readying themselves by improving their French. Prospective travellers who attended the Inns of Court or, from the later seventeenth century, the French-run schools and academies of London gained a managed introduction to metropolitan society and training in gentlemanly accomplishments, with a view to preparing them for time abroad. They undertook domestic travel, too, with the Continent in mind. Foreign travel was the highest point in a wider culture of educational mobility, within which families found preparations and alternatives. Each stage shaped the experiences of those who ventured abroad.

School, university, London institutions and domestic travel created a sense of 'home' against which young men could judge what they encountered across the Channel. This process is clearest among those who divided their lives and possessions between England and Ireland. The elder Robert Southwell bragged of his grandson, Sir Philip Perceval, who had attended London schools, Oxford and Lincoln's Inn before impending continental travel, that he was 'well stockt & furnisht with the learning of England'.[159] Sir Philip's uncle then sent him on an English tour, to afford him 'some Standard in your mind for the like things, when you observe them abroad' and, he told his tutor, to 'preserve that Value for his Country which is due unto it'.[160] These families consciously followed English practice: Sir Robert Southwell suggested that Sir Thomas attend Foubert's academy before travel, 'as many Englishmen doe', while Francis Parry told Sir John Perceval, who had lived in England from the age of 3, that a preparatory tour of his 'own country' would further 'the design to live like Englishmen'.[161] This design promised social advancement, as Robert Molesworth felt that 'neither the improvement of our children nor their genius' would suit an Irish education. Sir Alan Brodrick sent the younger Alan from Eton to Cambridge precisely because 'Whoever proposes to come to any thing considerable hereafter must

[159] BL, Add MS 46952, fo. 240: elder Robert Southwell to Sir John Champante, 7 Dec 1675.
[160] BL, Add MS 46953, fo. 103: Sir Robert Southwell to Sir Philip Perceval, 5 Jun 1676; NLI, MS 7863, fo. 34: same to Jean Gailhard, 5 Jun 1676.
[161] BL, Add MS 47025, fo. 42v: Francis Parry to Sir John Perceval, 24 Jun 1701.

make England the theatre he resolves to appear on'. He saw many 'little fellows' from Ireland who 'by only spending some of their time there make themselves thought to be men of consequence'.[162] Education in London and the south-east was bound up with ethnic self-identifications, as these families began a process of buying into the English elite that many would complete on the Continent.[163] That said, these differences encouraged them to voice a more widely held desire to lay foundations for foreign travel. Cohen observes the construction of the English gentleman in France, but this process began with the acquisition of shared knowledge, networks and experiences at home, for the Ingrams of Yorkshire and the Leighs of Gloucestershire as much as for those of Anglo-Irish background.[164] Young travellers ventured abroad as Englishmen but, rather than coming automatically, that sense of origin required years of domestic mobility.

Travel journals often begin as young men leave London for the coast, but that should not prevent historians from looking beforehand for explanations of how they understood their journeys. This chapter has embedded time abroad in longer courses of domestic and foreign mobility, adding a new dimension to ongoing efforts to take the educational aspects of travel more seriously. The contribution of these courses to British elite formation must, in turn, be seen as a whole. If elite formation can often seem impersonal, meanwhile, attention to family strategy puts faces to the process. Educational history usually focuses on major institutions, but their significance was determined by parents, guardians and pupils. Family choices also bring to light smaller institutions, especially the schools and academies of London, that demonstrate how far domestic educational mobility depended upon the transnational movement of French professionals. These 'mediators' helped to make the English capital an underappreciated aspect of elite British travel culture.[165] Finally, divisions between domestic and foreign travel have

162 NLI, microfilm P.3752: Robert Molesworth to Laetitia Molesworth, 29 Jul 1709; SHC, 1248/4/53: Sir Alan Brodrick to Thomas Brodrick, 27 Jul 1717.

163 Toby Barnard, *Making the Grand Figure: Lives and Possessions in Ireland, 1641–1770* (New Haven, CT, and London: Yale University Press, 2004), pp. 310–45; Richard Ansell, 'Educational Travel in Protestant Families from Post-Restoration Ireland', *Historical Journal*, 58:4 (2015), 931–58. On Protestant and Catholic noblemen, see Jane Ohlmeyer, *Making Ireland English: The Irish Aristocracy in the Seventeenth Century* (New Haven, CT, and London: Yale University Press, 2012), pp. 435–46.

164 Cohen, 'Grand Tour: Constructing the English Gentleman'.

165 Ansell, 'Foubert's Academy'; Gallagher, *Learning Languages*, pp. 14–54; Michael Green, 'Bridging the English Channel: Huguenots in the Educational Milieu of the English Upper Class', *Paedagogica Historica*, 54:4 (2018), 389–409. On mediators, see Stedman, *Cultural Exchange*, p. 259.

proved untenable, with 'home tours' developing as preparations for 'Grand Tours' as much as in opposition to them. The term 'Grand Tour' gives a monolithic impression of foreign travel that sits uneasily with the tailored domestic education outlined here, but Chapter 2 explains how families similarly fashioned continental voyages to suit their means, ambitions and obligations.

2

Finances, Social Standing and the 'Grand Tour'

Countless publications on the Grand Tour make clear that educational travel was an expensive undertaking. If something is so costly, there is always a point at which people must decide whether they can afford it and, if not, whether they can stretch to a cheaper approximation. For the last hundred years, though, most historians have presumed that the Grand Tour was what the rich did, a package that families could or could not afford. Where they have considered money, they list spending on artwork, food, accommodation and transport or trace developing lines of cash or credit.[1] This chapter places the expense of a recognisable Grand Tour more centrally, as a way for wealthy families to perform their standing and an adaptable ideal to those below them in society. Some undertook conspicuous Italian travel while others—often of lesser means—used alternative forms of continental education. Debates over whether young men *should* travel were academic by the late seventeenth century, even though they continued to take up a misleading amount of print. An enduring question for families was whether they *could* travel, with time abroad a major factor in educational strategy and household finance. What did this decision involve at a family level? And what does it reveal about the relationship between travel and social mobility, or indeed stability? This chapter finds answers in the letters and accounts of the Ingram, Rawdon, Brydges, Leigh and Molesworth families, who provide several generations of foreign education, subject to changing

[1] William Edward Mead, *The Grand Tour in the Eighteenth Century* (Boston, MA, and New York: Houghton Mifflin, 1914), pp. 3, 170–206; Black, *British Abroad*, pp. 83–110, especially p. 106. For hints of nuance, see G. E. Mingay, *English Landed Society in the Eighteenth Century* (London: Routledge & Kegan Paul, 1963), pp. 137–8, 141; Rosenheim, *Emergence of a Ruling Order*, p. 44; French and Rothery, *Man's Estate*, p. 145. For perspective on spending, see Antoni Mączak, *Travel in Early Modern Europe*, trans. Ursula Phillips (Cambridge: Polity Press, 1995), pp. 72–94.

means, ambitions and obligations. They reconnect travel to questions of social mobility, landed self-identity, generational change and elite formation.

Since noblemen and gentlemen were expected to live up to their standing, conspicuous travel was one way in which families could maintain or acquire pre-eminence. A journey through France and Italy helped the 18-year-old ninth Earl of Derby to restore the Stanley name, which had suffered from the execution of his royalist grandfather and the sequestration of ancestral lands in 1651. His father, the eighth earl, was imprisoned after an uprising in 1659 and met with mixed success in recovering estates after the Restoration. The ninth earl represented the fourth generation to undertake educational travel, dating back to the 1580s, and time abroad was essential to reviving the family reputation.[2] Derby married the Duke of Ormond's granddaughter immediately before his travels of 1673–5, during which Ormond worked 'to make his fortune fitt for his Quality at his Returne'. Noble life was cheaper abroad, but travel would do more than save money. It displayed family standing, while providing materials with which heirs could raise or sustain it. If Derby were to live 'in some proportion with his forefathers', the 'Improvement of his person' and the acquisition of 'parts sutable to his condition', including the French language, would be as important as 'the good Management of his fortune'. When crop failure and the Third Anglo-Dutch War hit revenues from the Isle of Man, Ormond allowed Derby £100 a month of his own credit, 'something more' than the Stanley estate would bear. He gambled that travel would, as the earl's tutor put it, help 'to Restablish a family almost lost'.[3]

Educational travel was widespread among the gentry as well as the nobility, even if few families could match the prestige of the Stanleys or Ormond's financial muscle.[4] When James Brydges, the Duke of Chandos, discovered that his eldest son was wildly overspending on the Continent,

[2] Leo Daugherty, 'Stanley, William, sixth Earl of Derby (bap. 1561, d. 1642)', *ODNB*, online edn, 2018 (https://doi-org/10.1093/ref:odnb/72296); Barry Coward, 'Stanley, James, seventh Earl of Derby (1607–1651)', *ODNB*, online edn, 2008 (https://doi-org/10.1093/ref:odnb/26274); John H. Rains, III, 'Stanley, William George Richard, ninth Earl of Derby (1655–1702)', *ODNB*, online edn, 2004 (https://doi-org/10.1093/ref:odnb/40803); Coward, 'The Social and Political Position of the Earls of Derby in Later Seventeenth-Century Lancashire', *Transactions of the Historic Society of Lancashire and Cheshire*, 132 (1982/3), 127–54, at 138.

[3] Bodleian, MS Carte 50, fo. 103: Ormond to James Forbes, 7 Aug 1673; fo. 109: same to same, 18 Sep 1673; fos 235–6: Ormond to Thomas Fairfax, 6 Feb 1674; 219, fo. 123: same to Mr Cholmondley, 11 Aug 1673; 243, fo. 93: Fairfax to Ormond, 17 Mar 1673/4; fo. 105: same to same, 20 May 1674.

[4] Ormond's annual income was £24,000: Toby Barnard, *Irish Protestant Ascents and Descents* (Dublin: Four Courts, 2003), p. 49.

he warned that 'if Nobody was to travell, but who spent after the rate of 2000£ per Annum ... there would be very few young Gentlemen go abroad'.[5] This was indeed an extravagant sum. Between the mid-seventeenth and mid-eighteenth centuries, the gentry and nobility developed a remarkable consensus that a sensible traveller would spend £300 a year. If historians continue to focus on the costliest journeys, confining their attention to those who saw Italy on their parents' money, they will indeed find fewer young gentlemen than went abroad. To extend the military metaphor implied by family strategy, the gentry used a range of tactics to secure continental experience. Parents and guardians delayed journeys to lessen pedagogical expenses; they rationed travel according to birth order, aptitude or even the inclination of sons; they borrowed money to keep young men abroad; they altered the destinations, routes and activities of an ideal Grand Tour, recognising that standing still was cheaper than remaining in motion; they shortened periods on the Continent, allowing time only for the Netherlands or France. Kin facilitated foreign travel, as they had domestic education, by contributing funds, connections and expertise, and kinship networks increasingly overlapped with the military and the diplomatic service, offering other routes to continental experience. These tactics stretched familiarity with the Continent beyond those who could, on the face of it, afford time abroad. They ensured that travel helped to make, as well as reflect, the social elite.

The Ingrams: Stability through Travel

The Ingrams, Viscounts Irwin of Temple Newsam, were of more recent vintage than the ancient Stanleys, but they similarly demonstrate the centrality of travel to the standing and self-image of a noble family. Financial ruthlessness and three marriages had allowed Sir Arthur Ingram, a London merchant who began his career in Turkey and Italy, to build an estate in Yorkshire and Lincolnshire worth £9,000 a year by his death in 1642. Other family members held property worth another annual £3,000. These sums allowed the Ingrams to join the longer-established Saviles at the top of Yorkshire society.[6] Rental income had fallen from over £5,500 in

[5] HL, mssST57, vol. 20, fo. 99: James Brydges to Charles Stuart, 6 Feb 1721/2.

[6] Anthony F. Upton, *Sir Arthur Ingram, c. 1565–1642: A Study of the Origins of an English Landed Family* (London: Oxford University Press, 1961); J. T. Cliffe, *The Yorkshire Gentry from the Reformation to the Civil War* (London: Athlone Press, 1969), p. 30; John P. Ferris and

1659 to £4,500 in the mid-1660s and settled around £4,000 in the 1670s and 1680s, before another slight fall to just over £3,500 at the turn of the eighteenth century.[7] These figures underestimate annual totals, as they ignore other sources of income, but they still place the Ingrams around the mid-point for spending power in the English peerage, despite the unusual fall in their rentals.[8] As a successful newcomer, Sir Arthur had secured his position through office-holding, competitive building and opulent living, but educational travel was also central to his family's longer-term establishment among the nobility.[9]

The Ingrams travelled in three successive generations between the 1650s and the 1710s, through traces of their early journeys are few. Two of Sir Arthur's grandsons, Henry, the future first viscount, and Arthur, left for France in mid-1655, in their mid-teens. If the age difference was not too great, noble brothers often travelled together throughout the early modern period, as even the wealthiest families recognised economies of scale.[10] Henry and Arthur may have travelled with Thomas Cordell, an experienced tutor of exiled royalists, but were certainly accompanied by George Townshend, whose wages of £100 a year turned into a generous annuity. Henry may have returned for marriage negotiations in 1657, with Arthur still abroad on £150 a year, but both were recalled from France in early 1660.[11] In the next generation, Henry's sons, Edward and another Arthur, travelled to France and Italy in the late 1670s and early 1680s.

Simon Healy, 'Ingram, Arthur (*c.* 1565–1642)', in Andrew Thrush and John P. Ferris, eds, *The History of Parliament: The House of Commons, 1604–1629* (Cambridge: Cambridge University Press, 2010).

7 WYAS, WYL100/EA/15/6: General Rental, 1659; WYL100/EA/15/12: Summary Rents from Yorkshire and Lincolnshire Properties, 1666; WYL100/EA/15/13: General Rental During Minority of Edward, 2nd Viscount Irwin, 1666–84; WYL100/EA/15/22: Yorkshire Rentals, 1688–96; WYL100/EA/15/23: Yorkshire Rentals, 1696–1707; WYL100/EA/15/24: Lincolnshire Rentals, 1697–1710.

8 Cannon, *Aristocratic Century*, pp. 129–32, though Viscount Irwin/Irvine was a Scottish title. Rents stagnated in the century after 1640: Rosenheim, *Emergence of a Ruling Order*, p. 49.

9 Stone and Stone, *Open Elite?*, pp. 303, 399.

10 For the evidence of passports, see John H. Ghazvinian, '"A Certain Tickling Humour": English Travellers, 1560–1660', D.Phil. thesis (Oxford, 2003), pp. 110–11.

11 WYAS, WYL100/C/20, fos 366–81: Thomas Ingram to Henry Ingram, [*c.* late 1650s]; WYL100/C/9A, George Townshend to John Roades, 4 Jul 1702; WYL100/EA/12/17/1: Stewards' vouchers for education, 1623–1744; *Calendar of State Papers, Domestic Series, [of the Commonwealth, 1649–60]*, 13 vols (London: Longman & Co. and Trübner & Co., 1875–86), viii, p. 575: pass for Thomas Cordell and Henry and Arthur Ingram to go to France, 19 Oct 1655; xiii, p. 561: pass for John Everard to France, to bring over Henry and Arthur Ingram, 7 Feb 1659/60; Geoffrey Smith, *The Cavaliers in Exile, 1640–1660* (Basingstoke: Palgrave Macmillan, 2003), pp. 75–6.

They took a younger brother from the neighbouring Stapylton family as their tutor. Again, little survives: they were at Saumur in mid-1678 and still abroad in late 1681, but only Edward was at Padua in May 1683.[12] Traces of these two generations emerge from the letters of another Edward, the fourth viscount, eldest son of the younger Arthur and the best documented traveller in the family. If travel had helped to establish the Ingrams over the late seventeenth century, it was necessary to maintain their standing by the time of the fourth viscount's voyage.

The family maintained its tradition of sending sons abroad together in 1704, two years after the third viscount's death. The fourth viscount crossed the Channel at 17 with his 16-year-old brother Rich and their tutor John Haccius, as well as Thomas Worsley, the 17-year-old son of one of Irwin's trustees. They settled in Delft and prepared for the University of Leiden, where they began a few months later, but soon departed the Netherlands after Irwin had fought a duel. He and Haccius spent several months in Germany but eventually set off for Italy, reaching Venice by May 1705. Rich and Worsley meanwhile returned home, the former directly and the latter after a separate period in Germany.[13] Irwin and Haccius were at Siena for four months and Rome for three, before returning to Siena and proceeding to Lucca. They probably abandoned plans to return to Rome, as they passed through Genoa, Florence and Bologna on the way back to Venice, which they had reached by January 1707. The journey home is less clear, amid reports that Irwin had turned back towards Florence, but he returned to England that autumn after three and a half years abroad.[14]

Plenty of evidence survives for the pressures that the fourth viscount's voyage placed on his estate, given the need to explain himself to those who held the purse strings. Irwin's letters between the ages of 17 and 20 chart a declining relationship with the trustees of his father's will, especially the elder

12 BL Add MS 46955B, fo. 44v: Lady Perceval to Sir Robert Southwell, 20 Jul 1678; WYAS, WYL100/EA/12/17/2: receipt for bill of exchange and letter of credit, 8 Sep 1681; Horatio F. Brown, ed., *Inglesi e Scozzesi all' Università di Padova dall' anno 1618 sino al 1765* (Venice: 1921), p. 239.

13 WYAS, WYL100/C/9A: John Haccius to John Roades, 18 Jun [1704]; Irwin to Dowager Viscountess Irwin, 19 Sep [1704]; Thomas Worsley to Roades, 5 Mar 1704/5; WYL100/C/9B: Haccius to [Roades], 16 Dec 1704; WYAS, WYL100/C/9B: Worsley to Roades, 30 Jun 1705; WYL100/C/10: Haccius to same, 9 Feb 1705; Richard Hey to Dowager Viscountess Irwin, 15 May 1705.

14 WYAS, WYL100/C/9A: Irwin to John Roades, 27 Aug 1705; same to same, 5 Dec 1705; same to same, 25 Feb 1706; WYL100/C/10: John Roades to Irwin, 8 Jan 1706/7; Thomas Worsley to Roades, 18 Jan 1706/7; WYL100/C/20, fo. 160: Irwin to Dowager Viscountess Irwin, 9 Apr [1706]; fo. 185: same to same, 9 Jun [1706]; fo. 242: same to same, 14 Aug [1706]; WYL100/EA/10: Thomas Palmer to Roades, 10 May 1707.

Thomas Worsley and John Roades, the estate steward. These men could do little to frustrate the viscount's desires as he neared his majority, however, particularly when they aligned with those of his mother. Townshend, another trustee, renounced 'the using of any Coessive power', recognising 'his Lordships great power which he knows now very well', even at 17. Worsley excused his 'ffreedome' in hoping that Irwin would be a good financial 'manager' and Roades's drafts are similarly servile.[15] The trustees could only try to persuade Irwin into prudence by appealing to the honour of his family, the precedent of his ancestors and the health of his estate. Irwin occasionally justified spending with reference to educational improvement, but he more often resorted to the language of personal honour, even when he invoked earlier generations. At an age at which adolescent noblemen and gentlemen were increasingly sensitive to such questions, Irwin saw the financing of his travels as a test of his independence and authority.

Expense had always been a central issue, informing the shape and timing of Irwin's travels. Though the Ingrams had no portions to find for daughters, the cost of educating eight boys had worried the third viscount before his death.[16] Soon afterwards, Townshend, then 80 years old, drew on his experience of travel to 'compute' its likely cost for the Ingram heir. The first viscount had spent just over £700 a year abroad in the 1650s, though Townshend warned that travel would now be 'more Chargable and more daingerus'. The trustees agreed that the sensible option, amid debts and the need to provide for his brothers, was first to send the fourth viscount for two years at Cambridge. Eventually, on the eve of Irwin's travels, Townshend reminded him that a speedy marriage upon his return would secure the estate, pay for his brothers and place him 'in as good a condition as any of his predecessors'.[17]

The 17-year-old initially moved carefully abroad, though Haccius warned the trustees that it would be 'necessary to live honourably'—that is, expensively—during Irwin's 'first step into the world'.[18] Roades, by contrast, begged Irwin not to spend too freely in the Netherlands, reminding him that his uncle had supported three servants and a tutor on £500 a year and that his father had spent under £800 in three years abroad, 'as appears by the

[15] WYAS, WYL100/C/9A: George Townshend to John Roades, 29 Apr 1704; [Thomas Worsley] to Irwin, 13 Nov 1704.

[16] *'A Little News from Horsham': Some 18th Century Letters from Hills*, ed. Susan C. Djabri (Horsham: Horsham Museum Society, 2002), p. 14.

[17] WYAS, WYL100/C/9A: George Townshend to John Roades, 4 Jul 1702; same to same, 7 Jul 1702; same to same, 29 Apr 1704.

[18] WYAS, WYL100/C/9A: John Haccius to John Roades, 3 Oct 1704.

Governors booke of Accounts now in my Custody'. If the young viscount kept similar records, they would prove useful to 'others herafter, as yor unkles & ffathers are to us now'. Although Irwin argued that his father and uncle were misleading precedents, 'since I never heard they had any thing but servants about them', Roades knew that they had taken Stapylton as their tutor.[19] Relations remained cordial, but the lines were already drawn between personal honour and financial prudence, and in the battle over family memory.

Irwin became more demanding as he left the Netherlands. He wanted his Dutch allowance of £600 a year to be raised to £800 for Italy, though he promised 'not to ask more in his Travels'.[20] Since the trustees recognised that the journey south would be more expensive, Worsley argued that 'the later you Travell the better'. He suggested that the viscount might gain more from the voyage at an older age, but 'a great reason at present' was 'the scarcity of money' occasioned by land purchases and lawsuits related to the estate.[21] Irwin nevertheless received the increased sum and more, knowing that his mother would not leave him 'seeing how I might improve and not having where withall to do it'.[22] The trustees' financial deliberations, which forced the viscount to wait in Germany rather than proceed to Venice, sowed the seeds of disharmony. When a Dutch banker hinted that overspending would see him recalled to England, Irwin was astounded that 'Trustees should pretend to so much Authority over a man of my age, as to stop all payments and command me so absolutely to come back, when they know I can't in honour do it.' Now 18, he began to wonder 'whither they will be my Enemies, or Friends'.[23]

Irwin found his answer in the Anglophone communities of Italy, where his sense of personal honour widened divisions with those at home. Further delays in funding soon forced him to leave Venice for Padua, having 'by my poverty expos'd my self to the whole English Nation here'. Fellow travellers, including the Duke of Shrewsbury and the Yorkshire gentlemen Thomas Frankland and Metcalfe Robinson, had, 'knowing me[,] wondered att my

19 WYAS, WYL100/C/9A: John Roades to Irwin, 16 Aug 1704; Irwin to Roades, [Aug 1704]; Roades to Irwin, 6 Sep 1704.

20 WYAS, WYL100/C/9A: John Haccius to John Roades, 28 Nov 1704; WYL100/C/9B: same to [same], 16 Dec 1704.

21 WYAS, WYL100/C/9A: [Thomas Worsley] to Irwin, 13 Nov 1704.

22 WYAS, WYL100/C/20, fo. 97: Irwin to Dowager Viscountess Irwin, [Dec 1704]. Irwin received £970 between June 1705 and April 1706: WYL100/C/10: Thomas Palmer to John Roades, 1 Aug 1706.

23 WYAS, WYL100/C/20, fo. 161: Irwin to Dowager Viscountess Irwin, [early 1705].

Circumstances', shocked 'to see me so miserable'.[24] Such claims were self-serving, calculated to play on concerns about the opinion of neighbours, but they also carried some truth. Robinson indeed told his father about Irwin, worrying about how 'this conduct or misfortune may change him', and the sentiment clearly circulated in Yorkshire.[25] The Ingrams and their employees remained attuned to the reports of fellow travellers, whether they gave 'pitifull' accounts or 'mighty advantageous Characters' of the young viscount.[26] Irwin told the same story of embarrassment from Rome, which he also left prematurely, 'not onely because it was too expensive a Place for me but also because several private Gentlemen lived much more like a Lord then my self'. He had been 'inferior in everything' to another Yorkshireman, John Bourchier, 'who has though but a Gentleman, a greater allowance than I'. Irwin claimed to be no 'lover of a vain show and Equipage' but stressed the need 'to live as a man of my Quality in a foreign Country (where it does not cost so much)', especially 'when I know I have where withall to live so in my own Country where it costs considerably more'. At present, however, Siena, Lucca and other 'little towns' were more 'fitted' to his purse than Rome.[27]

Irwin compared himself to his predecessors as well as his contemporaries, claiming that his small allowance had amazed 'every body, who knew my Unkle' in Rome. The second viscount had had 'not onely a Coach and six Horses; but also a Palace, open table, and all the great Company in Town'. His nephew, however, cut a different figure, 'though I am as capable to make as great a one as he did, having the same estate'. Roades was better placed to remember that the second viscount's travels had been 'fatall, for he contracted so much Debt upon himself by it, that after his return, he scarce ever enjoyed himself so long as he Lived'.[28] In truth, too, Irwin did not have the same estate, which had suffered from declining rents since the mid-seventeenth century.[29] Stories about the preceding few generations bolstered collective identity in elite families, but the Ingrams demonstrate that they could also generate friction, particularly when they concerned the expense

[24] WYAS, WYL100/C/9A: Irwin to [John Roades], 23 May [1705]; WYL100/C/20, fo. 163: same to Dowager Viscountess Irwin, 30 May [1705].

[25] WYAS, WYL150/6005/46: Metcalfe Robinson to Sir William Robinson, 21 May [1705].

[26] WYAS, WYL100/C/10: Thomas Palmer to John Roades, 19 Sep 1706; Irwin to Dowager Viscountess Irwin, 22 Jan 1706/7.

[27] WYAS, WYL100/C/9A: Irwin to John Roades, 5 Dec 1705; same to same, 25 Feb 1706; WYL100/C/10: same to Dowager Viscountess Irwin, 25 Feb [1706].

[28] WYAS, WYL100/C/9A: Irwin to John Roades, 5 Dec 1705; same to same, 12 Dec 1705; Roades to Irwin, 29 Mar 1706.

[29] The lack of inflation allows valid comparisons: Rosenheim, *Emergence of a Ruling Order*, p. 49.

of travel.[30] Where Roades had begged Irwin not to outspend his ancestors in the Netherlands, he and the viscount disagreed over the meaning of their precedent in Italy. The Ingrams saw themselves as a family who travelled, drawing on bequeathed texts and living memory, but opinion differed over how that identity should be maintained.

Irwin's frustrations in Venice and Rome threatened to break the relationship with his trustees, whom the 19-year-old now openly attacked. Although Worsley and Roades were 'so very carefull he should not spend a hundred Pound less or more, for fear of ruining himself', he reminded his mother that it was 'vain to tell me my estate wont bear it since I know well enough what I have'. Paying him quarterly, the usual method at school and university, rather than supplying a letter of credit, also left Irwin feeling 'like a child of Fifteen'.[31] He continued to complain on his way out of Italy. On his second stay in Venice, he met no 'English Gentlemen who travells with his man, but what has six or seven hundred Pounds a year, which is much more then I have, who must have more servants than that'.[32] These claims seem unlikely, given the consistency with which others mention £300 as a benchmark allowance, but they probably reflect Irwin's status anxiety as well as his hopes for financial gain. Far from Temple Newsam, his standing and his adolescent masculinity hinged upon his spending power among peers in Venice and Rome.

Where Irwin mobilised honour and precedent in defence of his authority and financial independence, the trustees tried to limit his spending by warning of its consequences for the estate. As the viscount settled into his first taste of Venice in mid-1705, Roades had warned that he was 'Just begining yor Lordshipps receipt of rents, but never mett with so generall an outcry of scarsity of money in my whole time'. He sent Irwin's quarterly £200 to Lucca the following year but complained of rains, flooding and plummeting corn and cattle prices in northern England, causing 'such sad times for Tenants to raise money … as the like was never known in my time'. The 'badness of the times' forced him to borrow in order to make up the delayed allowance.[33] Rental books reveal that income dipped in 1705, from the usual £3,500 to only £2,756, and struggled to recover over the next two

30 Henry French, 'The "Remembered Family" and Dynastic Senses of Identity among the English Gentry *c.* 1600–1800', *Historical Research*, 92:257 (2019), 529–46.

31 WYAS, WYL100/C/20, fo. 160: Irwin to Dowager Viscountess Irwin, 9 Apr [1706]; fo. 242: same to same, 14 Aug [1706].

32 WYAS, WYL100/C/10: Irwin to Dowager Viscountess Irwin, 22 Jan 1706/7.

33 WYAS, WYL100/C/9A: John Roades to Irwin, 23 Jun 1705; WYL100/C/10: same to same, 25 Jul 1706; same to same, 18 Sep 1706.

years.[34] From a financial perspective, Irwin's travels could not have come at a worse time. He laid out £3,700 in the course of his three and a half years abroad, however, of which £1,800 came in the final nine months.[35] The trustees reminded Irwin of his domestic revenue and expenditure, warning that he had 'sunk much of his Annuall Income'; indeed, £1,800 represented 80 per cent of the total rents for those nine months in 1707. In that year and the next, as Irwin spent freely during his second stay in Venice, Roades even squeezed an additional £313 from the Lincolnshire lands, beyond the biannual rental payments, for his employer's use abroad.[36]

Among themselves, the trustees lamented Irwin's adolescent excess. Worsley feared that the viscount had followed 'most of our English youth Abroad', who were 'Extravigent & lives very High'. In a letter to Roades, he punctured Irwin's defence of his spending: 'whear Lyes the Advantage? Will extravigent Liveing Abroad, make him more Easie or Increase his Intrest att Home? I [do not] doubt the Contrary. And am sure they're noe ffriends which Advise itt'. As Irwin became more assertive in Italy, Worsley felt that he was treating 'his Trustees like Slaves (if not Children)', using 'ffare or fowle meanes to have what he Pleases'.[37] Outwardly, though, the trustees were more conciliatory as the viscount neared his majority and sent any money that he requested. Roades could do little more than remind Irwin of the cost of expanding his estate and educating his brothers, warning him 'not to stretch too farr' abroad. Only then could he 'Live as Honorably & Great, as any of yor Ancestors' upon his return, though Roades wisely erased 'but not otherwise' from his draft.[38]

At root, these clashes revolved around how travel related to honour. The language of honour regularly justified opposed courses in the period, but Irwin and his trustees differed over whether to prioritise the honour of the traveller or his family.[39] Was it best to support a young man abroad, where, removed from the country houses and transgenerational deference of

[34] WYAS, WYL100/EA/15/23: Yorkshire Rentals, 1696–1707; WYL100/EA/15/24: Lincolnshire Rentals, 1697–1710.

[35] Christopher Gilbert, 'A Nobleman and the Grand Tour: Lord Irwin and Marco Ricci', *Apollo: The International Magazine of Arts*, 51 (1966), 358–63, at 363, though Irwin (b. 26 Dec 1686) did not turn 21 at Venice. He entered his '21th year of Age': WYAS, WYL100/C/10: John Roades to Irwin, 8 Jan 1706[/7].

[36] WYAS, WYL100/EA/15/24: Lincolnshire Rentals, 1697–1710.

[37] WYAS, WYL100/C/9B: Thomas Worsley to John Roades, 11 Aug 1705; same to same, 6 Oct 1706; WYL100/C/10: same to same, 4 Jan 1705/6.

[38] WYAS, WYL100/C/10: John Roades to Irwin, 8 Jan 1706[/7]; Thomas Worsley to Roades, 18 Jan 1706/7; same to same, 15 Feb 1706[/7].

[39] Thomas, *Ends of Life*, p. 148.

home, his reputation rested more than ever on conspicuous consumption? Or should a family resist lavish spending on travel, in order to secure its financial stability and thereby its capacity to live 'Honorably & Great' in the future? Irwin felt that travelling well was an obligation for someone of his income and background. His exposure to fellow travellers and foreign hosts led him to emphasise personal honour, although he sometimes justified himself as projecting family status. Indeed, his sensitivity to individual reputation drew him into duels in the Netherlands and possibly Italy. The trustees were more concerned with the health of the estate, as befitted their role, and less worried by delicate late adolescent masculinity. They saw the costs of travel as threats to family standing rather than opportunities for its assertion.

Who was right? More than is usually recognised, financial stability was essential for the continuity of the elite.[40] At the same time, though, wealth served little purpose unless it allowed landowners to confirm their social position. Contemporaries understood that the pressure to do so through consumption was particularly strong in Britain and Ireland. When Irwin met some Genoese noblemen in Lucca, 'The knowledge they have of England made them, I believe, take me for a very poor inconsiderable person, which I can assure you troubles me very much.'[41] Away from home, esteem depended entirely on the individual figure that Irwin could cut. In his telling, the Genoese knew that spending power and social standing went hand in hand among the English, and communities of young travellers—with whom they must have been so familiar—offered a clear illustration. The volatility of continental experience in the families examined below suggests a balance needed to be struck, between living up to one's status abroad and ensuring that later generations could do the same. Throughout the period, it remained a central question in family strategy.

Clashes on this theme were common because young men travelled at an age at which they wanted to assert their independent authority.[42] That was part of the point, of course, but it made disagreements inevitable. Sir Robert Southwell sent his nephew, Sir Philip Perceval, abroad at 20 with a 'general rule of frugality ... having Brothers and Sisters to provide for' and an English estate to buy.[43] Perceval, however, increasingly resented the influence of his tutor, Jean Gailhard, as he came of age in France, feeling 'free to doe what he pleases being master of his person and estate'. He bought fine

40 Rothery and Stobart, 'Inheritance Events and Spending Patterns'.
41 WYL100/C/20, fo. 242: Irwin to Dowager Viscountess Irwin, 14 Aug [1706].
42 French and Rothery, *Man's Estate*, pp. 141–2.
43 NLI, MS 7863, fo. 34: Sir Robert Southwell to Jean Gailhard, 5 Jun 1676.

clothes without consulting Gailhard, who reported that his pupil thought himself 'no child now but knowes without being told what company is fitt for him to keep and how to govern himself'. A servant, Thomas, broadcast that his master 'was resolved to spend and to live high, having an estate able to beare it'. Soon, Perceval engineered Gailhard's dismissal.[44] The Frenchman surely had his most recent pupil in mind when, the following year, his book denounced untameable young men 'passed 20', who 'think themselves wiser than their Teachers'.[45] Perceval's uncle astutely judged that Gailhard's successor, Alexandre de Rasigade, should act 'as a favourite & not as a Tutor', doing his best to keep expenses around £300 a year and no more than £400. Southwell recognised that his nephew was 'now in that pitch of Age, and other Circumstances, which usually make young men think themselves capable of governing others and a sort of affront, to think that they need to be governed'.[46] The assertion of elite male independence, not least by spending for an audience of peers, was also the crux of the issue between Irwin and his trustees. It surfaces in the correspondence of countless educational travellers.

Battles over spending were not simply generational, though. In fact, many parents and guardians followed Lady Irwin in recognising the social value of living well abroad. James Brydges even argued on behalf of his nephew, William Leigh, that an insufficient allowance would 'damp his mind' and 'drive him into actions mean & unworthy of him'. Brydges felt strongly that Leigh needed the standard £300 a year.[47] Tutors generally agreed, as a decent budget was not only in their own interests but also avoided clashes that would always end, as Gailhard learned, in favour of a socially superior charge. As such, they echoed the language of their pupils. Derby's tutor informed Ormond that his young nobleman did 'not intend to be so extravagant' as some of the English then living 'verie high' in Paris, but 'it is verie expedient he should live decently and handsomly, as a person of his qualite ought to doe'.[48] Ormond felt the same way, as many parents and guardians believed that honour demanded appropriate spending. Trustees

[44] BL, Add MS 46954A, fo. 274: Jean Gailhard to Sir Robert Southwell, 27 Jun 1677; 46955A, fo. 32v: Southwell to Sir Philip Perceval, 24 Jan 1677/8.

[45] Gailhard, *Compleat Gentleman*, ii, p. 20; John Gallagher, 'Language and Education on the Grand Tour of Sir Philip Perceval', in Helmut Glück, Mark Häberlein and Andreas Flurschütz da Cruz, eds, *Adel und Mehrsprachigkeit in der Frühen Neuzeit. Ziele, Formen und Praktiken des Erwerbs und Gebrauchs von Fremdsprachen* (Wiesbaden: Harrassowitz Verlag in Kommission, 2019), pp. 113–33, at p. 133.

[46] BL, Add MS 46954B, fo. 3: Sir Robert Southwell to Alexandre de Rasigade, 9 Jul 1677.

[47] HL, mssST57, vol. 4, fos 7–8: James Brydges to Theophilus Leigh, 1 Jun 1710.

[48] Bodleian, MS Carte 243, fo. 87v: James Forbes to Ormond, 7 Aug 1673.

and estate managers were more reluctant to endorse high expenses, given their task of maintaining the estate and, perhaps, their lesser attachment to family self-perceptions. As with tutors, however, there was a limit to how far they could control heirs who were increasingly sensitive to authority. The appeals by Irwin's trustees to family honour and estate finance were common among men who held diminishing influence as their employers aged and asserted themselves; they were frequently as ineffective.[49]

A recurring theme for those advising caution, from Southwell to Townshend, was the need to provide for younger brothers. The question of how to deal with them generated anxiety throughout the period: given that the heir might die, but that younger sons would otherwise need to support themselves through respectable work, how should they be educated, and at what cost?[50] As this chapter will show, answers varied on whether they should travel. It is not known if Sir Arthur Ingram's eldest grandson, Thomas, spent time abroad before his brothers crossed the Channel in the 1650s, but the family's solution then becomes clearer. In the late 1670s and early 1680s, and again from 1704, the eldest two sons travelled, the first more extensively than the second. Irwin cast his voyage of 1704–7 as a preformed whole, trusting from Düsseldorf that his mother would agree with the 'design of going to Italy, to finish what I have begun'. He rightly assumed that he would be denied nothing 'for my educations sake'.[51] The shape of travel was less defined for Rich, whose allowance in the Netherlands was under half that of his brother. Haccius complained about teaching them together, since their 'Education cannot be the same', and Irwin tired of lending Rich money to appear respectably at The Hague. The viscount still felt that his brother had 'grown a Pretty Gentleman, and will with Travelling, which is absolutely necessary for him, make a good Man', but neither their mother nor Rich's trustees were 'for gratifieing Extravagant Humors'. They rejected Rich's request for a higher allowance, despite the sizeable inheritance that he expected from his maternal grandfather, and agonised for months over whether he justified the expense of another £100-a-year tutor.[52] While Lady Irwin agreed that Rich should have 'a Maintenance

[49] Anthony Fletcher, 'The Ambition of a Young Baronet: Sir Thomas Isham of Lamport, 1657–1681', in Angela McShane and Garthine Walker, eds, *The Extraordinary and the Everyday in Early Modern England* (Basingstoke: Palgrave Macmillan, 2010), pp. 31–47.

[50] Thirsk, 'Younger Sons'; French and Rothery, 'Male Anxiety'; Muir, *Gentlemen of Uncertain Fortune*.

[51] WYAS, WYL100/C/20, fo. 158: Irwin to Dowager Viscountess Irwin, 8 Jan [1705].

[52] WYAS, WYL100/C/9A: Thomas Worsley to John Roades, 2 Jul 1704; John Haccius to same, 3 Oct 1704; Richard Thornton to Dowager Viscountess Irwin, 26 Feb 1704/5; WYL100/C/10:

and education suitable to his Estate and Quality', the trustees would not contemplate such an expensive option. Rich came back to London after only eight months abroad.[53] Friends encouraged him into the army, which offered 'Knowledge in the World', and he returned to the Continent in a military capacity a few months later. Rich was marching through the Netherlands in April 1705 as Irwin made his way to Italy, the pair having only left Leiden before Christmas.[54] Educational travel was desirable for Rich, as a second son, and would pay off when he inherited upon Irwin's death in 1714, but his family never committed to the expense in the way that they had for his elder brother.

Travel formed no part of educational planning for the third viscount's six other sons.[55] When the fourth, Henry, travelled to the Netherlands and France at 25 in 1716–18, he did so outside the bounds of family strategy. The journey was his own idea and he funded it from personal savings, where Irwin's time abroad had been a matter for the whole family. Their father, the third viscount, had left £2,000 in trust for each of his younger sons, a sum that had grown to £3,803 by 1719, with interest of £570 projected for the next three years. Henry spent £828 abroad, living beyond the interest on his fortune, a guideline for educational investment laid down in his father's will and called to mind by Rich, then fifth viscount. He relied, where necessary, on the latter's credit.[56] Henry's movements only emerge from financial records and a few incoming letters, but he spent his early months touring Holland and learning French. He travelled to the Loire, studying for six months at the academy in Angers before a tour of France. Lady Irwin was 'mightyly against' Henry's plans for Italy, remembering the fourth viscount's profligacy there, but Rich promised to meet any bills he might incur. In the end, Henry turned

---

Richard Hey to Thomas Kitchingman, 9 Jan 1705; WYL100/C/20, fo. 98: Irwin to Dowager Viscountess Irwin, 31 Oct [1704]; fo. 350: Dowager Viscountess Irwin to Rich Ingram, [late 1704/early 1705].

53 WYAS, WYL100/C/9A: Richard Thornton to Dowager Viscountess Irwin, 19 Dec 1704.

54 WYAS, WYL100/EA/12/17/5: Earl of Halifax to Dowager Viscountess Irwin, 19 Dec 1704; WYL100/C/9A: George Hollings to John Roades, 17 Apr 1705; WYL100/C/9B: Thomas Worsley to same, 26 Dec 1704.

55 Arthur trained as a lawyer: WYAS, WYL100/C/10: Dowager Viscountess Irwin to John Roades, 3 Jan [1710]. John and Charles joined the army: Charles Dalton, ed., *George I's Army, 1714–1727*, 2 vols (London: Eyre & Spottiswoode, 1910–12), i, p. 293; ii, pp. 272, 304. George joined the Church: WYL100/C/12: Dowager Viscountess Irwin to Henry Ingram, 2 Aug 1721. William was apprenticed to an Amsterdam merchant but has left no evidence of further travel: WYL100/C/20, fo. 178: William Ingram to Dowager Viscountess Irwin, [1720s].

56 WYAS, WYL100/C/20, fo. 340: Rich Ingram to Henry Ingram, 19 Oct [1718]; WYL100/EA/12/17/8: Accounts for Henry Ingram, 1716–20; TNA, PROB 11/488/419: Will of third Viscount Irwin, 24 Jun 1706 [died 1702]. The will estimates annual interest at 5 per cent.

northwards after Bordeaux and Montpellier, making a long stay in Paris before coming home.[57] The journey had strong educational components, despite Henry's advanced age, but his brother and mother perceived it as a luxury on which they were happy for him to spend his own money. Perhaps Henry, who enjoyed himself in the company of other Englishmen, could now afford something that he had missed in adolescence, when he had not had control over his inheritance. Non-essential travel had no place in family strategy, reflecting the early modern view that young people needed only as much education as prepared them for their life roles, but landowners such as the Ingrams always saw time abroad as a good thing.[58] For Henry's eldest brother, Irwin, it had also been essential to maintaining family standing.

The Rawdons: The Effects of Changing Means

The Rawdons, who began the seventeenth century among the Yorkshire gentry only 12 miles from Temple Newsam, demonstrate the effects that changing means could work on travel over several generations. George Rawdon was educated at schools in Bradford and York, not far from the family's small estate, and he crossed the Irish Sea as a young man in the mid-1620s, serving as private secretary to the first and second Viscounts Conway. Rawdon's links to the Conways, who were major landowners in Down and Antrim, helped him to develop his own Irish estate, centred on Lisburn. He married the second viscount's daughter, Dorothy, and continued to enjoy the patronage of his son, the first Earl of Conway. Rawdon held civil and military office in Ireland during the turbulent 1640s and 1650s, becoming a baronet in 1665, but never forgot the Conway family's 'kindness in raising me in the world from a low degree'.[59] Fragmentary accounts suggest that the Rawdons received around £700–£1,000 a year in the 1670s, an income that, at the top end, was enough to live as a gentleman of quality in contemporary

[57] WYAS, WYL100/C/11: Arthur Ingram of Barrowby to Henry Ingram, 25 Dec 1716; same to same, 11 May 1717; same to same, 21 Oct 1717; WYL100/C/12: same to same, 31 Mar 1718; WYL100/C/20, fo. 325: Rich Ingram to same, 28 Oct [1716]; fo. 326: same to same, 25 Jun [1717].

[58] Thomas, *Ends of Life*, pp. 18–20.

[59] R. M. Armstrong, 'Rawdon, Sir George, first Baronet (1604–1684)', *ODNB*, online edn, 2008 (https://doi-org/10.1093/ref:odnb/23177); Patrick Little, 'Rawdon, Sir George (1604–84)', *DIB* (https://www.dib.ie/biography/rawdon-sir-george-a7585); *CSPD, Charles II*, xix, p. 511: Sir George Rawdon to Conway, 15 Dec 1677.

Ireland.[60] Sir George and Lady Rawdon recognised the value of educational mobility for each of their sons, with Edward, John and Arthur all spending time in France during the 1670s, but travel stretched the family's means. They resorted to economisation, borrowing and patronage to secure foreign experience in this first generation, showing how a family of lesser means might fit educational travel to their spending power.

Correspondence is patchy for the time that the Rawdon brothers spent in France, but John arrived at Foubert's Parisian academy in mid-1674 and Edward joined him a year and a half later. Arthur, meanwhile, spent much of his childhood in France on account of a persistent lung problem that may have been tuberculosis. He had lived with the Conways in England from the age of 7, receiving early introductions to London, but doctors soon recommended the healthier climate of France.[61] Arthur crossed the Channel with a tutor in May 1671, aged 8, and planned to spend the winter at Montpellier, then renowned for medical expertise. They paused for a few months at Orleans, intending to improve his French, and toured provincial France before reaching their destination. All the while, Arthur continued his French and Latin.[62] The family considered sending him from Montpellier to Geneva, which Conway's well-travelled brother-in-law, John Finch, advised as a healthier summer home with good teaching and a 'great store of Nobility and Gentry', but the boy was instead at Saumur by late 1673. He seems to have made a long stay.[63] Arthur was back in England by March 1674, probably on account of his tutor's ill health, but he returned to France with a new man soon afterwards. They stopped in Paris on the way back to Montpellier, again on medical advice. Arthur finally moved to Geneva after 18 months, describing sights along the Rhône, and he stayed there for

[60] WYAS, Bradford, SpSt/5/4/1/1: Richard Rawdon, 'Irish Accompts', 1652–86; HL, mssHAF, box 20, fos 15, 46; box 22, fos 12, 34: Phelim O'Neill, Accounts for Sir George Rawdon, 1671–6; *CSPD, Charles II*, xvi, p. 225: Sir George Rawdon to Conway, 18 Apr 1674. For the £1,000 threshold, see Toby Barnard, *A New Anatomy of Ireland: The Irish Protestants, 1649–1770* (New Haven, CT, and London: Yale University Press, 2003), pp. 60–4.

[61] HL, mssHA 14486: Conway to Sir George Rawdon, 20 May 1670; *The Conway Letters: The Correspondence of Anne, Viscountess Conway, Henry More, and Their Friends 1642–1684*, ed. Marjorie Hope Nicolson, rev. Sarah Hutton (Oxford: Clarendon, 1992), pp. 325–6: same to same, 12 Dec 1670.

[62] HL, mssHA 14507: Conway to Sir George Rawdon, 15 Sep 1671; 15603: Arthur Rawdon to Conway, 19 Dec 1671; *Conway Letters*, pp. 338–9: Conway to Rawdon, 20 Jun 1671.

[63] HL, mssHA 14508: Conway to Sir George Rawdon, 17 Oct 1671; *CSPD, Charles II*, xvi, p. 37: Rawdon to Conway, 29 Nov 1673.

a year. He had spent six years abroad by the age of 14, when circumstances dictated that he return home rather than make a planned journey to Italy.[64]

How did the Rawdons afford this level of travel on an income of no more than £1,000 a year? Arthur's allowance alone was a quarter of that in 1674, and Edward's even more two years later, so expenses in France became the most important factor in the family economy. The Rawdons and the Conways had shared the cost of sending Edward and John to university and the arrangement persisted abroad, with the earl calling on Sir George to repay some expenses but declining to bill him for others.[65] Conway had recognised that John was 'not like to make a Scholler' and suggested France rather than Cambridge as 'more suitable to his inclinations', which tended towards the military. His college tutor later agreed that he had 'a Genius for this employment, and none to any thing els'. Conway estimated that a year with Foubert would cost about £170, little more than John's Cambridge expenses, and presented time in Paris as a more sensible investment.[66] Foubert's physical curriculum would, after all, offer good preparation for the army. Conway bought John a commission as he arrived in France, trusting that Sir George would 'breed him up to be fit for it', and arranged for his nephew to gain experience, after the academy, in an English brigade then in French service. Conway provided John with a sum for his personal maintenance in Paris, while Rawdon paid his academy fees.[67] Even so, Sir George hoped that John would make only 'one years stay att the Accademy for I am soe unprovided to manteyne them abrode and to live as I doe att home that I must take up or run in dett which I hate'.[68] When Edward too arrived in Paris, having first studied the law with a view to becoming 'a very considerable man', his father hoped that he would spend only three months abroad. Conway drew on his own memories of France to warn that Edward would 'know nothing, nor learne no Exercises'

[64] HL, mssHA 14523: Conway to Sir George Rawdon, 22 Jul 1673; 15264: Humphrey Maunsell to Dorothy, Lady Rawdon, 6 Jul 1674; 15604: Arthur Rawdon to Conway, [*c.* April 1676]; 15242: A. Magenis to Conway, 7 Apr 1677; *CSPD, Charles II*, xix, p. 133: Sir George Rawdon to same, 16 May 1677.

[65] HL, mssHA 14512: Conway to Sir George Rawdon, 22 Jun 1672; 14527: same to same, 6 Jun 1674; 14549: same to same, 23 Jun 1676.

[66] HL, mssHA 14526: Conway to Sir George Rawdon, 17 Apr 1674; 15373: Henry More to Conway, 5 Sep 1676.

[67] HL, mssHA 14531: Conway to Sir George Rawdon, 29 Sep 1674; 14535: same to same, 21 Nov 1674; 14543: same to same, 14 Dec 1675. Conway mentions Lord George Douglas, commander of the brigade: John Childs, *The Army of Charles II* (London: Routledge & Kegan Paul, 1976), pp. 244–50.

[68] HL, mssHA 15667: Sir George Rawdon to Conway, 2 Sep 1674.

in such a short stay. Sir George saw the logic but struggled to fund his sons, considering laying off servants, moving into a smaller home and selling land to keep them abroad. Conway eventually came to the rescue, promising to pay any expenses that Edward incurred above £200 a year as John left the academy, but even this sum represented a sacrifice for the Rawdons.[69]

Sir George and his brother-in-law similarly shared the burden of Arthur's upbringing, though it fell more on Conway. As the boy left for France for the first time, his father thanked the earl for his 'care and cost for that child that gives you so much trouble', though Sir George initially continued to participate. The 'hard winter' of 1673–4 stretched his means, however, as 'so sudden and general a destruction' of cattle and crops had not been known since the 1641 rebellion, and Rawdon received only a fifth of his £500 half-yearly rental that spring. He depended on Conway's credit until, a couple of years later, the earl took over all of Arthur's expenses.[70] Even contributing to the foreign education of three sons was more than the Rawdons could afford, of course. Sir George may have hated owing money but the Irish Statute Staple Books, a way of tracking contemporary lending, indicate that he borrowed around £2,300 between 1674 and 1679. The source covers most of the seventeenth century but Rawdon only appears as a debtor in these years, the precise period during which his sons left home for their upbringing.[71]

Some families rationed continental experience according to birth order, as many had with university. The Percevals, for instance, sent Sir Philip to France and Italy in the 1670s but his younger brothers, Robert and John, stayed at home. Robert treated a tour of southern Wales according to the principles of foreign travel, keeping a diary of 'everything I thought remarkable', but found within it 'many trivials things and very few worth a mans imprinting in his memory'. His description of Abergavenny, a town 'encompassed round about with the high hills (in likenes to Rome or Jerusalem)', suggests frustrated hopes of travel or a nice sense of irony. John similarly envied Sir Philip's voyage, wishing that he had been 'born to such

[69] HL, mssHA 14526: Conway to Sir George Rawdon, 17 Apr 1674; 14545: same to same, 11 Jan 1675/6; 14546: same to same, 14 Mar 1675/6; 14549: same to same, 23 Jun 1676; Sean Kelsey, 'Conway, Edward, Earl of Conway (*c.* 1623–1683)', *ODNB*, online edn, 2010 (https://doi-org/10.1093/ref:odnb/39681).

[70] HL, mssHA 14498: Conway to Sir George Rawdon, 21 Jan 1670[/1]; 14499: same to same, 15 Feb 1670[/1]; 14546: same to same, 13 Mar 1675/6; *CSPD, Charles II*, xi, p. 296: Rawdon to Conway, 3 Jun 1671; xvi, pp. 212–13: same to same, 28 Mar 1674; p. 225: same to same, 15–18 Apr 1674.

[71] Jane Ohlmeyer and Éamonn Ó Ciardha, eds, *The Irish Statute Staple Books, 1596–1687* (Dublin: Dublin Corporation, 1998), p. 280.

an Employment as might require a harmless ramble about some parts of the World', and he followed his elder brother 'on the globe & Mapps'. John internalised the older generation's reasoning, however, recognising that travel was 'an accomplishment butt onely fitt for elder Brothers; butt a hazardous & chargeable sort of Life by which noe Estate will be gott'.[72] Where the Ingrams tended to send two sons abroad, the Percevals used the same logic of future life roles to limit continental experience to their heir.

In the same decade as the Percevals, the Rawdons paid more attention to the 'genius' and even the inclinations of their sons. John, though a younger brother, travelled first to the Parisian academy, in accordance with his military bent, while Edward spent longer in England to acquire the legal knowledge more standard for a gentleman. Conway's willingness to fund Arthur's long periods abroad, despite his status as the youngest brother, owed much to ill health and an emotional attachment built up by brief fosterage. 'He is very fond of me', he told the boy's father, 'and therefore I have reason to take a great deale of care of him'. Arthur's travels also reflected Conway's feeling that, of the three brothers, 'if he lives will be the finest Gentleman'; 'if he recovers his health', he wrote elsewhere, 'he will be a man of extraordinary parts'.[73] Arthur's progress in French demonstrated the gains of travelling at an impressionable age—he spoke it 'as well as any man and writes it better'—and the boy himself felt some improvement in his health, but this approach inevitably came with risks.[74] Sir George had fought in Ireland during the 1640s, wary of any peace with Catholics, and proved a staunch supporter of the Established Church after the Restoration.[75] Predictably, he worried about the faith of his youngest son in France. Conway showed no such fears, even sending Arthur to Paris in the company of Catherine of Braganza's chaplain, whom he jokingly instructed not to convert the boy on the way. Sir George, by contrast, argued for moving Arthur to Geneva from an early stage, finding 'no danger there of Papistry'.[76] The recovery of the boy's health and the encouragement of his potential won out, however. Arthur's uncle and tutor agreed that he was, though a third son, 'likely to

72 BL, Add MS 46953, fo. 138: Robert Perceval to Sir Robert Southwell, 22 Jul 1676; 46955B, fo. 125: John Perceval to same, 23 Sep 1678.

73 HL, mssHA 14489: Conway to Sir George Rawdon, 28 Jun 1670; 14526: same to same, 17 Apr 1674; 14549: same to same, 23 Jun 1676.

74 HL, mssHA 14525: Conway to Sir George Rawdon, 31 Mar 1674; 15604: Arthur Rawdon to Conway, [Apr 1676].

75 Armstrong, 'Rawdon', *ODNB*; Little, 'Rawdon', *DIB*.

76 HL, mssHA 14499: Conway to Sir George Rawdon, 15 Feb 1670[/1]; *CSPD, Charles II*, xi, p. 553: Rawdon to Conway, 1 Nov 1671.

make as much use by beeing abroad as any one that ever traveld'.[77] The Percevals' £4,000 a year was four times the income enjoyed by the Rawdons, easily covering the £915 that Sir Philip had spent during two and a half years on the Continent in the 1670s, so decisions over whether to send one or several sons abroad were about more than money.[78] Families placed varying weight on birth order, talent and their sons' desires when striking a balance between means and ambitions.

Unfortunately, the Rawdons demonstrate that stretching to educational travel could be a risky choice. John joined the English brigade in Flanders directly from Paris, but soon died in the 'smart battelle' anticipated in his final letter home.[79] Perhaps his 'genius' had not been for the military after all. Conway had meanwhile grown frustrated with Sir George's penny-pinching and was planning to send Edward and Arthur on a three-year tour of France and Italy, possibly fulfilling his own abortive plans of Italian travel in the early 1640s. Conway blamed himself for John's death, however, and resolved never to 'attempt any thing of the like kinde any more'.[80] Sir George understandably favoured a subsidised Italian tour for his sons, urging Conway to rethink, but Edward also died before plans could be made. He suffered fatal wounds in a duel with Metcalfe Robinson's father, William, a fellow traveller in Paris.[81] Sir George quickly recalled his only surviving son from France, hoping that he would 'in some measure prove worthy of the love, charge and care that has been taken for his education'.[82] Word spread that the Rawdons had buried '2 hopeful gents … in one year', but Conway's

[77] HL, mssHA 14546: Conway to Sir George Rawdon, 14 Mar 1675/6.

[78] For Perceval's finances during travel, see BL, Add MS 46953, fos 190–1, 220; 46954A, fo. 206; 46954B, fos 33, 112–15v, 231–2v; 46955A, fos 176–83v; 46956A, fos 122–48v. Sterling calculated from exchange rates within. For the family's means, see Barnard, *Irish Protestant Ascents and Descents*, p. 44.

[79] HL, mssHA 15700: John Rawdon to Sir George Rawdon, 16 Jul 1676.

[80] HL, mssHA 14336: Edward, second Viscount Conway to Edward Conway (future first Earl), 30 Dec 1641; 14553: Earl of Conway to Sir George Rawdon, 19 Sep 1676; 14554: same to same, 4 Nov 1676.

[81] *CSPD, Charles II*, xviii, pp. 524–5: B. Magenis to Mr Magenis, 9 Feb [1677]; fo. 548: Sir George Rawdon to Conway, 10 Feb [1677]; BL, Add MS 4293, fo. 28b: Thomas Birch, Notes on Rawdon family [18th century]. '(Sir) Wm Robinson', the 'Yorkshire gentleman' who killed Edward Rawdon, was surely Metcalfe's father. He was 21, had recently left Cambridge and Gray's Inn and would inherit the baronetcy in 1690: Eveline Cruickshanks and Ivar McGrath, 'Robinson, Sir William, 1st Bt. (1655–1736) of Newby, Yorks.', in D. W. Hayton, Eveline Cruickshanks and Stuart Handley, eds, *The History of Parliament: The House of Commons 1690–1715* (Cambridge: Cambridge University Press, 2002).

[82] *CSPD, Charles II*, xix, p. 81: Sir George Rawdon to Conway, 7 Apr 1677.

ambition had paid off in the sense that a third son inherited with the benefit of a French education.[83]

Continental travel would wax and wane in the family for the next century, depending on financial circumstances. Conditions seemed to improve over the next two decades, either side of Sir George's death in 1684, with total income surpassing £1,200 in 1680–1 and a rental income of over £2,100 when Sir Arthur died on his 33rd birthday in 1695.[84] He had expected to be the main beneficiary of Conway's will, but his sudden exclusion, as well as substantial debts and the need to provide for two unmarried sisters, Brilliana and Dorothy, meant that 'there was not one sixpence of ready money left' during much of Sir Arthur's short adulthood, despite the improved rental.[85] Educational travel allowed relations to prove the earl's affection for his nephew and therefore, they hoped, that he had not been of sound mind when altering his will. It was 'wel k[n]owne', insisted Sir Arthur's aunt, Elizabeth Rawdon, that 'his onkle [Conway] sent him into france an[d] his oune chaplen with him an[d] al the care that could be'.[86] These arguments ended in failure, however, ensuring that the minority of the young Sir John, born in 1690, was dominated by creditors and legal fees. Little survives of his education, which probably took place at home, and there is no evidence that he travelled abroad, an unlikely course in the circumstances. Sir Arthur's widow, Helen, nevertheless managed the situation so effectively that annual rents rose to £3,100 in 1717 and almost £4,000 at Sir John's premature death in 1723.[87]

With healthier finances, the family could contemplate sending young men back to the Continent. Sir John's 18-year-old son and namesake undertook the most comprehensive journey that the Rawdons had yet attempted. Only a few letters survive, but he headed for the Loire in the spring of 1738, settling for several months of study at Blois. He made an autumn tour of France and wintered in the south of the country. Sir John crossed to Italy by sea, spent a month learning the language at Siena and passed the winter of 1739–40 in Rome. He visited Naples in the spring

83 BL, Add MS 4293, fo. 28b: Birch, Notes on Rawdon family.

84 HL, mssHAF, box 23, folder 22: Phelim O'Neill, Accounts for Sir George Rawdon, 1680–1; box 27, folder 4: Sir Arthur Rawdon, List of debts, 1694–5; John Bergin, 'Rawdon, Sir Arthur (1662–95)', *DIB* (https://www.dib.ie/biography/rawdon-sir-arthur-a7583).

85 Hayton, *Anglo-Irish Experience*, p. 179.

86 WYAS, Bradford, SpSt/6/1/13: Elizabeth Rawdon to Priscilla Rawdon, 6 Feb 1689.

87 Hayton, *Anglo-Irish Experience*, pp. 178–82, 184.

before returning home via Venice and France.[88] Italian experience helped to enhance the family's standing, alongside Sir John's unusual longevity, his rising income, his noted library and art collection, three noble wives and his elevation to the earldom of Moira in 1762.[89]

Unsurprisingly, Moira's two eldest sons returned to the Continent in 1773. The tenth Earl of Huntingdon, his brother-in-law by his third marriage, had housed Francis and John in London and overseen their education, since the Rawdons still lived predominantly in Ireland. Huntingdon had himself visited the Low Countries, France, Italy, Spain and Portugal over two journeys in the 1750s, leaving England again for France and Italy in 1771.[90] He considered placing Francis and John with the British ambassador in Paris, an old fellow traveller, or sending them to academies in provincial France or Turin before a tour of the Netherlands and Germany. Huntingdon ultimately summoned Francis (19) and John (16) to come through France in the summer of 1773 and meet him in Italy. He saw himself as an 'Uncle-Governor', introducing his nephews to his many contacts among the Italian nobility and showing them the sights on their way to Naples, where they settled to enjoy the hospitality of Huntingdon's friend, Sir William Hamilton, the British ambassador. John sailed back to Ireland after less than a year abroad, his uncle frustrated with his laziness, but Huntingdon sent Francis to Vienna by himself in the autumn of 1774. They were supposed to reunite in Paris and Berlin, but Francis left instead for America, in fulfilment of the army commissions that Huntingdon had been buying him since Oxford.[91]

A century after Conway had sent his nephews to France, the Rawdons turned again to a well-connected maternal uncle with greater continental experience to arrange educational travel. They did so under similar terms: Moira funded his sons in England and allowed Francis £300 a year on the Continent, with Huntingdon covering further expenses with 'unlimited credit'; meanwhile, the boys' uncle surely spent more than the £110 bill

[88] PRONI, D2924/1/44: Sir John Rawdon to Thomas Prior, 23 Apr 1738; 46: same to same, 22 Nov 1738; 47: same to same, 20 Jul 1739; NYCRO, ZON13/2/28: Frederick Frankland to Thomas Worsley III, 26 Mar 1740; John Ingamells, *A Dictionary of British and Irish Travellers in Italy, 1701–1800* (New Haven, CT, and London: Yale University Press, 1997), pp. 800–1.

[89] Rosemary Richey, 'Rawdon, John (1720–93), 1st Earl of Moira', *DIB* (https://www.dib.ie/biography/rawdon-john-a7586).

[90] Joan Coutu, *Then and Now: Collecting and Classicism in Eighteenth-Century England* (Montreal: McGill-Queen's University Press, 2015), pp. 127–54.

[91] PRONI, D2924/1/80: Francis Hastings, Earl of Huntingdon to Earl of Moira, 25 Jun 1768; D2924/1/158–203: Francis, Lord Rawdon to Earl and Countess of Moira, 1767–75; T3765/M/2/14/53–106: Huntingdon to Countess of Moira, 1766–75.

he sent Moira for John's whole time abroad.[92] As with Conway and Arthur Rawdon, the strong bonds created by childhood fosterage moved Huntingdon to support onward travel. He was 'so wrapt up in my nephews' that, for Francis's journey, he was ready to 'open my purse with the same cordiality as if he was my own son'.[93] Huntingdon had fathered an illegitimate child, Charles, on his first visit to Paris but refused to marry, grooming Francis to inherit his titles and Leicestershire estate. He accordingly planned his eldest nephew's travels while he was at Oxford, reasoning that 'as he is to represent my family as well as his father's I wish he should have an education suitable to his situation'.[94] Like Conway, Huntingdon sought parental approval and some funding for plans that he worked out alone, thanking Moira for the 'obliging resignation' of his sons 'to my care'. The Rawdons still worried about costs, as they had done in the 1670s, but Huntingdon now encouraged them to spend what was necessary to sustain their noble status abroad, calling for investment in 'this critical period of their education'.[95] Cynically, perhaps, Huntingdon may have been encouraging the Rawdons to subsidise his own family strategy, as an unmarried uncle enhancing the reputation of his substitute son. Either way, Francis and John needed a sizeable allowance because 'their dress &c must be adapted to their situation & rank', though the elder brother went to America before he could 'make that figure (in the several courts which he was to have visited) so indispensible for a man of his rank'.[96] Huntingdon could now, like the Ingrams, appeal to family precedent, arguing that Moira was 'well acquainted with the style of living amongst the first company abroad'.[97]

Plenty had changed in elite travel culture between the 1670s and the 1770s, as is clear from Huntingdon's multiple journeys, his long residence in Naples and the British women who populate his letters from Italy.[98] Within that culture, however, the principles of male educational travel had remained remarkably consistent.[99] What had changed for the Rawdons were

92 PRONI, T3765/M/2/14/97: Huntingdon to Countess of Moira, 30 May 1774; 100: same to same, 22 Oct 1774.

93 PRONI, T3765/M/2/14/82: Huntingdon to Countess of Moira, 18 Sep 1771; 101: same to same, 1 Feb 1775.

94 PRONI, T3765/M/2/14/80: Huntingdon to Countess of Moira, 20 May 1771; Coutu, *Then and Now*, p. 134.

95 PRONI, T3765/M/2/14/92: Huntingdon to Countess of Moira, 21 May 1773.

96 PRONI, T3765/M/2/14/94: Huntingdon to Countess of Moira, 22 Oct 1773; 102: same to Earl of Moira, 1 Feb 1775.

97 PRONI, T3765/M/2/14/96: Huntingdon to Countess of Moira, 22 Feb 1774.

98 For British women in Italy, see Sweet, *Cities and the Grand Tour*, pp. 23–64.

99 French and Rothery, *Man's Estate*, pp. 137–83.

the financial means and social connections that they could bring to bear on the undertaking. So too had their obligation to do so, as they rose from Yorkshire gentlemen to Irish noblemen and heirs to an English earldom. Demographic chance complicated the family's early investments in travel, for which they relied on cost-cutting, borrowing and patronage, but the need to 'appear abroad becoming their rank', as Huntingdon put it, shaped the experiences of later, more elevated Rawdons.[100]

The Brydges Family: Advancement through Travel

If travel demanded that noble families live up to their position, it also allowed opportunities for such newcomers as the Brydges family to secure their progress. James Brydges had roots in the Gloucestershire and Herefordshire gentry, with his father, a baronet, becoming eighth Baron Chandos when the main line failed in 1676.[101] James left Oxford to travel at 19 in 1692, spending most of his two years abroad at the academy of Wolfenbüttel. He made another trip to Hanover later in 1694 but soon returned to London, where he pursued a role in government. Brydges's efforts truly bore fruit in 1705, when he became Paymaster of the Forces Abroad, a position that would transform the wealth and standing of his family. Links to the electoral court at Hanover, created through travel, smoothed his rise to the earldom of Carnarvon in 1714 and the dukedom of Chandos in 1719.[102]

Brydges used his post as army paymaster to amass a personal fortune of around £700,000, a phenomenal sum in contemporary terms. He represents a striking example of the opportunities for personal enrichment offered by the greater military commitment to the Continent after the 1688 Revolution, as he gambled on sieges, exploited exchange rates, deducted percentages from army pay and provided the troops with financial services. Brydges's carefully built network, alongside outright bribery, allowed him to escape investigation for corruption on a massive scale.[103] He invested plenty of the proceeds in land, claiming by 1727 that his son was heir to some £17,000

[100] PRONI, T3765/M/2/14/93: Huntingdon to Countess of Moira, 1 Jul 1773.

[101] P. G. M. Dickson and J. V. Beckett, 'The Finances of the Dukes of Chandos: Aristocratic Inheritance, Marriage, and Debt in Eighteenth-Century England', *Huntington Library Quarterly*, 64:3/4 (2001), 309–55, at 311–12.

[102] HL, mssST 57, vol. 1, fos 9–21: James Brydges, letters to contacts in Germany, 1693–5.

[103] Dickson and Beckett, 'Finances of the Dukes of Chandos', 313–14; Aaron Graham, 'Auditing Leviathan: Corruption and State Formation in Early Eighteenth-Century Britain', *English Historical Review*, 128:533 (2013), 806–38.

a year in rent, which various offices raised to an annual £22,000, though a yearly rental income of £12,000 may have been more accurate.[104] Brydges recognised that he needed to spend in order to secure his standing in the highest reaches of society. His programmes of building, collecting, artistic patronage and conspicuous consumption are well known, centred on his palatial new residence at Cannons and several London projects.[105] An equally significant, but so far overlooked, investment in the educational travels of his sons and wider family met the same end of establishing his social position.

Brydges's new wealth allowed him to send his two sons, John and Henry, on lavish continental journeys. They had mostly prosecuted their comprehensive domestic travels together, but their paths diverged with John's final trip through the West Country in April and May 1721. That June, the 18-year-old left for the Netherlands with a tutor, an army officer and an experienced travelling servant. John studied for six months at Leiden and toured Holland and Flanders before moving to an academy in Paris. He came back to England briefly but returned for a tour of France and Germany, taking in Geneva, Vienna and Hanover. He was home by January 1724, after two and a half years abroad.[106] John's travels had been long in the planning, but Henry's voyage emerged as a solution to an immediate problem. Brydges had intended a 'quite Different' upbringing for his second son, hoping that he would 'become a man of Learning'. He entered Cambridge as a student on £200 a year, rather than as a nobleman like his brother, who had followed his 'Inclinations & Pleasure' at Oxford. Henry was supposed, as a younger brother, to 'raise his Fortunes in the world' through study, but Brydges soon found that he had not been following the plan. Henry had 'most shamefully misemployed and perfectly lost' his six months at Cambridge, regularly sleeping into the afternoon.[107] Brydges's means allowed him to plot educational strategies on a broad canvas: since an English university had failed, he would send Henry 'beyond sea & endeavour to reclaim him in one of those abroad'. Leiden had worked well for John, so Henry set off for the Netherlands with the same army officer in October

104 HL, mssST57, vol. 30, fo. 160: James Brydges to Andrew Robinson, 26 Jul 1727; Dickson and Beckett, 'Finances of the Dukes of Chandos', 321.

105 Susan Jenkins, *Portrait of a Patron: The Patronage and Collecting of James Brydges, 1st Duke of Chandos (1674–1744)* (Aldershot: Ashgate, 2007); Aaron Graham, 'Connoisseurship, Consumption, Company, and James Brydges, First Duke of Chandos, 1705–13', *Huntington Library Quarterly*, 80:4 (2017), 539–57.

106 HL, mssST57, vols 18–23: letters by James Brydges, 1721–4.

107 HL, mssST57, vol. 23, fo. 336: James Brydges to Captain Marchand, 9 Mar 1723/4; vol. 24, fos 24–5: same to Robert Jenkin, 13 Apr 1724; fos 81–2: same to same, 9 May 1724; fos 175–6: same to same, 29 Jul 1724; fo. 217: same to same 22 Aug 1724.

1724, aged 16.[108] He also toured the Low Countries and northern Germany but moved to a different French academy, at Angers, after a year at Leiden. Henry probably toured France and Lorraine but certainly visited Brussels and Hanover before returning to England in August 1727, just under three years after he had set out.[109]

Family strategy shaped how Brydges sent his sons to the Continent. He followed many parents and guardians in stratifying the experience of travel but, where others restricted access to the Continent by birth order, Brydges sent both of his sons abroad with differing budgets and priorities. Although Henry had been supposed to work harder at home than his elder brother, essential aspects of John's travels became optional for a second son: their father was 'by no means against' Henry frequenting assemblies at The Hague, for instance, a vital test of John's conduct and standing. Where Brydges criticised John's projected annual spending of £2,000, he recycled the line for Henry with suitable adjustments, arguing that 'if the Case was such as that no Gentleman could travell under the Expence of 1000£ a year there would be a much smaller Number go abroad than there does'.[110] Itineraries, too, were shaped by dynastic concerns. John had been destined to travel from an early age but ventured no further than the Low Countries, France and Germany, staying within reach as his father tried to arrange his marriage at home. Brydges had been thinking of securing Henry a military commission and was planning to let him tour France and Italy before joining the army in Flanders, but John's death from smallpox only three years after his return left Henry's future 'quite altered'.[111] Like the early Rawdons, Brydges's decision to send his second son abroad paid off upon his inheritance. There can be few more straightforward illustrations of travel's dependence on family strategy than that of Henry waiting in France to hear whether his sister-in-law had had a boy or a girl, an event that would dictate whether he could continue to Italy. John's widow, Catherine, gave birth to a daughter, so Henry returned to sit in parliament.[112] Not only did family circumstances initially mean that Italy was more feasible for a

[108] HL, mssST57, vol. 24, fos 188–9: James Brydges to Robert Jenkin, 4 Aug 1724; fo. 271: same to Walter Senserf, 5 Oct 1724.

[109] HL, mssST12, vol. 3, fos 287–96: accounts for Lord Henry's travels; ST57, vols 26–30: letters by James Brydges, 1725–7.

[110] HL, mssST57, vol. 25, fo. 28: James Brydges to Andrew Robinson, 20 Nov 1724.

[111] HL, mssST57, vol. 29, fo. 129: James Brydges to Andrew Robinson, 21 Dec 1726; fo. 253: same to same, 21 Mar 1727; fo. 278: same to same, 30 Mar 1727.

[112] HL, mssST57, vol. 30, fo. 56: James Brydges to Andrew Robinson, 9 Jun 1727; fo. 151: same to same, 23 Jul 1727.

younger brother, but matters at home ensured that, for all their father's wealth, neither John nor Henry went further than France and Germany.

Changing spending power nevertheless shaped travel over three generations. Though there are few traces of James Brydges's time at Wolfenbüttel, he later implied that 'crampt' finances had made it 'hardly possible' for him to live like a gentleman there.[113] He offered far more to his sons. It was unusual even among the nobility for brothers to travel separately, as it was inevitably more expensive, but John and Henry both combined periods of institutional learning with lengthy tours, accompanied by tutors, companions and servants. Their father hoped that the improvements of travel would allow John 'to make that figure which will be expected from one of your Rank'.[114] Brydges had suffered huge losses on the South Sea Company—perhaps £200,000—in the months before he sent his eldest son abroad, however, and he made several ineffectual attempts at retrenchment.[115] When John spent £1,000 in only six months abroad, a sum that 'must have been squandred away in idle & trifling expences', he warned that even £1,500 a year would be more than he could spare and 'much greater than any eldest son ought to expect'. John compared his spending with that of other travellers, a common tactic, but his father protested that 'the Gentlemen you instance … are three of as considerable Estates as any in the Kingdom'. Brydges could think of many more who 'never travelled at such a rate', encouraging John to live 'in the most frugal manner that is consistent with reputation' in Paris. A few years later, he would use near-identical phrasing with Henry.[116] As John prepared to celebrate Easter in Rotterdam, Brydges advised that he should make a donation for the church organ 'so as not to outdo any other English Gentleman, who hath paid that compliment before him, nor yet to be the least Contributor to such a work'. Brydges was wary of playing the upstart, an accusation to which his ascent and lifestyle readily exposed him. Caught between avoiding ostentation and living up to his dukedom, he wanted his sons 'to observe a proper Medium' abroad.[117]

Financial decline soon took such decisions out of the family's hands. Brydges compounded his losses on the stock market with a string of poor

113 HL, mssST57, vol. 4, fos 7–8: James Brydges to Theophilus Leigh, 1 Jun 1710.

114 HL, mssST57, vol. 18, fo. 345: James Brydges to John Brydges, 3 Jan 1721/2.

115 Dickson and Beckett, 'Finances of the Dukes of Chandos', 318–19.

116 HL, mssST57, vol. 20, fo. 99: James Brydges to Charles Stuart, 6 Feb 1721/2; fos 140–1: same to same, 9 Mar 1721/2; vol. 22, fos 87–8: same to same, 15 Oct 1722; vol. 28, fo. 21: same to Andrew Robinson, 25 Mar 1726.

117 HL, mssST57, vol. 20, fos 140–1: James Brydges to Charles Stuart, 9 Mar 1721/2; Dickson and Beckett, 'Finances of the Dukes of Chandos', 327, 332.

choices about land, building, investment and lifestyle during the 1730s, dying with debts of around £83,000 in 1744. Henry owed at least £49,000 when he succeeded to the dukedom, making a combined £132,000. Sales of land and property followed, with Cannons stripped and sold in 1747. Serious cutbacks left the second duke on a rental income of only £1,600 a year by 1759 but still saddled with debts.[118] In this context, Henry's only son, another James, travelled less ambitiously in 1750–2. The 19-year-old future third duke spent less than a year and a half abroad, making tours of western Germany, Paris and the Loire Valley between winters at the University of Göttingen and Orleans. Though Marquis of Carnarvon, he travelled without a tutor and studied among gentlemen, as had his grandfather, leaving behind the retinues employed by the previous generation.[119] Spending on travel contracted alongside the other ways in which the family had looked to establish their position, from land and buildings to elections and artistic patronage. Cannons House was demolished in 1753, months after the future third duke had returned from the Continent. He considered moving to Florence later in life, an option that would have lowered his living costs, but there had been no question of an educational voyage as far as Italy.[120] If travel held a central place in family strategy and estate finance, the availability of funding, for the Brydges family as for the Ingrams and Rawdons, in turn shaped the timing, route, duration and purpose of journeys.

The Leighs: Travel through Patronage

John and Henry Brydges did not have to look far for examples of young men who travelled on under £1,000 or £2,000 a year. Despite their father's wealth, his rapid rise had left them with many cousins among the modest gentry.[121] If kin networks were central to domestic educational mobility, Brydges's letter-books show how relations of lesser means might similarly use them to undertake foreign travel that was otherwise beyond their reach. Brydges's educational patronage is clearest for the children of his eldest

[118] Dickson and Beckett, 'Finances of the Dukes of Chandos', 327–38.
[119] HL, mssSTB, box 12, folder 53: James Brydges II, Letter-book, 1750–1; mssST109, vol. 1: Brydges, Travel diary for Germany and France, 1751.
[120] Dickson and Beckett, 'Finances of the Dukes of Chandos', 337, 344. Finances improved between the 1760s and 1780s but had already shaped the future third duke's travels: 339–45.
[121] Rosemary O'Day, 'Matchmaking and Moneymaking in a Patronage Society: The First Duke and Duchess of Chandos, *c.* 1712–35', *Economic History Review*, 66:1 (2013), 273–96, at 275.

sister, Mary, and Theophilus Leigh of Adlestrop. Leigh had travelled himself in the late 1660s, but the journey has left few traces. He remembered, in the middle of a family dispute over budgets abroad, that he had spent less than £200 a year and had travelled with one servant 'till I found I could stand on my own legs', suggesting that the Leighs had barely afforded travel in his generation.[122] He found himself in a similar position with his six surviving sons and four daughters, a tally that severely stretched family means, particularly after Mary died in her late thirties. It had 'always' been Leigh's 'inclination to let my eldest son see something of the world ... in order to make him a man as well as a Gentleman', but he relied upon Brydges to send William to the Continent.[123] Their experience makes clear that travel was never limited to those who could afford it.

William Leigh travelled between 1710 and 1713. The chance arose at short notice for him to visit Italy with Viscount Cornbury, only son of the Earl of Clarendon, and the family seized the opportunity for social advancement. Brydges's new wealth and reputation furnished the money and bills of exchange required for an immediate departure, as the party crossed to the Netherlands in May 1710. Cornbury and Leigh travelled with a tutor and three servants, two for Cornbury and one for Leigh.[124] They passed through the Netherlands, Germany and Bohemia, taking in Hanover, Berlin, Prague and Vienna, before heading down to Venice and six months in Rome. The party toured Italy and stopped in Geneva and Frankfurt on the way back to the Low Countries.[125] They stayed in Utrecht from January 1712 in order to observe the peace negotiations, but Cornbury turned homewards in June, leaving Leigh to explore Flanders. Brydges secured his nephew a passport for France before the end of hostilities, as the position of army paymaster eased travel in practical as well as financial terms, and William stayed in Paris for nine months. He spent another month along the Loire before coming home in November 1713, after three and a half years abroad.[126]

122 HL, mssST58, vol. 7, fo. 56: Theophilus Leigh to James Brydges, 13 Nov 1710. He had travelled after Oxford and Middle Temple: SBT, DR671/60: Draft inscriptions for Theophilus Leigh's monument, 1727.

123 HL, mssST58, vol. 5, fo. 272: Theophilus Leigh to James Brydges, 14 May 1710.

124 SBT, DR18/20/18b: Theophilus Leigh to James Leigh, 10 Jun 1710.

125 SBT, DR671/59a: William Leigh, Travel journal, 1710–11; DR18/20/18b: William Leigh to Theophilus Leigh, 23 Sep 1711; same to same, 17 Oct 1711; same to same, 10 Nov 1711; same to same, 3 Dec 1711; same to same, 20 Dec 1711.

126 SBT, DR18/20/18b: Theophilus Leigh to William Leigh, 19 Jan 1711/12; William Leigh to Theophilus Leigh, 29 Jun 1712; same to same, 21 Sep 1712; same to same, 13 Oct 1712; same to same, 18 Jan 1713; same to same, 20 Sep 1713; HL, mssST57, vol. 9, fo. 265: James Brydges to same, 23 Nov 1713.

The journey spiralled beyond the modest expectations with which Theophilus Leigh had sent his son abroad. He had hoped that his own travel budget of £200 a year would be 'fully sufficient' as 'a great sum out of my small estate, who have so many more to provide for, & lye under difficulties more than enough'. Brydges had helped with the costs of departure but Leigh hoped to 'stretch to supply' his son for the rest of his travels. The original plans comprised the 'nearest cut through Germany into Italy', which would take a year and a half, a period in Flanders and, peace permitting, another six months in France. This addition would require more money, but Theophilus hoped that the basic journey would cost little more than £350.[127] The figure was always unrealistic, as Cornbury's tutor, Dr Hobart, knew from experience that a traveller who 'comes off at the years end for 300£ is a good Husband'.[128] In the event, William spent perhaps three or four times his father's estimates in his three and a half years abroad. The issue was quickly apparent, such that Theophilus was already considering recalling him in the summer of 1711. By May 1712, at the mid-point of the voyage, he found that 'my son has put me to more than double the charge I design'd'. William had in fact already spent almost £800 by the beginning of the year and is unlikely to have slowed his spending.[129] He certainly gave no 'good reason' for his long period of sightseeing and socialising in Paris, which 'especially at this juncture, is none of the cheapest Parts of the world'.[130] Despite his hopes, Theophilus relied on Brydges's money and credit throughout the voyage. Beyond allowances from his father, William was over £400 in his uncle's debt by mid-1711, had added at least £150 more by late 1712 and continued to draw on him in France the following year.[131] Theophilus assessed the damage after his son's return, recognising that William's travels, which had been twice their intended length, had made 1713 'a time of more than ordinary expence to me', particularly in the form

[127] HL, mssST58, vol. 5, fos 272–3: Theophilus Leigh to James Brydges, 14 May 1710; vol. 6, fos 37–8: same to same, 3 Jun 1710; vol. 7, fos 55–7: same to same, 13 Nov 1710. In the 1720s, the Leigh estate was no more than £1,300 a year: SBT, DR18/31/884a: Typescript copy of Mary Leigh, 'History of the Leigh Family of Adlestrop', 1788, p. 71.

[128] SBT, DR18/20/18b: William Leigh to Theophilus Leigh, 26 Dec 1710.

[129] HL, mssST58, vol. 8, fo. 241: Theophilus Leigh to James Brydges, 9 Jul 1711; vol. 11, fo. 251: same to same, 19 May 1712; SBT, DR18/20/18b: William Leigh to Theophilus Leigh, 8 Feb 1712.

[130] HL, mssST57, vol. 9, fos 63–4: James Brydges to William Leigh, 3 Apr 1713; fos 199–200: same to same, 23 Sep 1713.

[131] HL, mssST57, vol. 5, fo. 116: James Brydges to Theophilus Leigh, 3 Jul 1711; vol. 9, fos 145–6: same to William Leigh, 28 Jul 1713; SBT, DR18/20/18b: William Leigh to Theophilus Leigh, 17 Oct 1711; same to same, 14 Jun 1712; same to same, 13 Oct 1712.

of debts to Brydges. Most worryingly, William's voyage had eaten into the portions intended for his sisters. Theophilus saw that the only way to repair the estate, which had 'been 200 years in the family, & was once a much greater', would be to find 'some good match' for his eldest son, a prospect that travel itself advanced by furnishing connections and cachet.[132] Time abroad could be a high-stakes investment. William's marriage to Mary Lord and her 'Considerable Portion' in 1722 could not alleviate debts worsened by the South Sea Bubble, however. After Theophilus's death, William lived more cheaply at Utrecht between 1726 and 1732, while Brydges oversaw the repayment of his creditors.[133]

During his travels of 1710–13, William Leigh had worried about the burden placed on his family but reached for an old excuse. While there were 'at present a great many of English abroad', there was 'not one of them but what have spent much more then I have'.[134] It is notable that Leigh identified a large community of travellers even during the War of the Spanish Succession, but the standard budget of £300 a year makes it hard to agree with his claim. More convincing was the argument that he had 'not been master of my self', tied to Cornbury's decisions and expenses.[135] Some noblemen subsidised those with whom they travelled, but Cornbury devised a system in which William was always 'at Half … so that If the Reconing comes to 30s. I pay ten: which is just even, my Lord having Four in company with him and I two, myself included'.[136] Less fair, perhaps, was the need to live up to Cornbury's status. Although William could not 'handsomely' have avoided such expenses, he promised his father that travelling alone would allow him to suit his 'motions & stay in places & way of living in proportion to my purse'.[137] Theophilus agreed that Cornbury had led his son into extravagance, spending six weeks 'idly & expensively' at Berlin and touring Germany 'to a great expence & little purpose'. He looked forward to 'a thorough reformation' after William parted with Cornbury in the Netherlands.[138] Even then, however, expenses remained high. Theophilus

132 SBT, DR18/20/18b: Theophilus Leigh to James Brydges, 23 Oct 1714.

133 HL, mssST57, vol. 28, fo. 77: James Brydges to Andries Pels, 23 Apr 1726; fo. 312: same to William Leigh, 6 Sep 1726; SBT, DR671/61–67: William Leigh, Diaries while resident in the Netherlands, 1726–32. Mary and their children returned to England in 1730.

134 SBT, DR18/20/18b: William Leigh to Theophilus Leigh, 26 Dec 1710.

135 HL, mssST58, vol. 10, fo. 266: William Leigh to James Brydges, 22 Mar 1712.

136 SBT, DR18/20/18b: William Leigh to Theophilus Leigh, 30 May 1710.

137 SBT, DR18/20/18b: William Leigh to Theophilus Leigh, 8 Feb 1712; HL, mssST58, vol. 10, fo. 266: same to James Brydges, 22 Mar 1712.

138 HL, mssST58, vol. 7, fos 55–9: Theophilus Leigh to James Brydges, 13 Nov 1710; vol. 12, fos 53–4: same to same, 6 Jun 1712.

consented to additional time in the Low Countries and France, unable to resist his son's promises to 'spend my time to the best advantage', nor Brydges's arguments that 'the Charge & Pains you have been at in his Education' would be 'very well bestowd'.[139] The problem had been the level at which the Leighs had engaged in travel culture, as a gentleman's estate was running up against the kind of travel thought suitable by a nobleman and his tutor—and also, perhaps, the kind that Brydges wanted for his own sons. William nevertheless promised 'to make a return in some measure proportionable' to his father's investment, and Theophilus still hoped that 'the great Expences of his 3 year's Travail may turn to good Account'.[140]

The Leighs served as a model for Brydges's patronage, coming of age just as the proceeds of his position as army paymaster rolled in. He soon extended subsidies to other nephews. In 1714, Brydges sent John Chamberlayne, the eldest son of his third sister, Emma, to the Low Countries and Paris with money and introductions. He encouraged the 20-year-old to travel in style, reminding him that 'I did not send you abroad with an intention you should want.' Chamberlayne was planning to tour France, a suggestion with which his uncle agreed, but he died of a fever in the capital that summer.[141] As always, travel was a risky investment. Brydges promised to repair the loss by the 'carefull education of those who remain', supporting John's brothers at Westminster, funding one at Oxford and helping the other into the army.[142] Brydges's patronage had age limits, however. Captain Alexander Jacob, the son of his second sister, Elizabeth, secured his commission through his uncle's influence and used service in the Low Countries to see Paris in 1718.[143] He asked Brydges about his itinerary but was told that he had, at 25, reached 'years of discretion'; his uncle would have 'nothing at all to do with giving any advice'. Brydges offered £100 towards any debts accrued in Paris, all he could 'possibly spare', but denied Jacob the sums lavished on

139 HL, mssST57, vol. 4, fo. 243: James Brydges to Theophilus Leigh, 23 Dec 1710; mssST58, vol. 10, fos 236–7: William Leigh to James Brydges, 12 Jan 1711/12.

140 SBT, DR18/20/18b: William Leigh to Theophilus Leigh, 13 Oct 1712; Theophilus Leigh to James Leigh, 18 May 1713.

141 HL, mssST12, vol. 1, fo. 75: James Brydges, Accounts with brokers, 1711–14; mssST57, vol. 10, fos 117–18: Brydges to John Chamberlayne, 10 May 1714; fos 168–9: same to same, 2 Aug 1714; fos 173–4: same to Edmund Chamberlayne, 10 Aug 1714; 'Lady Chandos' Register', ed. C. H. Collins Baker, *The Genealogists' Magazine*, 10:8 (1948), 255–64, at 264.

142 HL, mssST57, vol. 10, fos 173–4: James Brydges to Edmund Chamberlayne, 10 Aug 1714; vol. 16, fos 193–4: same to same, 16 Jun 1719; vol. 30, fo. 302: same to same, 9 Oct 1727; *Cassandra Brydges*, p. 185: Cassandra Brydges to Emma Chamberlayne, 8 Dec 1724.

143 HL, mssST57, vol. 6, fo. 257: James Brydges to James Leigh, 8 Apr 1712.

younger nephews.[144] As Henry Ingram found at the same age, relations who supported educational travel for the family's sake might see it as a personal luxury only a few years later.

Brydges was driven to invest in family travel by love and self-interest.[145] Although he had not developed emotional bonds with his nephews through fosterage, as had Conway and Huntingdon, his support for their travels relied on his responsibility towards their mothers. Ties of early modern siblinghood extended beyond marriage, with brothers and sisters offering each other practical and emotional support in adult life. Heirs felt obliged to ensure the welfare of married sisters and often provided for their children, with an emphasis on consanguineal kin enduring well into the eighteenth century.[146] Brydges indeed offered greater financial help to the sons of his sisters, whether living or dead, than to those of his younger brother, Henry, whose family barely appears in his letter-books. Notably, Sir Robert Southwell and the earls of Conway and Huntingdon had all arranged English education and foreign travel for their sisters' sons. Brydges may have drawn distinctions of birth here, too, as he only funded an Italian journey for William Leigh, the eldest son of his eldest sister. Brydges, Southwell, Conway and Huntingdon used their wealth and connections to develop a patriarchal sense of self that dependents learned to flatter. All four men offer suggestions of a 'true patriarchy', in which authority held despite the presence of a sister's husband.[147] Their letters contain acknowledgements from nephews and sisters that they had played a 'fatherly' role, and resignations by fathers themselves of authority over their sons.[148] Brydges and Theophilus Leigh reconciled themselves to the fact that the latter's debts were unlikely to be repaid in his lifetime, but Leigh's obedience to his brother-in-law offered

144 HL, mssST57, vol. 15, fos 115–16: James Brydges to Captain [Alexander] Jacob, 21 Jan 1717/18.

145 O'Day, 'Matchmaking and Moneymaking', 292: the same balance explains a package of education and marriage offered to female relations.

146 Ruth Perry, *Novel Relations: The Transformation of Kinship in English Literature and Culture, 1748–1818* (Cambridge: Cambridge University Press, 2004), pp. 108–9, 111, 153; Amy Harris, *Siblinghood and Social Relations in Georgian England: Share and Share Alike* (Manchester: Manchester University Press, 2012), pp. 56, 67–8, 117, 122–3; Bernard Capp, *The Ties that Bind: Siblings, Family, and Society in Early Modern England* (Oxford: Oxford University Press, 2018), pp. 2, 33, 53–4.

147 Rosemary O'Day, *The Family and Family Relationships, 1500–1900: England, France and the United States of America* (Basingstoke: Macmillan, 1994), pp. 71–7, 87–93.

148 BL, Add MS 46952, fo. 169: Lady Perceval to Sir Robert Southwell, 19 Sep 1675; 46955A, fo. 71: Sir Philip Perceval to same, 3 Mar 1678; *CSPD, Charles II*, xi, p. 553: Sir George Rawdon to Conway, 1 Nov 1671; HL, mssST58, vol. 3, fos 200–1: Theophilus Leigh to James Brydges, 16 Jan 1708[/9]; vol. 4, fo. 29: James Leigh to James Brydges, 13 May 1709.

alternative gains. He followed 'whatever Commands' Brydges might 'think fitt' for his children and hoped that his sons would 'justly dutyfully & gratefully make up' his debts in adulthood, though William's flight to Utrecht ensured that they would not do so in cash.[149] The obligation of heirs towards sisters and their children tended to come with a reciprocal expectation of deference and service.[150] Intertwined with his blood ties to the sons of his sisters, Brydges recognised the power of educational patronage to create a network of upwardly mobile dependents.

Travel through Institutions

Seen from the other side, of course, Brydges's relations tapped into his wealth and status to travel beyond their means. For the Leighs and the Chamberlaynes, he helped to fulfil long-held parental desires to send sons abroad. If kin networks allowed families to take advantage of educational opportunities at home and abroad, parents also made use of the expanding British presence on the Continent. The later seventeenth century saw rising public interest in European affairs and, after the 1688 Revolution, an interventionist foreign policy directed against France.[151] The restored Stuarts had made increasing use of envoys, meanwhile, and the diplomatic corps undertook longer and more frequent missions by the 1680s and 1690s. The number and significance of posts rose after the revolution, with the Continent at the centre of diplomacy even as Britain became an imperial power.[152] The army was also a European institution: it primarily fought on the Continent, especially between 1689 and 1756, and British soldiers fought European enemies alongside European allies and auxiliaries.[153] Before then, too, detachments of the kind with which John Rawdon died in 1676 had fought for neighbouring states.[154] The English, Scottish and Irish armies had shrunk at the Restoration, leaving few opportunities for a career, but expansion under James II intensified after the revolution, bringing

[149] HL, mssST58, vol. 3, fos 200–1: Theophilus Leigh to James Brydges, 16 Jan 1708[/9]; vol. 6, fos 37–8: same to same, 3 Jun 1710.

[150] Capp, *Ties that Bind*, pp. 36, 143; Harris, *Siblinghood*, p. 127.

[151] Claydon, *Europe and the Making of England*, pp. 125–219; Steve Pincus, *1688: The First Modern Revolution* (New Haven, CT: Yale University Press, 2009), pp. 305–65.

[152] Helen Jacobsen, *Luxury and Power: The Material World of the Stuart Diplomat, 1660–1714* (Oxford: Oxford University Press, 2012), p. 3; D. B. Horn, *The British Diplomatic Service, 1689–1789* (Oxford: Clarendon, 1961), pp. 13–16.

[153] Conway, *Britain, Ireland and Continental Europe*, p. 266.

[154] Childs, *Army of Charles II*, pp. 162–83.

veterans home and opening up paths for younger men. From 1689, the wars against France took most of this growing army to Flanders and sent the navy, the recipients of longer-term investment, overwhelmingly to the Mediterranean.[155] Few historians set continental travel alongside diplomatic and military engagement, but families used Britain's European institutions to achieve various kinds of educational journey.[156]

Diplomacy and educational travel existed, to some extent, in a virtuous circle. While contemporaries believed, as they had done since the sixteenth century, that a continental voyage offered good training for ambassadorial employment, diplomats in turn enabled journeys. Across the early modern period, envoys and ambassadors took young gentlemen abroad in their retinues. Some of these companions intended to begin a similar career themselves, while others profited from an opportunity to travel.[157] Connections to a diplomat often depended upon kinship or neighbourhood, as officials followed the common practice of merging public and private networks.[158] Sir Robert Southwell, for instance, continued his role as an educational fixer over two decades of ambassadorial work, which took him to Lisbon twice between 1665 and 1668, to Brussels in 1671 and to Brandenburg in 1680–1. His father-in-law, Sir Edward Dering, turned these postings to the advantage of his children. Dering's 18-year-old eldest son, also Edward, joined the second trip to Portugal, touring southern Spain with Southwell's assistant, Francis Parry, who had recently served as a travelling tutor. Little survives of the journey, but Edward also accompanied Southwell home through France.[159] Another of Dering's sons, Charles, went to Brandenburg at a similar age, his father 'taking this opportunitie to let him see this part of Europe under so good a guide'.[160] Southwell later reflected on his postings in a family history addressed to his son, assuring him that had

155 Roger B. Manning, *An Apprenticeship in Arms: The Origins of the British Army 1585–1702* (Oxford: Oxford University Press, 2006), pp. 400, 403, 407, 443; Conway, *Britain, Ireland and Continental Europe*, pp. 240–5.

156 Conway, *Britain, Ireland, and Continental Europe*, is one exception.

157 Brennan, ed., *Origins of the Grand Tour*, pp. 10–12, 17–19; Horn, *British Diplomatic Service*, pp. 123–30; Mori, *Culture of Diplomacy*, pp. 152–3. For the same phenomenon among the Dutch and Flemish, see Verhoeven, *Europe within Reach*, pp. 56–61.

158 Aaron Graham, *Corruption, Party, and Government in Britain, 1702–1713* (Oxford: Oxford University Press, 2015).

159 BL, Add MS 35100, fo. 27: Edward Dering and Francis Parry to Sir Robert Southwell, 16 Apr 1669; TNA, SP78/127, fo. 34: Southwell to [Joseph Williamson], 3 Aug 1669. Parry had travelled through France, Italy, Spain and Germany in 1665–8: Bodleian, MS Eng. Lett. c 328/3, fos 477–598.

160 *The Diaries and Papers of Sir Edward Dering, Second Baronet, 1644 to 1684*, ed. Maurice F. Bond (London: HMSO, 1976), p. 117.

he 'beene borne in some of these adventures, or of age enough in the latter ones; I had not fail'd to have taken you with me'.[161] Once Edward was old enough, in the 1690s, his father twice used Whitehall contacts to send him to the Low Countries with government officials. The young man shadowed his superiors, explored the region and visited courts in which 'my name had many friends'. Sir Robert had intended the trips as preparation for a diplomatic posting, although Edward soon began an administrative career.[162] Whether or not young men later pursued ambassadorial work, diplomatic retinues offered opportunities for travel, supervision and instruction.

The Molesworth family, from County Dublin and Yorkshire, used diplomatic networks to achieve a spread of educational travel, though their fragmentary papers offer few glimpses of their means. Robert Molesworth often pleaded poverty in search of patronage, but he continued to buy land and improve his estates throughout supposedly troubled periods. Molesworth nevertheless portrayed himself as a Whig martyr, with family finances seeming to fluctuate as he and his sons gained or lost office under various ministries.[163] Investment in the South Sea Company also makes gauging their wealth more complicated. John, the eldest son, enjoyed a 'paper fortune' of £100,000 in the summer of 1720, but almost all of it disappeared when the bubble burst. Robert lost a more modest two-thirds of his £2,000 investment, which he had probably made in order to provide for his children.[164] He and Laetitia had a 'very numerous' family, listing eight sons and three daughters in 1709, and gave the boys a 'very chargeable' education 'for the publick service'.[165] This course stretched the family budget, as did the financial pressures on diplomats, who often received their pay late or not at all, with Robert serving as envoy to Denmark in 1689–92 and John

[161] Bristol Archives, 45317/16/2/1: Sir Robert Southwell, Narrative of Southwell family history, 1700, p. 30.

[162] Beinecke, Osborn MS b66: [Edward Southwell], Holland voyage, 1691; BL, Add MS 38015, fo. 389: Sir Robert Southwell, 'Instructions to my son Edward Southwell when in Holland', 5 Jan 1690/1; Koninklijke Bibliotheek, MS 133c.6: Edward Southwell to Sir Robert Southwell, 31 Aug 1696; Katharine Fremantle, 'A Visit to the United Provinces and Cleves in the Time of William III: Described in Edward Southwell's Journal', *Nederlands Kunsthistorisch Jaarboek*, 21 (1970), 39–68.

[163] D. W. Hayton, 'Molesworth, Robert (1656–1725)', in Hayton, Cruickshanks and Handley, eds, *House of Commons 1690–1715*.

[164] Patrick Walsh, *The South Sea Bubble and Ireland: Money, Banking and Investment, 1690–1721* (Woodbridge: Boydell, 2014), pp. 79, 81, 91, 102–3, 105.

[165] BL, Add MS 61639, fo. 4v: [Robert Molesworth], Autobiographical fragment, [1709]. Only nine of an original 17 children were living in 1721: NLI, microfilm P.3753: Robert Molesworth to John Molesworth, 15 Aug 1721.

posted to Florence and Turin in the 1710s and 1720s.[166] If these postings consumed Molesworth finances, however, especially as John's arrears of pay stacked up, they also offered opportunities for educational travel.[167]

As with the Ingrams and the Leighs, little survives of the Molesworths' first forays on the Continent. Robert recalled 'severall voyages into forreign countreys' made between 1684 and 1688, 'allwayes takeing Holland in my way'.[168] His travels must also have included Italy, as he proved familiar with its climate and prices in later life, while his published work claims experience of France and Germany.[169] His contemporary rental income of £2,800, a rare detail, would indeed have permitted extensive travel.[170] Molesworth ventured abroad as a husband and father in his late twenties, though, modelling the adult travel that he would advocate in print.[171] His journeys to the Netherlands may have been visits to his brother-in-law, then serving in the Dutch army, but they were also, perhaps, attempts to gain the patronage of William of Orange. If so, they paid off after the 1688 Revolution, when Molesworth was sent to Copenhagen.

Molesworth provided ambitious travel to his two eldest sons in the final years of the seventeenth century, taking a similar approach to the Ingram and Brydges families. John and Richard were around 19 and 18, travelled with a tutor, lived on £300 a year, probably saw France and Italy and returned to England in mid-1700, though little else survives of the journey.[172] After a period at Middle Temple, the two men left again 'to travel for some years' in 1703. They served on the Continent as aides-de-camp to the Duke of Marlborough, with Richard beginning a long military career but John leaving after a year, probably for a diplomatic retinue in Hanover.[173] Both sons

166 Horn, *British Diplomatic Service*, pp. 46–75.

167 NLI, microfilm P.3752: Robert Molesworth to Laetitia Molesworth, 23 Feb 1711/12; BL, Add MS 61639, fo. 120–1v: same to [Earl of Sunderland], 6 Mar 1714/15.

168 BL, Add MS 61639, fo. 3: [Molesworth], Autobiographical fragment; TNA, SP44/336, fo. 201: pass for Robert 'Moldsworth' to go beyond seas, 10 Aug 1685.

169 NLI, microfilm P.3753: Robert Molesworth to John Molesworth, 30 Jun 1720; same to same, 20 Oct 1722; Molesworth, *Account of Denmark*, pp. 214, 239.

170 Hayton, 'Molesworth', in Hayton, Cruickshanks and Handley, eds, *House of Commons 1690–1715*.

171 Molesworth, *Account of Denmark*, sig. A4v.

172 John and Richard were in London in February 1697, they came home through France in May 1700 and John had visited Florence before 1703: NLI, microfilm P.3752: Robert Molesworth to Laetitia Molesworth, 23 Feb 1697; same to same, 17 May 1700; Le Comte Lorenzi to [John Molesworth], 8 Jun 1703; HMC, *Report on Manuscripts in Various Collections, Vol. VIII* (London: HMSO, 1913), p. 221: Robert Molesworth to Laetitia Molesworth, 24 May 1700.

173 *Letters of Eminent Men, Addressed to Ralph Thoresby, F.R.S.*, 2 vols (London: Henry Colburn and Richard Bentley, 1832), i, p. 444: Robert Molesworth to Ralph Thoresby, 12 Apr 1703;

crossed the Channel regularly in the next couple of years, spending up to an annual £700 between them. Richard remained with the army and John probably continued to travel in diplomatic retinues, as he was hopeful of his own posting in 1707.[174] John planned to take a tutor and one or two of his younger brothers, including Coote, aged only 9, of whom he was 'fondest', but the job did not materialise.[175] He finally received a posting to Florence three years later, again planning to take Coote, but in the end he left his brother to prepare at a French school in London. Plans changed, however, and Coote did not come to Florence. John might have taken another brother, Bysse, but Robert decided that the 11-year-old should 'stay & study till he be fit for the university, to make either a parson or a Lawyer of'.[176] Although Richard used breaks from the army to meet John twice in Italy between 1712 and 1714, travelling back from the second visit through France, John was unable to use his Tuscan post to widen educational travel in the family.[177]

John left Florence in 1714, but efforts continued to secure time abroad for his brothers. Where the Ingrams had seen travel by sons beyond the second-born as a luxury, the Molesworths spread foreign experience to Coote and Bysse, their sixth and seventh sons. They travelled in different capacities to their elder brothers, however. Coote studied at Leiden between 1717 and 1718, 'still with a view to being employed in publick affairs', and the 20-year-old recognised that his father had favoured him over his other younger sons. John, who visited Leiden on his way to take the waters at Spa, reported that 'Coote had left a very good Character there, both for his Conduct, Capacity, and Application'; he was likely to get 'the greatest Reputation' and 'the best fortune of us All'. In 1718, the favour of his parents and eldest brother carried Coote from Leiden to six months of French and 'Genteel Exercises' at Geneva.[178] Meanwhile, Bysse's future changed at 20 when he secured a position with the second Baron Carteret on his

NLI, microfilm P.3752: same to Laetitia Molesworth, 7 Aug 1704; TNA, SP98/21: John Molesworth to [Thomas Wentworth, Lord Raby], 20 Dec 1704.

[174] HMC, *Various Collections, Vol. VIII*, p. 233: Robert Molesworth to Laetitia Molesworth, 12 Apr 1705; TNA, SP44/390, p. 205: pass for John Molesworth to Holland, 6 Jul 1705; SP44/391, p. 114: pass for Richard Molesworth to Holland, 24 Apr 1705.

[175] NLI, microfilm P.3752: Robert Molesworth to Laetitia Molesworth, 11 Jun 1707.

[176] NLI, microfilm P.3752: Robert Molesworth to Laetitia Molesworth, 10 Jun 1710; same to same, 6 Nov 1710.

[177] William Molesworth, 'John Molesworth (1679–1726) as a Patron of Art: Complacence, Connoisseurship and Commissions', M.Litt. thesis (Trinity College Dublin, 2010), pp. 58, 70.

[178] NLI, microfilm P.3753: Coote Molesworth to John Molesworth, 6 Feb 1718; John Molesworth to Laetitia Molesworth, 3 Oct 1719; Coote Molesworth, Family history, in Mayo, 'Molesworth's *Account of Denmark*', pp. 219–20.

year-long diplomatic posting to Sweden from late 1719. The connection probably emerged from Robert's long-standing cultivation of Tory patrons, despite his radical Whig politics. He would later complain that Carteret had indulged his son, allowing him to run up debts of £360, but he recognised that competition for places was high. Carteret's papers indeed contain letters from several correspondents seeking to place young men in his retinue.[179] When Bysse fell ill, Robert accordingly sent Coote back from Geneva to the Netherlands, ready to replace his brother should he die in Sweden. Bysse recovered, so Coote spent another year at Leiden before returning to London and entering Middle Temple in 1720.[180]

Later that year, a new opportunity arose when John was posted to Turin. Coote, then 22, wrote to his mother in the hope that she would use her 'interest with my brother to procure me leave to accompany him', raising and refuting all objections to his leaving the Inns of Court for Italy. He could avoid an expensive London lifestyle, though an envoy's brother 'must make some little figure'. He could take his law books with him and spend the mornings in study. In the afternoons, John could introduce him 'to the polite world', giving him 'some notion of a Court, and the turn of a Gentleman', which would help to 'throw off that bashfullness and awe which I have foolishly contracted'. These improvements would make him a better lawyer, but Italian 'might be of use to me, in case my attempts in the Law shou'd be vain'. The academy at Turin, meanwhile, offered lectures in civil law and, should the family afford it, training in riding, fencing and dancing. Finally, he would provide company for his sister-in-law 'in a strange country'. Coote was 'sure my good Brother had not thought on these advantages, or he wou'd not have said No when it was asked him whether I went with him'.[181] Lady Molesworth's intercession obviously worked, as Coote accompanied John to Turin.

The arrangement held for two years, during which Coote received an unspecified allowance from his parents. He did not attend the academy, but unofficially acted as his brother's secretary.[182] Importantly, however, Coote

179 NLI, microfilm P.3752: Bysse Molesworth to John Molesworth, 25 Jul 1719; P.3753: John Molesworth to Laetitia Molesworth, 3 Oct 1719; Robert Molesworth to [John Molesworth], 27 Feb 1721/2; BL, Add MS 22512, fo. 87: Carteret to Mr Tilson, 19 Mar 1719/20; 22514, fo. 9: same to Mr Jackson, 3 Oct 1720; D. W. Hayton, 'Molesworth, Robert, first Viscount Molesworth (1656–1725)', *ODNB*, online edn, 2008 (https://doi-org/10.1093/ref:odnb/18901).

180 Coote Molesworth, Family history, in Mayo, 'Molesworth's *Account of Denmark*', pp. 219–20.

181 NLI, microfilm P.3753: Coote Molesworth to [Laetitia Molesworth], [1720].

182 NLI, microfilm P.3753: Robert Molesworth to John Molesworth, 19 May 1721; Coote Molesworth, Family history, in Mayo, 'Molesworth's *Account of Denmark*', pp. 219–20.

was not in Turin for diplomatic training. John admitted that his brother could not study the law 'with the same Application as he might at the Temple', but he read 'every day something relating to his future profession, & in the mean time improves in some Accessory qualifications perhaps not intirely foreign from it'. Here, John included 'French (which he is Master of)' and 'a little Italian', alongside classical and legal knowledge. Coote lacked 'impudence', a 'commodity' that he would acquire at the 'refin'd Court' of Savoy, but further study at Middle Temple would allow him to 'make a figure in Chancery'.[183] Coote's time in Italy was effectively ended by Carteret in 1721, when the new Secretary of State for the South instructed John to take 22-year-old Bysse, who had returned from Sweden the previous year, and 'breed him up' for a government position. The family could not give Bysse much money in Turin, with John reassuring his father that it was not 'requisite he should appear fine', but Robert nevertheless spent twice Bysse's annual allowance on transport and clothing 'good enough for my 7th son'. John was unable to support two brothers in Turin, so Coote had to return to London.[184] Little survives of the journey, but Coote later remembered that he had 'obtained permission to visit the most remarkable towns in Italy; by which means he was twice at Rome, at Naples etc. etc. & came home through France'.[185] Bysse served as John's secretary until at least 1724, when Carteret, by then Lord Lieutenant of Ireland, promised to find him a position.[186] At the latest, Bysse must have left Italy with John in 1725.

The Molesworth approach to educational travel shows some stratification by birth order, but also left room for individual 'genius'. Time abroad was essential for the two eldest brothers, John and Richard, as it would soon be for Irwin and Rich Ingram, and later for John and Henry Brydges. None of these families adopted the rigidity of the Percevals, who afforded travel to Sir Philip alone. The Molesworths even offered scope for negotiation by brothers beyond the eldest two: where Henry Ingram had waited until he could fund himself, Coote Molesworth persuaded his family to pay for his stay in Turin and a tour of Italy. As with generations of Rawdons, emotional bonds help to explain journeys by younger brothers, since John was 'fondest' of Coote and their mother argued for his travels. Like Conway

183 NLI, microfilm P.3752: John Molesworth to Robert Molesworth, 10 Dec 1721.

184 NLI, microfilm P.3752: John Molesworth to Robert Molesworth, 10 Dec 1721; P.3753: Robert Molesworth to [John Molesworth], 27 Feb 1721/2.

185 Coote Molesworth, Family history, in Mayo, 'Molesworth's *Account of Denmark*', pp. 219–20.

186 HMC, *Various Collections, Vol. VIII*, p. 376: John Lekeux to [John Molesworth], 17 Oct 1724.

with Arthur Rawdon, they also sent Coote abroad because they believed that he would make the most of it. In both cases, journeys by younger sons added to those of their elder brothers, rather than replacing them. Decisions over travel weighed emotional and economic concerns, but some families mobilised networks and resources to spread continental experience down the birth order without compromising the travels of their eldest sons. The Molesworths had sent John, Richard and a tutor on a lengthy journey, but they also afforded continental experience to Coote and Bysse through the diplomatic corps.

Families similarly took advantage of the army and navy. War instinctively seems bad for travel: in the traditional telling, peace with Spain opened the Continent to British travellers from 1604, conflict prevented journeys at intervals throughout the next two centuries and the French Revolutionary and Napoleonic Wars ended the Grand Tour for good.[187] Conflict certainly shaped itineraries, with Sir John Perceval avoiding France during the War of the Spanish Succession and William Leigh awaiting peace in order to enter the country. Attention to family strategies, however, reveals more complicated relationships between educational travel and the military. Travellers consistently volunteered to fight in the Low Countries and Germany between the late sixteenth and mid-eighteenth centuries, either in preparation for a military career or as looser training in the martial side of elite masculinity.[188] Initially, young men served in foreign armies or, like John Rawdon, in English units in the service of other states. After 1689, however, the British military presence on the Continent more easily afforded travellers a taste of army life. Those who travelled during the War of the Spanish Succession, for instance, socialised with officers, toured camps and considered temporary service.[189] From the other side, families used the military, like the diplomatic corps, as a route to further travel. Daniel Dering, whose brothers had joined Southwell's ambassadorial retinues, was with the navy in Spain by his mid-twenties. As the threat of war with France declined in 1679, he hoped for parental permission to visit Florence and Lucca, since his ship would soon sail for Genoa and Livorno.[190] Such opportunities proliferated with the greater commitment to the Mediterranean after 1689.

187 For more nuance, see John Towner, 'The English Tourist and War, 1500–1800', in Richard Butler and Wantanee Suntikul, eds, *Tourism and War* (London: Routledge, 2013), pp. 49–63.
188 Roger B. Manning, *Swordsmen: The Martial Ethos in the Three Kingdoms* (Oxford: Oxford University Press, 2003), pp. 103–38; Goldsmith, *Masculinity and Danger*, pp. 75–109.
189 WYAS, WYL150/6005/12: Metcalfe Robinson to Sir William Robinson, [*c.* Mar 1704]; Goldsmith, *Masculinity and Danger*, pp. 83, 94.
190 Kent Archives, U1713/C1/53: Daniel Dering to Sir Robert Southwell, 11 Nov 1679.

The second Arthur Rawdon, already a naval officer at 17, landed in Italy for the winter of 1739–40 and accompanied his brother, Sir John, home through France.[191] Increasingly, the British armed forces offered chances to merge educational and military travel.

James Brydges, as army paymaster, exploited military networks to facilitate travel in his family. He used contacts from wartime finance to send money to sons and nephews abroad during the 1710s and 1720s, continuing his lucrative practice of merging official and private connections.[192] Jacob Senserf of Rotterdam and Andries Pels of Amsterdam received bills of exchange, extended credit and forwarded letters for William Leigh, John Chamberlayne, John Brydges and Henry Brydges. In turn, James Brydges secured a pass to England for Senserf's son, Walter, 'on his ramble'.[193] Brydges delved deeper for his sons, as Richard Cantillon extended credit in Paris and Pels provided money and introductions in France, Germany and Vienna.[194] The army also provided at least three companions for John and Henry during domestic and foreign travel.[195] Their father is a classic example of the fact that Britain's deepening military engagement with the Continent offered the chance for financial gain. Less well known are the opportunities it provided to acquire social and cultural capital through travel.

When Brydges sent Theophilus Leigh's heir, William, to Italy, he had already placed his second son, James, abroad in the army. In 1708, Brydges found 15-year-old James a commission with the Huguenot Colonel La Fabrèque's's regiment of French dragoons, serving with the British in Portugal, and promised that the young man might 'depend upon me for making your fortune'.[196] Brydges bought his nephew promotions beyond his

[191] PRONI, D2924/1/41-3: Sir John Rawdon to Thomas Prior, 20 Jul 1739; *Joseph Spence: Letters from the Grand Tour*, ed. Slava Klima (Montreal: McGill-Queen's University Press, 1975), p. 284: Spence to Mirabella Spence, 1 Jun 1740.

[192] Graham, *Corruption, Party, and Government*, pp. 183–227. For Senserf, Cantillon and Pels, see pp. 103–4, 113, 198, 200, 204, 218.

[193] HL, mssST12, vol. 1, fo. 75, Brydges's accounts with Pels, 1714; mssST57, vol. 7, fo. 46: James Brydges to William Leigh, 3 Jun 1712; vol. 18, fos 277–8: same to Charles Stuart, 18 Sep 1721; vol. 19, fo. 95: same to Jacob Senserf, 27 Jan 1721[/2]; vol. 24, fo. 271: same to Walter Senserf, 5 Oct 1724; mssST58, vol. 10, fos 75–7: Jacob Senserf to James Brydges, 8 Dec 1711.

[194] HL, mssST57, vol. 19, fo. 113: James Brydges to Andries Pels, 11 Jul 1721; vol. 20, fo. 141: same to Charles Stuart, 9 Mar 1721/2; vol. 22, fo. 256: same to Captain Lancaster, 3 May 1723.

[195] Captains Marchand and Robinson during their domestic travels, Captain Lancaster abroad with John and Captain Andrew Robinson (presumably the same) abroad with Henry.

[196] HL, mssST57, vol. 3, fo. 46: James Brydges to James Leigh, 14 Sep 1709; Charles Dalton, ed., *English Army Lists and Commission Registers, 1661–1714*, 6 vols (London: Eyre & Spottiswoode, 1892–1904), vi, p. 264.

years, wrote recommendations to his commanding officers and sent money to cover his movements across western Europe.[197] Captain Leigh stayed with La Fabrèque for two and a half years, served briefly as major in a Spanish regiment, spent a few months as lieutenant colonel in an English regiment in Catalonia, crossed to Livorno and rode non-stop for 12 days to begin an equivalent position in the Low Countries. He took over as colonel in 1712, aged only 19, but the regiment was disbanded the following year.[198] When James died of consumption in 1714, Lady Chandos, who had hosted him during his domestic education, mourned her grandson as 'a young man that gave early promises of his being an honour & advantage to his family'.[199]

Theophilus Leigh thought of William's travels and James's military career in similar terms. Brydges had given 'my second son an occasion of travelling abroad', just as he would allow his eldest to 'see something of the world too'. Both received from their uncle a 'more improving & manly Education'.[200] Adolescent service abroad allowed James to lose his 'Childishness', to harden himself against discomfort, to 'learn somwhat' and to make connections, particularly to Tory officers of his father's 'church-party'. Theophilus sent him away with standard warnings against wine and women, and James later suspected that Brydges had chosen La Fabrèque because 'in the french Regiments they live more sober than in the strictest university'.[201] A Francophone regiment indeed offered educational opportunities for such a young officer. Theophilus insisted that his second son improve his French, 'which I would have you by noe meanes to neglect'. Portuguese and Spanish would come 'of necessity', leaving time for 'those Qualifications, with many others which attend a Travailler, will fit you for other good Employments as well as the Military'.[202] Allied success in Spain might allow James the opportunity of 'seeing some part of that kingdom', where he should continue

197 SBT, DR18/20/18b: James Brydges to Sir Robert Rich, 22 Jul 1711; same to Duke of Argyll, 22 Jul 1711; James Leigh to Theophilus Leigh, 13 Nov 1711; HL, mssST58, vol. 10, fos 237–8: William Leigh to James Brydges, 26 Feb 1711/12.

198 SBT, DR18/20/18b: Theophilus Leigh to James Leigh, 16 Apr 1711; James Leigh to Theophilus Leigh, 4 Sep 1711; same to same, 27 Jun 1712; Theophilus Leigh to James Leigh, 6 Jul 1713.

199 SBT, DR18/20/18b: Elizabeth, Lady Chandos to William Leigh, 23 Jan 1713/14.

200 HL, mssST58, vol. 5, fo. 272: Theophilus Leigh to James Brydges, 14 May 1710; SBT, DR18/20/18b: same to James Leigh, 10 Jun 1710.

201 HL, mssST58, vol. 3, fo. 200: Theophilus Leigh to James Brydges, 16 Jan 1708/9; SBT, DR18/20/18/47: same to James Leigh, 4 Jul 1709; DR18/20/18b: same to same, 25 Sep 1710; same to same, 11 Dec 1710; James Leigh to Theophilus Leigh, 4 Sep 1711. Warnings were appropriate, as family tradition later held that 'rich wines' and 'conviviality' in Portugal had brought on James's consumption: DR18/31/884a: Leigh, 'History of the Leigh Family', p. 51.

202 SBT, DR18/20/18/48: Theophilus Leigh to James Leigh, 10 Dec 1709.

'reading & learning a Constant account of all forreign as well as domestique transactions', which would be 'profitable for yor instruction hereafter'.[203] Theophilus asked James to write to him in French, finding grammatical errors that his son explained by the fact that he had learned through speaking. After his transfer to an English regiment, though, James hoped to keep his fluency 'by reading French Books'.[204]

The Leigh brothers met in the Low Countries in 1712, when William was returning from Italy and James taking up his new commission. They briefly travelled together, spending time in Rotterdam, Antwerp, Ghent, Bruges and Brussels, the latter 'inferiour to very few towns' that James had seen. Theophilus equally hoped that he would help William to 'see what he can, of the Army' in Flanders.[205] After several years apart, William found that his brother had made 'very great improvements' not only as a soldier but also 'in all the other good Qualities' that made 'a Compleat Gentleman'. James spoke excellent French, with 'a greater command of words and a better accent' than William could hope to acquire 'for some time'. He was 'very obliging' and 'Courtious', too, possessing a good knowledge of 'manners, which of all sciences are the most usefull'.[206] Perhaps William exaggerated for their father, but the intended outcomes of James's years abroad in the military were identical to those that his elder brother had sought on a more leisurely journey to Italy. Although William and James Leigh would usually appear in separate books on the Grand Tour and the army, from a family perspective, the exploitation of kinship ties and military networks allowed two sons to travel for their education.

Whose Strategy Was Family Strategy?

This book often uses a shorthand of families acting as one, but relations could hold competing visions of how travel might serve personal or family interests. Fathers and uncles tended to exercise authority, not only because they held the purse strings but also because men such as Sir Robert Southwell

[203] SBT, DR18/20/18b: Theophilus Leigh to James Leigh, 25 Sep 1710.

[204] SBT, DR18/20/18b: Theophilus Leigh to James Leigh, 16 Apr 1711; James Leigh to Theophilus Leigh, 28 Jul 1711; Theophilus Leigh to James Leigh, 1 Sep 1711; James Leigh to Theophilus Leigh, 13 Nov 1711.

[205] SBT, DR18/20/18b: William Leigh to Theophilus Leigh, 29 Jun 1712; Theophilus Leigh to James Leigh, 21 Jul 1712; William Leigh to Theophilus Leigh, 21 Sep 1712; James Leigh to same, 2 Oct 1712.

[206] SBT, DR/18/20/18b: William Leigh to Theophilus Leigh, 30 Jul 1712.

and the Earl of Huntingdon had personal experience of educational travel. Sometimes, as with James Brydges and Theophilus Leigh, wealthy and well-travelled uncles even outranked fathers. Decisions emerged not as patriarchal decrees, however, but from unequal negotiations among family members.[207] They dealt with two issues: whether a young man should travel and, if so, what he should do on the Continent. The first was often settled early, allowing long-term planning through prior stages of education, but the second saw ongoing debate.

Mothers continued to influence the education of their sons, in England and on the Continent, for all that elite male upbringing was an incremental removal from home.[208] Sir George Rawdon sent Edward, John and Arthur to France after 'discourse with their mother', something that must have shaped many strategies where both parents survived.[209] In these conversations, sons often saw their mothers as advocates. Coote Molesworth went to Laetitia first, hoping that she would use her 'interest' to support his plans for Italy. Even where negotiations were with trustees rather than a father, Irwin applied the pressure to his mother, presuming that she would argue for his educational needs. Both sons were correct, and both mothers successful, but they suggest that women's authority derived from relationships with men.[210] Even Lady Perceval, who was in France and had consulted educators there, failed to convince her brother, Sir Robert Southwell, to send her youngest son to the 'brave Academy' of Saumur. Conscious of her weak position, she made clear that her eldest, Sir Philip, also favoured the plan. Southwell still thought his sister vulnerable to the arguments of 'airy Frenchmen', however, and kept John in England.[211] Women's influence on travel varied, but it depended on the disposition of husbands, brothers and even sons.

What about the influence of sons? It is important to distinguish here between 'genius' and 'inclination', two words that appear regularly in discussions of travel. Parents, guardians and tutors identified a young man's 'genius' on the basis of his personality and academic performance rather than his desires. When they allowed 'genius' to shape a son's future, as landowners could only do for younger brothers rather than heirs predestined to inherit, it was often to maximise resources rather than indulge vocations. This left room for travel by younger sons such as John and Arthur Rawdon.

207 Barker, *Family and Business*, pp. 13–14, 78–81.
208 French and Rothery, *Man's Estate*, p. 83.
209 *CSPD, Charles II*, xvi, p. 213: Sir George Rawdon to Conway, 28 Mar 1674.
210 French and Rothery, *Man's Estate*, p. 68.
211 BL, Add MS 46955A, fo. 110: Lady Perceval to Sir Robert Southwell, 13 Apr 1678; 46956B, fo. 127: Southwell to John Perceval, 24 Sep 1678.

As in the long-term plans of James Brydges or the will of Sir Bourchier Wrey, however, many fathers insisted that an heir would spend time abroad with no thought to his character.[212] Seventeenth- and eighteenth-century parents usually paid some attention to the 'inclinations' of their sons, but filial input was limited on such a large investment as travel.[213] Again, it is worth distinguishing between whether a young man should go abroad and what he should do there. The first question was often resolved without sons. Whether time on the Continent was long foreseen, especially in families with a developing tradition, or depended upon identifiable 'genius', the older generation tended to know best. Even sons who had already inherited had difficulty affecting whether younger brothers could travel, with Sir Philip Perceval unable to bring John abroad and Irwin failing to keep Rich there. Although Coote Molesworth convinced his parents to let him go to Italy, he had been preparing for some form of travel from an early age. Parents could not prevent young men such as Henry Ingram from going abroad once they were of age and could fund themselves, but they could deny the chance to travel as part of family strategy.

If sons had little say over whether they went abroad before adulthood, those who travelled had more influence over what they did on the Continent. Slow communications allowed plenty of scope—it was easier to beg forgiveness than ask permission—and friction was common, as between Irwin and those at home. The viscount's trustees had decided when he should travel but they could do little, beyond pleading, to control his itinerary and spending on the Continent. The development of mastery over such questions was one of the aims of travel, however, and parents often built choice into itineraries as opportunities to exercise judgement. Just as Coote Molesworth gained permission for an Italian tour after his stay in Turin, so the younger Thomas Worsley could extend his time at Leiden into several months in Germany, including Hanover and Hamburg.[214] Tutors managed the independence of many travellers, another source of friction with increasingly assertive young men such as Sir Philip Perceval, but the authority of those judged able to travel alone was unchallenged. Sir William Robinson left 'the method of your travails' up to 21-year-old Metcalfe's 'own discretion', though he recommended 'some stay in Holland' as a 'necessary'

212 HL, mssEL 9084: Sir Bourchier Wrey's will, 16 Aug 1694: only his younger sons were to be bred according to 'their respective genius and inclinacons'.
213 Thomas, *Ends of Life*, p. 33.
214 WYAS, WYL100/C/9A: Thomas Worsley to John Roades, 24 Mar 1704/5; same to same, 8 Apr 1705; WYL100/C/9B, John Haccius to [same], 16 Dec 1704; Worsley to same, 30 Jun 1705; NYCRO, ZON13/1/274: Richard Hey to Worsley, 27 Oct 1705.

step before Italy.[215] Alan Brodrick received no parental instructions at 22, taking some advice from his uncle and great-aunt in London but mostly deciding for himself on the 'most convenient' route.[216] The practicalities and objectives of travel afforded greatest influence to those who were already on the Continent.

Conclusion

This chapter has placed educational travel in its essential contexts of family strategy, estate finance and landed identities. The Ingrams demonstrate how travel could become central to the standing and self-image of a noble family over several generations. It was possible to travel more cheaply, so the high cost of journeys in prominent families emerged from the need to live up to their position abroad. For the Rawdons, meanwhile, changing levels of financial and social capital shaped fluctuating engagement with travel over a century. If the Ingrams and later Rawdons felt that their elevated position obliged them to travel well, the Brydges family saw time abroad as an opportunity to secure recent advancement. As their wealth ebbed away, however, later Brydges travellers crossed the Channel with less ambitious intentions. The more modest Leighs, like the first generation of Rawdons, relied upon borrowing and the patronage of kin to travel beyond their means. With the growing British institutional presence on the Continent in the late seventeenth and eighteenth centuries, the Molesworths and the Leighs used military and diplomatic networks, overlapping with those of kin in office-holding families, to broaden their foreign experience. Of the travellers in this chapter, only elder brothers in the Perceval, Ingram and Molesworth families saw Italy on their parents' money. William Leigh, younger Molesworths and later generations of Rawdons instead relied upon patronage, kinship and institutional networks to venture that far. Many others, including the early Rawdons, younger Ingrams and—for all their wealth—the entire Brydges family, never went to Italy. Beginning with family practice, rather than working backwards from the Italian Grand Tour, reveals more diverse experiences.

Different forms of educational travel emerged from the various ways in which families balanced their means, ambitions and obligations. They faced their own pressures: while the Ingrams argued with employees and

[215] WYAS, WYL150/6002/51: Sir William Robinson to Metcalfe Robinson, 20 Mar [1704].
[216] SHC, 1248/6/137: Alan Brodrick II to Sir Alan Brodrick, 2 Feb 1724/5.

friends over memories of ancestral travel and the fuzzy demands of honour, the Brydges family and the early Rawdons sought—at different levels—to make the most of unprecedented outlays on continental education. In all instances, travel became a central issue in family strategy. A key question was how to divide foreign experience among brothers. Parents variously stressed birth order, aptitude or inclination, though they tended to prioritise travel by the eldest one or two sons. On the perennial issue of how best to train younger brothers, families struck a balance between wasting money on inappropriate journeys and risking that an heir might die abroad or soon after his return. As a result, second sons were the likeliest of younger brothers to travel, as they were to attend university, though they frequently crossed the Channel in a lesser capacity than the eldest son.[217] Estate finances were crucial in these decisions, as were the possibilities offered by kinship and institutional connections, but families also weighed affection for individuals and estimations of their abilities, such that even younger brothers might travel. Wealthy and experienced uncles occupied a prominent position, as they had in domestic educational mobility, and their priorities and emotional attachments carried significant sway alongside or instead of those of fathers. Mothers also held an important role, though one that must often be inferred, in deciding whether sons would travel. The preferences of young men themselves, meanwhile, more often shaped activities abroad than whether they crossed the Channel in the first place. Within the family, competing financial, emotional and reputational concerns led some to embrace the Italian Grand Tour but others to seek out adaptations, approximations and alternatives.

Travel served as a barometer for the changing wealth and self-image of landed families over generations, while also providing an outlet for their hopes for advancement or fears of decline. It offered one avenue for the investment of financial and social capital, in the hope of securing or enhancing family standing. In these ways, it was perhaps no different from other flexible methods, such as building, through which landowning families could buy into an elite identity as far as their budgets would allow.[218] This flexibility meant that far more young men could travel than could contemplate a journey of several years to Italy, spreading first-hand knowledge of the Continent deep into the gentry. Like other forms of consumption, travel did not merely reflect the social standing of its participants; it was crucial in helping to

[217] Wallis and Webb, 'Education and Training'.

[218] Stephen Hague, *The Gentleman's House in the British Atlantic World 1680–1780* (Basingstoke: Palgrave Macmillan, 2015).

make it. Chapter 5 will return to this question, asking how far investments in travel actually paid off. Educational voyages were not simply conspicuous consumption, though, as parents sent sons abroad to acquire real skills and connections that would contribute to family standing at home. Chapters 3 and 4 investigate these themes of learning and networking abroad.

3

Learning Abroad

This book has so far examined why families participated in educational travel, how young men prepared for journeys and how their parents afforded them; but what did travellers do once they were abroad? Historians tend to follow some contemporary theorists in seeing travel as an 'invisible academy' or an informal supplement to domestic education, giving an impression of peripatetic learning by osmosis.[1] This tendency is reinforced by the decision in most Anglophone studies to describe virtually any instance of elite travel during the seventeenth and eighteenth centuries as a 'Grand Tour', a term that implies constant movement. In fact, educational travel combined the experience of moving through foreign countries with that of making extended stays in specific towns.[2] It was during these crucial periods of 'settlement' that travellers learned most of the knowledge for which their families had sent them abroad. The prevailing emphasis on mobility misrepresents the experiences of those who blended such periods with tours and ignores those who could only afford settlement. As a result, most work hugely underestimates gentry familiarity with the Continent and, above all, with French culture.

With one exception, the families under discussion here did not describe the whole period that a young man spent abroad as a 'Grand Tour'. Historians often trace the term to Richard Lassels's posthumous *Voyage of Italy* (1670), written during the 1650s, but Lassels's single use of the phrase refers to the benefits of making 'exactly the Grand Tour of France, and the

[1] For example, Cesare De Seta, 'Grand Tour: The Lure of Italy in the Eighteenth Century', in Wilton and Bignamini, eds, *Grand Tour*, pp. 13–19, at p. 13; also Bignamini, 'Grand Tour: Open Issues', p. 31. For contemporary printed opinion, see Gallagher, *Learning Languages*, p. 160.

[2] Jean Boutier, 'Le Grand Tour des gentilhommes et les academies d'éducation pour la noblesse: France et Italie, XVIe–XVIIIe siècle', in Rainer Babel and Werner Paravicini, eds, *Grand Tour: Adeliges Reisen und Europäische Kultur vom 14. bis zum 18. Jahrhundert* (Ostfildern: Thorbecke, 2005), pp. 237–53, at p. 253; John Stoye, 'The Grand Tour in the Seventeenth Century', *Journal of Anglo-Italian Studies*, 1 (1991), 62–73, at 63.

Giro of Italy'.[3] This narrower meaning was more familiar to contemporaries. Travellers to France since the 1630s had consolidated patterns established by the previous generation, with some touring the country after a period of settlement and others making a less ambitious 'petit tour'. In the same decade, Englishmen consistently began venturing far enough beyond Venice and Florence to represent a 'giro d'Italia'.[4] More generally, from Sir Philip Perceval, who crossed the Channel a few years after Lassels's publication, to William Leigh in the 1710s, to the younger James Brydges in the 1750s, travel correspondence uses the word 'tour' to refer to bounded periods of heightened mobility. From Angers, Perceval and Jean Gailhard planned 'a short tour of nine or tenne days' along the Loire, considered a 'tour' of Brittany and referred to their two-month journey around France as 'the Grand Tour', to be followed by travel into Italy and Germany. Perceval's relations employed similar usages.[5] William Leigh 'made a Tour in North Holland' from Amsterdam, 'made the Grand Tour of Germany', left Rome 'to make the Tour of Italy', decided against 'any great Tour in Switzerland', made 'an agreable Tour' of three weeks in Flanders and came home after 'a Tour upon the Loire'.[6] With his 'Business' at the University of Göttingen almost finished, the younger James Brydges proposed 'a Tour' of two or three weeks, savouring the prospect of writing 'more entertaining Letters'.[7] Tours, whether 'grand' or small, mostly fitted within longer periods abroad. As Brydges's distinction between business and pleasure implies, they may not have been the most important aspects of those periods.

When correspondents wanted to indicate a young man's entire time abroad, they instead referred to his 'travels'. To return to the same examples, George Perceval hoped to visit Sir Philip before he 'quits English ground in order to his travels', Sir Robert Southwell habitually wrote of his nephew's 'Travels' (always capitalised) and a 1742 family history recalls when Perceval

[3] Lassels, *Voyage of Italy*, sig. A6r; Jean Boutier, 'Inventing the "Grand Tour": An Historiographical Success Story, between Anticipation and *Nachleben*', paper delivered at the *Art of Travel* conference, NUI Galway, 7 November 2016. For tracing the Grand Tour to Lassels, see Redford, *Venice and the Grand Tour*, pp. 10–12; Sweet, *Cities and the Grand Tour*, p. 23. On Lassels, see Edward Chaney, *The Grand Tour and the Great Rebellion: Richard Lassels and 'The Voyage of Italy' in the Seventeenth Century* (Geneva: Slatkine, 1985).

[4] Stoye, *English Travellers Abroad*, pp. 17–46, 117–33, 281–321.

[5] BL, Add MS 46954A, fo. 45: Jean Gailhard to Sir Robert Southwell, 7 Mar 1676/7; fo. 61v: Lady Perceval to same, 16 Mar 1676/7; fo. 158**: Southwell to Gailhard, [mid-May 1677]; fo. 274v: Gailhard to Southwell, 27 Jun 1677.

[6] SBT, DR18/20/18b: William Leigh to Theophilus Leigh, 14 Jul [1710]; same to same, 11 Jul 1711; same to same, 3 Dec 1711; same to same, 21 Sep 1712; same to same, 19 Jul 1713; HL, mssST58, vol. 7, fo. 57: Theophilus Leigh to James Brydges, 13 Nov 1710.

[7] SBT, DR18/20/18: James Brydges II to Barbara Blackley, 6 May 1751.

'was upon his Travels'.[8] Cornbury's tutor hoped that William Leigh would 'do himself a great deal of service & credit by his travels', while his father stretched 'to supply him, during his travels'.[9] And after his time at Göttingen and his tour through parts of Germany and France, the younger James Brydges began to think of home, 'however agreable my Travels have been to me'.[10] The families under discussion offer only one instance of using 'Grand Tour' to describe a whole period abroad, and even then in the process of rejecting it. In 1716, Sir Alan Brodrick saw little purpose in his stepson, Arthur Hill, 'making the grand Toure (as we have learned to call it)', and kept him in Ireland.[11] There are two points worth making about Brodrick's formulation, beside the fact that he is referring to educational travel in its entirety. First, Brodrick believes the term to be a new one, almost 50 years after Lassels. How had he 'learned' to use it in this way, in a family that had not travelled before the 1720s, and would many contemporaries have agreed? Only two books published in English before 1801 use 'Grand Tour' in their titles, by William Bromley (1692) and Thomas Nugent (1749).[12] Second, Brodrick's broad use of the term coexists with narrower senses of 'tour'. Seven years later, after a month at Hanover and three in Flanders, the younger Alan knew that he should continue his travels but not 'which Tour' his father wanted him to make. Later, he left Venice to 'make the best tour I can', before returning a few months later. On his way home, however, Alan referred to 'the whole Tour I have made since I left England'.[13] Fifteen years earlier, Theophilus Leigh had complained in similar terms that William had seen only Germany, Italy and France in three years abroad, enough time 'to make the whole Tour almost of Europe'.[14] Both Brodrick and Leigh still saw the need to distinguish between bounded and all-encompassing 'tours'.

[8] BL, Add MS 46951B, fo. 52: George Perceval to Sir Robert Southwell, 3 Nov 1674; 46953, fo. 68: Southwell to Lady Perceval, 20 May 1676; fo. 102v: same to Sir Philip Perceval, 5 Jun 1676; NLI, MS 7863, fo. 34: same to Jean Gailhard, 5 Jun 1676; Anderson, *Genealogical History*, ii, p. 365.

[9] SBT, DR18/20/18b: William Leigh and Dr Hobart to Theophilus Leigh, 26 Dec 1710; HL, mssST58, vol. 6, fos 37–8: Theophilus Leigh to James Brydges, 3 Jun 1710.

[10] SBT, DR18/20/18: James Brydges II to [Barbara Blackley], 15 Feb 1752.

[11] SHC, 1248/3/396: Sir Alan Brodrick to Thomas Brodrick, 10 Nov 1716.

[12] Boutier, 'Inventing the "Grand Tour"'; William Bromley, *Remarks in the Grande Tour of France & Italy* (London: 1692); Thomas Nugent, *The Grand Tour* (London: 1749). Confirmed by British Library's *English Short Title Catalogue* (estc.bl.uk).

[13] 1248/5/346v: Alan Brodrick II to [Martha Courthope], 30 Dec 1723; 1248/6/63: same to Sir Alan Brodrick, 4 Sep 1724; fo. 82: Thomas Brodrick to Alan Brodrick II, 6 Nov 1724; fo. 302: Alan Brodrick II to Sir Alan Brodrick, 11 Sep 1725.

[14] HL, mssST58, vol. 7, fo. 78: Theophilus Leigh to James Brydges, 27 Nov 1710.

The tendency to think of educational travel as a mobile Grand Tour is reinforced by the standard recourse to diaries and journals, whether printed or manuscript, as evidence of time abroad. These sources only offer a partial picture, as young men usually kept them during tours and put them aside during periods of settlement. The younger James Brydges studied at Göttingen for five months before he began his diary, which he kept during local tours and a longer journey to Paris and the Loire. He left it unfinished as he settled for another winter at Orleans.[15] Parents and guardians called on travellers to keep diaries as they began mobile periods, rather than as they crossed the Channel. When Sir Philip Perceval entered Italy, after almost a year of study at Angers, Sir Robert Southwell advised him to gather 'a good Stock of materials' from 'Sharpe Observation' and to fix 'the most memorable points, by writing in your Diary'.[16] Only as John Brydges thought of leaving France for a tour of Germany, nearly two years after he had begun at the University of Leiden, did his father recommend that he 'keep a Journall of his Travells & put down every thing he meets with that is remarkable'.[17] Some scholars express understandable frustration with monotonous travel diaries, 'a tragedy for the historian'.[18] They are less disappointing when they are seen as evidence for a specific part of a young man's travels, rather than the key to his encounter with the Continent.

If diaries and journals are usually silent on settled learning, they risk leaving out those who attended foreign institutions but could not afford lengthy tours. The experiences of these young men are vastly underplayed, but they are arguably the most important—surely the most popular—element of elite travel culture. Of 11 British students who were at Foubert's academy in Paris between 1674 and 1677, by no means the entire attendance for that period, only one appears in the matriculation records of the University of Padua, which had become a de facto visitors' book.[19] There were 37 British and Irish students at Göttingen in its first 20 years, from 1735 to 1755, but a maximum of five appear in John Ingamells's dictionary of eighteenth-century travellers to Italy, of whom two possibilities went there a decade or

[15] HL, mssST109, vol. 1: James Brydges II, Travel diary.

[16] BL, Add MS 46955A, fo. 59: Sir Robert Southwell to Sir Philip Perceval, 11 Feb 1677/8.

[17] HL, mssST57, vol. 22, fo. 208: James Brydges to Charles Stuart, 21 Mar 1722/3.

[18] Mączak, *Travel in Early Modern Europe*, p. 185.

[19] Viscount Cornbury, father of Leigh's companion: Brown, ed., *Inglesi e Scozzesi*, p. 168; D. W. Hayton and Henry Lancaster, 'Hyde, Edward, Visct. Cornbury (1661–1723)', in Hayton, Cruickshanks and Handley, eds, *House of Commons 1690–1715*.

more after the German university.[20] Of 76 Englishmen who attended the academy at Angers between 1755 and 1775, at a rate of up to 12 a year, those who may appear in Ingamells are always in a minority.[21] These ratios are only a rough guide: the incomplete list for Foubert's academy is compiled from research on the family side, in the absence of institutional records, and many must have gone to Italy without visiting Padua; later, Ingamells surely misses many who undertook Italian tours. The figures nevertheless suggest that settled learning in France, the Netherlands or Germany did not have to precede travel to Italy. The suspicion is borne out in family strategies, where French, Dutch and German institutions offered options for families who might not have been able to afford Italian travel and for the younger sons of wealthier parents.

This chapter recognises that settled learning was the bedrock of encounters with the Continent. It thereby advances what John Gallagher has called 'a history of educational travel with the education put back in'.[22] These efforts build on work by Cohen and Boutier, two of the few to analyse the educational role routinely claimed for travel. Discarding the shorthand of an 'invisible academy', Boutier finds significant institutional learning.[23] More recently, Gallagher has traced the 'everyday pedagogies' of travel, emphasising language acquisition and private tuition.[24] Beyond the beguiling simplicity of the Grand Tour, foreign education was as complex as domestic upbringing. On the Continent, as at home, families pieced together courses according to their means, ambitions and obligations. They created a wide range of experience that flourished below the level of eye-catching Italian tours, much of it centred on institutions.

Recent developments owe much to a new openness to international comparison and collaboration, as scholars increasingly see educational travel as a common practice among European elites, though one with national inflections.[25] Those working on continental subjects have long

[20] Gordon M. Stewart, 'British Students at the University of Göttingen in the Eighteenth Century', *German Life and Letters*, 33:1 (1979), 24–41; Ingamells, *Dictionary*, pp. 419, 506, 816, 915: George Hobart, Robert Robinson, James Grant, William Sutherland, Richard Sutton.

[21] Olivier Raguenet de Saint-Albin, 'Livre des pensionnaires et des externes à l'Académie d'équitation d'Angers au XVIIIe siècle (1755–1790), d'après un document inédit', *Revue de l'Anjou*, 68 (1914), 161–78, 339–60; Ingamells, *Dictionary*.

[22] Gallagher, *Learning Languages*, p. 161.

[23] Cohen, 'Grand Tour: Constructing the English Gentleman'; Boutier, 'Grand Tour des gentilhommes'.

[24] Gallagher, *Learning Languages*, p. 158; Gallagher, 'Language and Education'.

[25] Sweet, Verhoeven and Goldsmith, 'Introduction', in Sweet, Verhoeven and Goldsmith, eds, *Beyond the Grand Tour*, pp. 7–8.

emphasised academies and universities. Usage varied, with French and Florentine travellers attending less often than Germans and other northern Europeans, while the rise and decline of certain options moved along national chronologies. Dutch use of French academies fell in the late seventeenth century, for instance, just as their popularity rose among Bohemian families.[26] Anglophone studies have tended to rush over settlement in favour of a mobile Grand Tour, but were travellers from Britain and Ireland really the odd ones out? This chapter argues that they were not. It explores what travellers went abroad to learn and where they settled to do so, before placing tours in context. It concludes by considering what a more appropriate balance between settlement and mobility means for the study of educational travel and the cultural history of the elite.

What They Went Abroad to Learn

What were educational travellers supposed to learn? It can be tempting to fall back on nebulous ideas such as polish or refinement, as manuscript and printed sources often deal vaguely with 'those Improvements that are to be acquired by Travelling into Foreign Kingdoms'.[27] Family correspondence provides greater insight because it places generalisations alongside evidence of action. James Brydges, for instance, saw travel as an opportunity for his sons to improve their 'naturall Parts', but his letters and financial records trace John's and Henry's activities on the Continent with some precision.[28] They, and others of their standing, crossed the Channel in search of the 'parts' that made a 'compleat gentleman', to quote the title given to several conduct manuals in the period, including one by Gailhard.[29] From the mid-seventeenth to the mid-eighteenth centuries, the most prominent building

[26] Leibetseder, 'Across Europe'; Verhoeven, *Europe within Reach*, pp. 61–70; Eva Chodějovská and Zdeněk Hojda, 'Abroad, or Still "at Home"? Young Noblemen from the Czech Lands and the Empire in the Seventeenth and Eighteenth Centuries', in Sweet, Verhoeven and Goldsmith, eds, *Beyond the Grand Tour*, pp. 83–107, at pp. 87, 90.

[27] [Pullen], *Vindication*, p. 32.

[28] HL, mssST57, vol. 21, fo. 34: James Brydges to John Brydges, 22 Jul 1722.

[29] Henry Peacham, *The Compleat Gentleman Fashioning Him Absolute in the Most Necessary & Commendable Qualities Concerning Minde or Bodie that May Be Required in a Noble Gentlema[n]* (London: 1622); Gailhard, *Compleat Gentleman*. Also, the English titles of older works in translation: Baldassare Castiglione, *The Courtier: or, The Complete Gentleman and Gentlewoman* (London: 1729); Baltasar Gracián, *The Compleat Gentleman: or a Description of the Several Qualifications, both Natural and Acquired, that are Necessary to Form a Great Man* (London: 1730).

blocks were the French language and the physical 'exercises' of fencing, dancing and horse-riding, with many also learning some civil law. These forms of knowledge allowed young men to make the most of a continent where few spoke English—indeed, travellers tended to learn their exercises in French—but they also prepared them to impress back at home.[30]

Between 1650 and 1750, and for some time afterwards, the most important skill that travellers pursued was fluency in French. The language rose to dominance among European vernaculars and a fashionable position in British culture between the mid-seventeenth and the late eighteenth centuries, a period that roughly coincides with the involvement of gentry families in continental educational travel. The overlap was not perfect: in 1659, Robert Southwell arrived in France unable to converse in the language, which he understood only 'in a Book'. He 'could never venture to Speak' until, after five months in the country, he overheard 'other passengers talk French eternally' on the stagecoach to Lyon.[31] Southwell was not alone at the time, as many fell back on Latin while they built up their spoken French.[32] This inability to communicate would have embarrassed his descendants, however, as French became a European 'lingua franca' in the late seventeenth and eighteenth centuries, shared among courtly and learned circles across the Continent.[33] In Britain, its cachet had risen from Restoration Francophilia to the point where James Brydges could argue, in the 1720s, that French had 'become so universall that one can hardly live without it'. The language was, he claimed, 'almost as necessary as one's Mother Tongue'.[34] Brydges meant necessary at home, as much as abroad, and Chapter 5 will consider why this was so.

From the late seventeenth century until the late eighteenth, French was usually the most important skill—sometimes the only one—to which parents and travellers referred in correspondence.[35] As William Leigh crossed the Channel in 1710, he promised his father that he would 'improve myself as much as I can, by letting no opportunity slip of rendring me master of the French tongue', mentioning no other criteria by which to judge his

[30] Gallagher, 'Language and Education', p. 118.

[31] BL, Egerton MS 1634: Le Grand, 'Life of Southwell', fo. 10.

[32] Arturo Tosi, *Language and the Grand Tour: Linguistic Experiences of Travelling in Early Modern Europe* (Cambridge: Cambridge University Press, 2020), pp. 127–8, 172–4.

[33] Peter Burke, *Languages and Communities in Early Modern Europe* (Cambridge: Cambridge University Press, 2004), pp. 85–8.

[34] HL, mssST57, vol. 18, fo. 345: James Brydges to John Brydges, 3 Jan 1721/2; vol. 21, fo. 34: same to same, 22 Jul 1722; Lambley, *Teaching and Cultivation*, pp. 361–400.

[35] Conway, *Britain, Ireland and Continental Europe*, p. 203.

time abroad. Theophilus Leigh would agree with his son throughout his travels, referring to French as 'that necessary qualification'.[36] A few years later, James Brydges similarly mentioned only French when he urged his eldest son to master 'those Improvements which are so necessary for a young Gentleman', and thereby to impress a potential bride and her family. When John needed to return home within six weeks, in the context of a second match, Brydges urged him to devote the remaining time to his French.[37] From Arthur Rawdon in the 1670s to the Leigh brothers in the 1710s, sons wrote letters home in French to demonstrate their progress in this most important aspect of travel.[38]

To some degree, French rose at the expense of Italian, which asserted a rival claim on European elites until the mid-seventeenth century.[39] In a late sixteenth-century moment of 'Italophilia', the language enjoyed 'prestige status' in England. Travellers settled in Florence or the cheaper option of Siena to learn the purest Tuscan dialect, though Rome also had advocates by the mid-seventeenth century.[40] Soon afterwards, however, Italian was overtaken by the rise of French and the expansion of educational provision in France. Some still studied Italian before entering the country, such as Sir Philip Perceval, or took lessons in Rome or Siena, as did Sir John Rawdon as late as 1739, not least because of their lack of prior knowledge.[41] By the late seventeenth and eighteenth centuries, though, travellers mostly acquired the language in order to speak with locals rather than to bring it home. The same conclusion explains the emergence of German by the mid-eighteenth century, as travellers increasingly frequented Vienna and other courts as part of their social training.[42] Alan Brodrick in the 1720s and the younger James Brydges in the 1750s both began German from scratch while abroad, but

[36] SBT, DR18/20/18b: William Leigh to Theophilus Leigh, 30 May 1710; Theophilus Leigh to William Leigh, 19 Jan 1711/12.
[37] HL, mssST57, vol. 18, fos 344–5: James Brydges to John Brydges, 3 Jan 1721/2; fo. 364: same to Charles Stuart, 19 Jan 1721/2.
[38] HL, mssHA 15603: Arthur Rawdon to Conway, 19 Dec 1671; 15604: same to same, [*c.* Apr 1676]; SBT, DR18/20/18b: James Leigh to Theophilus Leigh, 28 Jul 1711; William Leigh to same, 16 Feb 1712.
[39] Tosi, *Language and the Grand Tour*, pp. 181–2.
[40] Gallagher, *Learning Languages*, pp. 66, 165–7. Robert Southwell learned 'the Tone, and cadence of the Language' by paying a Florentine chorister to read him Guicciardini's *History of Italy*: BL, Egerton MS 1634: Le Grand, 'Life of Southwell', fo. 10.
[41] PRONI, D2924/1/47-8: Sir John Rawdon to Thomas Prior, 20 Jul 1739; Gallagher, 'Language and Education', pp. 118–21.
[42] Sarah Goldsmith, 'The Social Challenge: Northern and Central European Societies on the Eighteenth-Century Aristocratic Grand Tour', in Sweet, Verhoeven and Goldsmith, eds, *Beyond the Grand Tour*, pp. 65–82.

Brodrick soon abandoned 'so extreamly difficult' an undertaking and Brydges found the language 'harder than any, I ever yet attempted'.[43] Between 1650 and 1750, those who learned Italian and German mostly did so in order to facilitate travel, or as an optional extra. None of the family papers used here contain letters home in Italian or German, partly because they could not assume the same readership as French in an era when correspondence circulated among family members.[44] Equally, they surely felt less pressure to demonstrate progress in secondary languages. Travellers spent longer acquiring French, which allowed them to engage with elites across Europe and, just as importantly, to impress a Francophone elite at home.

French declined as an aspect of gentlemanly education in the final decades of the eighteenth century as part of a retreat from what the English increasingly saw as overly effeminate, French-style politeness. Elite men turned towards plainer forms of manliness that were supposedly epitomised by their native tongue.[45] By this point, amid a 'crisis in language learning', they began to display a 'monolingual complacency' in their interactions with the Continent.[46] Young men still travelled through Europe in the nineteenth century, alongside journeys to the eastern Mediterranean and the expanding empire, and they still used both kinds of travel to develop elite male independence.[47] Indeed, European travel had never exclusively focused on acquiring politeness, providing opportunities too for developing and displaying 'hardy' masculinities.[48] Into the nineteenth century, however, fewer young men used France and Italy as behavioural models or sources of essential knowledge.

Between the mid-seventeenth and mid-eighteenth centuries, young men travelled for bodily as well as intellectual formation. They made no division between formal and informal training or essential and ornamental knowledge, learning physical and mental skills in the same settings.[49] The acquisition of French was not simply a question of verbal fluency, after all,

[43] SHC, 1248/6/302: Alan Brodrick II to Sir Alan Brodrick, 11 Sep 1725; HL, mssSTB, Box 12, Folder 53: James Brydges II to George Brydges, 18 Feb 1751.

[44] Gary Schneider, *The Culture of Epistolarity: Vernacular Letters and Letter Writing in Early Modern England* (Newark: University of Delaware Press, 2005), p. 22. Apart from one letter to Dudley Loftus, the Dublin linguist (BL, Add MS 38015, fo. 27), Robert Southwell's Italian correspondence was exclusively with Italians.

[45] Cohen, 'Manliness, Effeminacy and the French'.

[46] Tosi, *Language and the Grand Tour*, pp. 164–7.

[47] French and Rothery, *Man's Estate*, pp. 160–81.

[48] Goldsmith, *Masculinity and Danger*.

[49] Cohen, 'Grand Tour: Constructing the English Gentleman', 243.

but also of appropriate gestural codes.[50] Gailhard urged Sir Philip Perceval that, 'being come abroad to improve himself', he should settle until 'he had made some progresse in his exercises, the language customs of the countrey and bettered his intellectual parts'.[51] This emphasis on 'language customs' ensured that few travellers employed French tutors without also taking on a dancing master, who taught pupils how to bear themselves and greet others.[52] The goal was not only to speak French but to do so in the best company.

These considerations fed into the broader development of the body as a social symbol. Elites have set themselves apart through bearing since the ancient world, with differences of form and emphasis, but the Renaissance notion that bodily training could perfect noble birth spread throughout early modern Europe.[53] The role of gesture in social distinction may have been particularly strong in Britain and Ireland, where landowners lacked the legal supports enjoyed by many of their continental counterparts, and especially among the ill-defined gentry, who needed to perform their status in daily life.[54] Landed families built bodily formation into the education of their sons, emphasising riding, fencing and dancing, which taught graceful and confident movement. In a job application to Ormond, one Huguenot tutor described these exercises as 'very necessary' for a young man 'of family'.[55]

Notions of 'civil' and 'polite' bearing reached England from sixteenth-century Italy, with its Renaissance models of the courtier, and seventeenth-century France, as its royal court achieved cultural pre-eminence. Families accordingly saw travel to these countries as an opportunity to draw from the fonts of civility and politeness. France supplanted Italy in the second half of the seventeenth century and for most of the eighteenth, combining cultural prestige with expanding pedagogical provision. Indeed, many Frenchmen founded institutions for riding, fencing and dancing explicitly to avoid Italian travel.[56] Deportment, like language, was best learned through imitation as well as instruction, and a lengthy

50 Gallagher, *Learning Languages*, p. 196.

51 BL, Add MS 46954A, fo. 57: Jean Gailhard to Sir Robert Southwell, 21 Mar 1677.

52 Ellery Schalk, *From Valor to Pedigree: Ideas of Nobility in France in the Sixteenth and Seventeenth Centuries* (Princeton, NJ: Princeton University Press, 1986), p. 190.

53 Thomas, *In Pursuit of Civility*, pp. 35–45.

54 John Walter, 'Gesturing at Authority: Deciphering the Gestural Code of Early Modern England', in Michael J. Braddick, ed., *The Politics of Gesture: Historical Perspectives* (Oxford: Oxford University Press, 2009), pp. 96–127, at pp. 101, 113.

55 NLI, MS 2493, fo. 57: Peter Drelincourt, 'About education of men of family', [1670s].

56 Schalk, *From Valor to Pedigree*, pp. 177–201; Mark Motley, *Becoming a French Aristocrat: The Education of the Court Nobility, 1580–1715* (Princeton, NJ: Princeton University Press, 1990), pp. 123–67.

period in France offered deep immersion.[57] Physical training informed the timing of travel as well as destinations, as young men ideally learned their exercises in their mid- to late teens, early enough to shape stature and bearing. The Earl of Orrery's tutor recommended, soon after his fifteenth birthday, that it was time to start fencing and dancing, which 'would help him to grow tall & strong'.[58] Beyond the nobility, financial, educational and moral considerations tended to push travel into the late teens and early twenties.

Families discussed bodily formation as a crucial aspect of travel. In the 1670s, the main question surrounding the Rawdon brothers' stay in France was how to balance the family's means with the time needed to learn their exercises, an essential step in making them 'acceptable to the world'.[59] The importance of riding, fencing and dancing was clearest to travellers unable to acquire them. Orrery followed two generations of Boyles in pursuing such learning in France, but the 1685 Revocation of the Edict of Nantes disrupted his plans.[60] He worried that he knew 'none of my exercises yett', after four years abroad, and would 'be asshamed to return home' without them.[61] William Leigh's movements in 1710–13 were restricted by his dependence on Cornbury, but he fitted his education around the whims of his noble companion. Leigh found French tuition in Berlin, taking advantage of its large Huguenot community, and hoped to 'advance' in 'two other improvements' when he came to Rome, 'to wit Fencing & Riding the great Horse'.[62] Contemporaries knew that linguistic and bodily formation were central to successful travel; otherwise, efforts to deliver such curricula in England would not have presented themselves as preparations or substitutes for time abroad. To date, however, there has been little work in English on the universities and academies in which most travellers pursued such knowledge.

Behind French and exercises, travel correspondence frequently mentions civil law. Sir Robert Southwell insisted that Sir Philip Perceval devote himself 'intirely, for some months' to the subject at Angers, though the younger Edward Southwell could choose whether to tour Germany or 'Sitt down

[57] Gallagher, *Learning Languages*, pp. 197–8.
[58] NLI, MS 36, fo. 587a: Germain Colladon to Dowager Countess of Orrery, 25 Jul 1686.
[59] HL, mssHA 14526: Conway to Sir George Rawdon, 17 Apr 1674; 14545: same to same, 11 Jan 1675/6.
[60] Ohlmeyer, *Making Ireland English*, p. 441.
[61] NLI, MS 36, fo. 724: Earl of Orrery to Dowager Countess of Orrery, 5 Feb 1688/9.
[62] SBT, DR18/20/18b: William Leigh to Theophilus Leigh, 30 May 1710; same to same, 6 Sep 1710.

some time and read some Civil Law' on his way home from Italy.[63] He made the tour, but the subject was again essential by the time that the third Edward Southwell travelled in the 1750s. His mother, Katherine, listed it among the advantages of moving to Geneva, with friends assuring her that 'it was the most usefull knowledge' her son could obtain.[64] Civil law was of limited use in England but contemporaries such as John Aubrey, writing in the 1680s, saw it as training in abstract reasoning.[65] In a classical curriculum, its status as a Roman inheritance surely enhanced such perceptions. Sir Robert Southwell similarly argued that civil law furnished 'fit Maxims of General Reason', offering Perceval 'materials to enter into Manly Conversation'. Sir Philip's grandfather, the elder Robert Southwell, had never been 'more earnest for any thing', since civil law would leave him 'ripe & compleat to appeare in any company & in any discourse'.[66] It seems unlikely that the subject motivated much travel, though, as parental opinion varied on its importance. Civil law was often taught well at institutions with provision for French and exercises, so it may have provided an official focus while travellers pursued other aims. Lectures also offered the chance to maintain Latin. More directly, civil law prepared young men for conversation and observation in western continental Europe, where it shaped most legal systems. Finally, some parents may simply have saved the subject until travel because foreign institutions offered better teaching. Civil law was, Katherine Southwell concluded, 'the most usefull Study for an Englishman, tho, like all other things that are good for us, the most neglected here'.[67]

Travelling Tutors

As is already clear, the skills for which families sent young men abroad required settled learning. The location, duration and timing of settlements varied according to means and priorities, as families entered the kind of

[63] NLI, MS 7863, fo. 34: Sir Robert Southwell to Jean Gailhard, 5 Jun 1676; BL, Add MS 47031, fo. 120v: Edward Southwell to Sir John Perceval, 17 Mar 1725/6.

[64] Bristol Archives, 45317/2/5/2/23: Katherine Southwell to Edward Southwell III, 27 Jul 1758.

[65] Brian P. Levack, 'Law', in Tyacke, ed., *History of the University of Oxford: Volume IV*, pp. 559–68, at p. 568.

[66] BL, Add MS 46953, fo. 103: Sir Robert Southwell to Sir Philip Perceval, 5 Jun 1676; 46954A, fo. 28: elder Robert Southwell to same, 3 Feb 1676[/7]; fo. 83v: Lady Perceval to Sir Philip Perceval, 30 Mar 1677.

[67] Bristol Archives, 45317/2/5/2/31: Katherine Southwell to Edward Southwell III, 25 Aug 1758.

mixed economy in which they participated at home. On the Continent, as in London, educational entrepreneurs competed to attract the patronage of elite families. Settlement at certain towns or institutions was a consistently significant element of educational travel, however. Some, such as Edward and John Rawdon in 1670s Paris, spent their whole time abroad in an institutional setting. Even for those who went on to lengthy tours of Italy, settlement at a French, Dutch or German establishment took up a decent slice of their time on the Continent: Sir Philip Perceval was at Angers for 11 months and Viscount Irwin intended to stay at Leiden for at least a year. Although wide-ranging surveys deal separately with the Grand Tour and foreign institutional learning, these options were closely related in family strategies, as counterparts for some and alternatives for others.[68] For young men of ample means, the relationship between settlement and tour was often mediated by travelling tutors or governors. Indeed, for those who could afford his hefty wages, a tutor held great sway over itineraries, places of settlement, forms of learning and depth of engagement with specific establishments. Tutors shaped the continental experiences of those who could afford them.

Tutors are everywhere in Grand Tour historiography, providing much of its written evidence through published travel accounts and manuscript letters home, and they feature prominently in definitions of the phenomenon.[69] At the same time, however, historians rarely examine tutors directly, leaving their role at once underexplored and overestimated.[70] The tendency to call them 'bear-leaders', a term that does not appear in any of the archives discussed here, hides the extent to which these men were supposed not only to be experienced travellers and moral guardians but also skilled educators.[71] Sir Robert Southwell told Sir Philip Perceval that he had recruited 'a French gentleman', Gailhard, 'to governe the course of your Travels; As where to

[68] Conway, *Britain, Ireland and Continental Europe*, pp. 137–41; Brauer, *Education of a Gentleman*, p. 187.

[69] Redford, *Venice and the Grand Tour*, p. 14; Justin Stagl, *A History of Curiosity: The Theory of Travel 1550–1800* (Chur: Harwood Academic Publishers, 1995), p. 73.

[70] Exceptions include Jean Boutier, 'Compétence internationale, émergence d'une "profession" et circulation des savoirs: le tuteur aristocratique dans l'Angleterre du XVIIe siècle', in Maria Pia Paoli, ed., *Saperi a confronto nell'Europa dei secoli XIII–XIX* (Pisa: Edizioni della Normale, 2009), pp. 151–79; María Dolores Sánchez-Jáuregui, 'Educating the Traveler: The Tutors', in Sánchez-Jáuregui and Wilcox, eds, *English Prize*, pp. 88–97; Green, 'Bridging the English Channel'.

[71] The families under discussion here use 'tutor' or 'governor'. Historians' tendency to use 'bear-leader' is encouraged by Pier Leone Ghezzi's 1720s print, which takes the title literally: Wilton and Bignamini, eds, *Grand Tour*, pp. 102–3. The *Oxford English Dictionary*'s first example is from 1743 (www.oed.com).

goe; how long to stay; what things to learne; what acquaintance to make; and how to keep the Ballance of Expences even'.[72] A tutor acted as a 'director of studies', planning learning, recruiting masters for subjects, supervising progress, overseeing finances and offering a point of contact to parents.[73] He also did some teaching himself. At the same time, tutors were responsible for behaviour. When the Earl of Derby argued with his tutor, James Forbes, in France, Ormond feared that a rupture would risk 'great disadvantage & danger to his person & honour'.[74]

That tutors fulfilled a multifaceted role is clear from their job requirements. When Southwell engaged Gailhard, 'upon recommendation of our friends', the Frenchman had a back catalogue of writings on Italy, experience of leading several Englishmen abroad and a draft of his forthcoming *Compleat Gentleman*. Gailhard used the manuscript as a teaching aid, to judge by similar passages in Perceval's commonplace book, and he sent it to the printers as soon as he returned to London.[75] The Ingrams employed John Haccius with references from a fellow Yorkshire Whig, William Thompson, whose son had travelled with him over several years. Thompson introduced Haccius as 'a German born of good parentage, who were so carefull of his Education that he has all his Exercises and Languages in a great Perfection, and is well versed in most Sorts of Learning'. Just as importantly, he had 'seen a great deal of the World' and could introduce Irwin 'into the best Companys abroad'.[76] References to Haccius's birth and Gailhard's gentility were consciously made. Not only would these men set patterns of behaviour, but they also had to convince a social superior who was nearing adulthood, far from home, with few means to compel obedience. As Southwell advised Alexandre de Rasigade, who arrived on the Loire after Perceval had fallen out with Gailhard, they often had to 'governe as a favourite & not as a Tutor'. Southwell employed Rasigade on the strength

72 BL, Add MS 46953, fo. 102v: Sir Robert Southwell to Sir Philip Perceval, 5 Jun 1676.

73 Sánchez-Jáuregui, 'Educating the Traveler', pp. 91–2.

74 Bodleian, MS Carte 50, fo. 111: Ormond, 'Instructions for Mulys', 17 Nov 1673.

75 BL, Add MS 46953, fo. 264: Sir Robert Southwell to Sir Philip Perceval, 14 Jun 1677; 47113: Sir Philip Perceval, Commonplace book, *c.* 1669–77, fos 14–17v; TNA, SP44/25, fo. 154: Licence for Sir Thomas Grosvenor to travel with John Edesbury and 'Gaillard', 7 Apr 1670; *CSPD, Charles II*, xvi, p. 369: pass for William Peyto to travel with 'John Gailhard', 26 Sep 1674; Jean Gailhard, *The Present State of the Princes and Republicks of Italy* (London: 1668); Gailhard, *The Present State of the Republick of Venice* (London: 1669); Gailhard, *Compleat Gentleman*, ii, sigs A2r–A4r; Boutier, 'Compétence internationale', pp. 158–61.

76 WYAS, WYL100/C/9A: William Thompson to [John Roades], 18 Jul [1702]; TNA, SP44/346, fo. 461: Passes for William Thompson and John Haccius to go to Holland, 13 Nov 1696.

of previous travels with English gentlemen and provided references for the Frenchman's later voyages.[77]

Several families recruited native French-speaking Protestant tutors, who were fashionable in the decades after the Restoration.[78] Gailhard and Rasigade accompanied Sir Philip Perceval; Ormond recruited a succession of Huguenots, including Maximilien Misson, to take relations abroad in the 1670s and 1680s; Orrery brought a Swiss Protestant, Germain Colladon.[79] Their popularity as domestic tutors rested on the cachet of French culture, as, aside from better French, they offered the same methods and curricula as English counterparts.[80] When it came to travel, however, Huguenots boasted knowledge and connections. Some had attended the academies or universities that families had in mind, and Perceval even stayed briefly with Rasigade's family in France.[81] These advantages backfired in the few years after 1685, when the Revocation of the Edict of Nantes eroded their networks and reduced their mobility. In the longer term, Huguenot refugees lost their attraction as they assimilated into host communities.[82] They had always met resistance, with Lassels presenting scare stories about 'Governours that were Strangers' in the hope that parents would engage Englishmen like himself.[83] He would have been happy with James Brydges, who sent his sons abroad with countrymen in the 1720s. John employed Dr Charles Stuart, who came recommended as a travelling tutor, and Captain Lancaster, an army officer. Henry, the younger son, took only a soldier, Captain Andrew Robinson, a man of sense and experience who had 'found the time to apply himself not only to polite Learning but to the harder & more intent study of mathematicks'. For this, Brydges paid him an impressive £150 a year.[84] Whether or not he was French, the ideal tutor was a well-born man with knowledge of travel, fluency in the right languages, expertise in academic subjects and physical exercises, careful judgement and an easy authority.

[77] BL, Add MS 46954B, fo. 3: Sir Robert Southwell to Alexandre de Rasigade, 9 Jul 1677; fo. 63: Rasigade to Southwell, 28 Aug 1677; 46956A, fo. 14: same to same, 27 Jan 1679.
[78] Boutier, 'Compétence internationale', pp. 167–72.
[79] Ansell, 'Educational Travel', 945–6; Ansell, 'Reading and Writing Travels'.
[80] Michael Green, 'A Huguenot Education for the Early Modern Nobility', *Proceedings of the Huguenot Society*, 30:1 (2013), 73–91, at 91.
[81] BL, Add MS 46954B, fo. 271v: Sir Philip Perceval to Lady Perceval, 3 Jan 1678; Boutier, 'Compétence internationale', p. 168.
[82] Green, 'Huguenot Education', 91.
[83] Lassels, *Voyage of Italy*, sig. E3r.
[84] HL, mssST57, vol. 17, fo. 274: James Brydges to John Brydges, 25 Nov 1720; vol 24, fo. 271: same to Walter Senserf, 5 Oct 1724; vol. 27, fos 235–6: same to Andrew Robinson, 24 Jan 1725/6.

The complexity of the role encouraged its professionalisation in the second half of the seventeenth century, distinct from its domestic counterpart.[85] But if tutors grew with educational travel, they had an active role in the process, using print to weave themselves into the ideal. Some of the most influential writers on travel or education in the seventeenth and early eighteenth centuries—Robert Dallington (1605), James Howell (1642), Richard Lassels (1670), Obadiah Walker (1673), Jean Gailhard (1678), Maximilien Misson (1691), John Locke (1693), Joseph Addison (1705), John Breval (1726) and Edward Wright (1730)—had worked as travelling tutors. They helped to create the role in print. Lassels called for virtue, honour, fortitude, tact, prudence, gentility of birth and breeding, scholarly expertise, experience of travel and even good looks.[86] Gailhard followed his criteria a few years later, including the need to be 'well shaped', and criticised parents for whom 'cheapness is the prevailing qualification'.[87] Unsurprisingly, given their authorship, these foundational texts include a heroic vision of the tutor as an indispensable feature of educational travel. Many families bought into this vision, but others did not. Given the salaries due to celebrated tutors, some waited until their sons no longer needed supervision. Sir John Perceval (22), Alan Brodrick (21) and the younger Edward Southwell (19, almost 20) all set off for Italy without a tutor. Others enrolled sons at academies or universities, using institutions as alternative sources of discipline. The Rawdon brothers, for instance, took only servants to Foubert in Paris. As a significant factor in any travel budget, families arrived at varying interpretations of the tutorial role.

Tutors were prominent in thinking about educational travel, even if not universally adopted, and they placed settled learning at the heart of the ideal. Howell thought a traveller should 'retire to some University about the Loire', covering his studies at length, and urged him not to 'stirre' from Siena until he could speak Italian. Lassels advised travel in the mid-teens, to leave enough time for two or three years in Italy and three more in France. More reasonably, Gailhard called for a nine- or 10-month 'settlement' in France, a first step on which 'all agree', in order 'to learn the Language, Fashions, and Exercises'. He devoted a substantial portion of his treatise to the topic. All three men advocated kinds of learning—languages, riding, fencing and dancing—that required long periods in one place.[88] Their

85 Boutier, 'Compétence internationale', p. 157.

86 Lassels, *Voyage of Italy*, sig. E2v.

87 Gailhard, *Compleat Gentleman*, ii, pp. 7–16.

88 James Howell, *Instructions for Forraine Travell* (London: 1642), pp. 25–47, 102; Lassels, *Voyage of Italy*, sigs E6v–I1r; Gailhard, *Compleat Gentleman*, ii, pp. 33–58, at p. 33.

views drew on real-life pedagogy. When Perceval wanted to leave Angers, Gailhard reminded him that 'those who travell according to rules' settled in one place 'for a competent time, proportionable to that one is to be abroad'. Arguing that a 'rowling stone gathers no mosse', a line that he would reuse in print, he listed English noblemen who had done the same.[89] Huguenots inevitably favoured French establishments, not least because some were former students. Cases must have been rare of tutors convincing parents 'that all good breeding was in that poore towne, where their wifes were breeding children', but Lassels's suggestion of cosy relationships with academies along the Loire rings true.[90] After years of travel, tutors built up reusable networks, as the repeated engagement of the same men amplified their preferences.[91] Tutors placed particular models of settled study at the heart of their package.

John Haccius, for example, brought Irwin and his party to the Netherlands in 1704 for a period of secluded learning. They set up in Delft, where the viscount, Rich and Thomas Worsley spent three months preparing for the University of Leiden. Like many in the preceding century, they learned French in the Netherlands, home to a Francophone elite and a large Huguenot community.[92] The group immersed themselves in a French Protestant household, practising the language with their hosts and enjoying their meat cooked 'after the french fashion'. The viscount and Worsley began 'to chatter in French very prittily' and Rich developed a good pronunciation, though he spoke 'not very much'. After a while, Irwin began to agitate for the move to Leiden, recognising that he could speak 'now French enough to enter into the World'.[93] The young men prioritised the language, taught by Haccius himself over 'several hours a day', and received masters for music, drawing, dancing and fencing. For the latter, they used a man previously employed by the Earl of Hertford, whose tutor in mathematics Irwin had inherited at Cambridge. Haccius even found 'a dutch Man to instruct them to understand the Latin dutch prononciation', preparing them for lectures at Leiden. They broke their studies to visit The Hague and the 'famous' palace at Honselaarsdijk, but Haccius assured Irwin's trustees that, 'if you should

89 BL, Add MS 46954A, fo. 57: Jean Gailhard to Sir Robert Southwell, 21 Mar 1677; Gailhard, *Compleat Gentleman*, ii, p. 34.

90 Lassels, *Voyage of Italy*, sigs E3r–E3v.

91 Sánchez-Jáuregui, 'Educating the Traveler: The Tutors', p. 92.

92 C. D. van Strien, *British Travellers in Holland during the Stuart Period: Edward Browne and John Locke as Tourists in the United Provinces* (Leiden: Brill, 1993), p. 100; Burke, *Languages and Communities*, p. 86.

93 WYAS, WYL100/C/9A: John Haccius to John Roades, 18 Jun [1704]; Irwin to same, 8 Aug 1704; WYL100/C/9B: same to Dowager Viscountess Irwin, 15 Jul 1704. Bobinau, 'the French dyer in Delft', was likely an exiled Protestant, given his profession.

see the House we are in, you would think it an University and an Academie att the same time'.[94]

Universities

It was soon time for Irwin's group to begin at Leiden. Continental universities had long attracted Englishmen as part of European student mobility, with Leiden and Padua the most popular institutions between the late sixteenth and mid-seventeenth centuries.[95] Padua fell away as part of the decline of Italian options for settled learning, which suffered from the educational shift towards French language and culture, but Leiden adapted to retain its position among new or newly prominent institutions in France, the Netherlands and Germany.[96] Foreign universities featured variously in family strategy because, just as at home, young men attended with different ends in mind. Leiden attracted British and Irish students of varying means for a range of religious, political and academic reasons, and it acquired a reputation for excellence in medicine and law. In the few decades before 1740, the Dutch university may have offered competition to Oxford and Cambridge.[97] Others used Leiden as a finishing school, taking advantage of extracurricular tuition in French, history and geography as well as, from 1704, a riding school offering paid options for fencing, dancing, drawing and geometry.[98] The university stood among a host of institutions that encouraged elitism over the seventeenth century, providing tailored tuition to the sons of wealthy families. Some even specialised as 'tourist universities'.[99] For these students, who often made shorter stays, foreign universities complemented Oxford and Cambridge, rather than rivalling them. Leiden, for instance, served different purposes for Coote Molesworth over three periods of study. He attended

[94] WYAS, WYL100/C/9A: John Haccius to [John Roades], 26 Jan 1702/3; same to same, 18 Jun [1704]; same to same, 26 Aug 1704; WYL100/C/20, fo. 190: Irwin to Dowager Viscountess Irwin, 18 Jun [1704].

[95] Daniela Prögler, *English Students at Leiden University, 1575–1650* (Farnham: Ashgate, 2013), pp. 27–52; Jonathan Woolfson, *Padua and the Tudors: English Students in Italy, 1485–1603* (Cambridge: James Clarke, 1998).

[96] De Ridder-Symoens, 'Mobility', p. 438.

[97] Conway, *Britain, Ireland, and Continental Europe*, pp. 137–40; Prögler, *English Students at Leiden*, pp. 136–41, 226–32; Esther Mijers, *'News from the Republick of Letters': Scottish Students, Charles Mackie, and the United Provinces, 1650–1750* (Leiden: Brill, 2012), pp. 67–99; Mijers, 'Irish Students in the Netherlands, 1650–1750', *Archivium Hibernicum*, 59 (2005), 66–78.

[98] Mijers, *News from the Republick of Letters*, p. 69; Verhoeven, *Europe within Reach*, p. 66.

[99] De Ridder-Symoens, 'Mobility', pp. 428, 432–3.

'the lectures of Civil Law, Belles Lettres, and Philosophy' in 1717–18 and returned for another year after six months of French and 'Genteel Exercises' in Geneva, preparing himself for employment 'in publick affairs'. He came back to Leiden in a different capacity in 1727, studying medicine over several years, and eventually served as an army physician in Minorca.[100] The same families, even the same people, used foreign universities in different ways.

Irwin's party treated Leiden as a 'tourist university'. They found lodgings on the Rapenburg, a prestigious location 'according to the Dutch fashion', and spoke with 'nothing but fine Gentlemen'. Irwin hired masters for music, fencing and French, fitting them around daily lectures on natural law. There he met with 'very Learned Professors, who without being so great Pedants as ours, know more'. Rich meanwhile received tuition from Haccius and a drawing master, appearing to have 'allmost left all his Childish manners'.[101] Like other privileged foreigners at Leiden, the young men exploited the town's location for visiting The Hague and touring the surrounding region.[102] Irwin's party frequented 'the best Company' in the capital and recruited 'four very pritty Gentlemen' for a trip to Utrecht and the palace of Het Loo, though weather prevented the second leg. They planned the journey out of 'Curiosity', but Irwin was disgusted by the 'Nastiness' of the 'so much fam'd City of Utrecht', the 'least of any I ever saw'.[103]

The trustees had planned to keep Irwin at Leiden for at least a year, but he ruined these plans by fighting a series of duels, one of them in the fencing school. The group felt obliged to leave after only two months.[104] While Rich and Worsley made their separate ways home, Irwin brought forward his journey to Italy in order to escape the fallout. Haccius framed the move as a continuation of settled learning, rather than the beginning of a more mobile period: Irwin was 'resolved to go to Italy and there to Siena, a University, where he may study and learn the language'. The viscount repeated this justification and reassured those at home of the 'extraordinary' masters there. Irwin followed outdated practice by studying Italian at Siena

[100] NLI, microfilm P.3753: Coote Molesworth to [Richard Molesworth], [1727]; Coote Molesworth, Family history, in Mayo, 'Molesworth's *Account of Denmark*', pp. 219–20; Edward Peacock, *Index to English-Speaking Students who have Graduated at Leyden University* (London: Longmans, Green & Co., 1883), p. 69.

[101] WYAS, WYL100/C/9A: Irwin to Dowager Viscountess Irwin, 19 Sep [1704]; Thomas Worsley to John Roades, 23 Sep 1704; John Haccius to same, 3 Oct 1704.

[102] Mijers, *News from the Republick of Letters*, pp. 99–105.

[103] WYAS, WYL100/C/9A: John Haccius to John Roades, 3 Oct 1704; WYL100/C/20, fo. 98: Irwin to Dowager Viscountess Irwin, 31 Oct [1704].

[104] WYAS, WYL100/C/9A: Thomas Worsley to John Roades, 1 Nov 1704; WYL100/C/9B: anonymous letter, 29 Dec 1704.

for three months, but Roades, his estate manager, knew that the town had 'so great a Character in the world for improvement'.[105] This reputation helped Irwin and Haccius to put a positive spin on the disarray into which their plans had fallen.

Academies

If Leiden and Siena were traditional choices, newer options for institutional learning had established themselves in the decades before Irwin's travels. French educators founded academies throughout the seventeenth century, hoping to prevent travel to Italian universities by teaching noble accomplishments at home. In the heyday of such institutions, between 1600 and 1680, there were seven or eight academies in Paris and between 18 and 20 in the provinces. They prioritised riding, dancing, fencing and military mathematics (often known as 'fortification'), but they also offered languages, history, civil law and other subjects.[106] These institutions soon attracted foreign students, as neighbouring elites held similar fears about Italy and appreciated a closer alternative, to the point that French academies became a shared aspect of western European travel culture.[107] French, German and Dutch students outnumbered those from Britain and Ireland throughout the seventeenth century, though the two decades after the Restoration saw a marked increase in British students at academy towns such as Saumur.[108] At the same time, scarce official documents show a flurry of new foundations in the French provinces.[109] Academies grew together with educational travel, their pedagogical flexibility and institutional discipline suiting the varying demands of parents who wanted sons to live in or out, to make long or short stays and to study a range of subjects at their own pace.[110]

105 WYAS, WYL100/C/9A: Irwin to John Roades, 27 Aug 1705; Roades to Irwin, 22 Sep 1705; WYL100/C/9B: John Haccius to [Roades], 16 Dec 1704; WYL100/C/20, fo. 163: Irwin to Dowager Viscountess Irwin, 30 May [1705].

106 Schalk, *From Valor to Pedigree*, pp. 177–201; Motley, *Becoming a French Aristocrat*, pp. 123–67.

107 Verhoeven, *Europe within Reach*, pp. 62–6; Leibetseder, 'Across Europe', 437–9.

108 Willem Frijhoff, 'Étudiants étrangers à l'académie d'équitation d'Angers au XVIIe siècle', *Lias*, 4 (1977), 13–84; Philippe Chareyre, 'Les protestants de Saumur au XVIIe siècle, religion et société', in *Saumur, capitale européenne du protestantisme au XVIIe siècle* (Fontevrault-l'Abbaye: Abbaye royale de Fontevraud, 1991), pp. 27–70, at pp. 53, 56–7.

109 Archives Nationales, Paris, O/1/915/1-2: 'Académies Etablies dans les Provinces'.

110 Boutier, 'Grand Tour des gentilhommes', pp. 238, 247–8, 252.

Academies clustered in Paris and along the Loire. Both areas offered prestige varieties of French and a good supply of masters for all kinds of learning, but they were otherwise different propositions. While Orleans, Saumur, Angers and Blois were renowned for the purity of their spoken language, accepted wisdom held that Paris was better for exercises.[111] Many parents and authors nevertheless thought that travellers should avoid a lengthy stay in the capital so early in a journey. Gailhard claimed that Parisians were impatient with foreigners, their city expensive and full of 'debauched' Englishmen, arguing that a traveller needed linguistic and cultural awareness 'before he be ripe for Paris'. The Loire towns were 'seated in a pleasant and plentiful Country, their language is well spoken amongst people of some fashion, good and able Masters of Exercises are found in them, and the people thereof are very kind and civil to strangers'. Sir Philip Perceval, who spent almost a year at Angers, was at least the second traveller whom Gailhard had settled in the town, which his book would recommend as the best choice.[112] Besides educational advantages, the Loire towns were cheaper than Paris and had long been a stronghold of French Protestantism.[113] Some young men nevertheless agitated for a return to the capital, through which most had passed on the way to the Loire. Derby's tutor understood his aversion to Saumur, 'a very malancholly place' with 'little to be learn't', and Perceval would not allow his uncle to think him 'in any way fond of' Angers 'or the Company in it'.[114]

The convenience and cachet of Paris endured, with many families, even the pious and precarious Rawdons, settling their sons in the capital. Sir George Rawdon followed Conway's recommendations, based on the precedents of other Irish landowners, and surely welcomed his promise that 'they are very strict and severe in those Academyes'. Arthur's tutor reassured Lady Rawdon that Foubert, then master of an establishment in the Faubourg Saint-Germain, was 'himselfe a protestant, and so I think the greatest part of the gentlemen are in that house'.[115] A list of students and teachers there in 1668 shows that Foubert's academy was indeed popular with the Huguenot

[111] Gallagher, *Learning Languages*, pp. 162–4; Boutier, 'Grand Tour des gentilhommes', p. 251.
[112] HL, mssHA 3339: Jean Gailhard to seventh Earl of Huntingdon, 8 Jun 1670; Gailhard, *Compleat Gentleman*, ii, pp. 34–5.
[113] Gallagher, *Learning Languages*, pp. 162–3.
[114] Bodleian, MS Carte 243, fo. 93: Thomas Fairfax to Ormond, 17 Mar 1673/4; BL, Add MS 46954A, fo. 156: Sir Philip Perceval to Sir Robert Southwell, 23 May 1677.
[115] HL, mssHA 14526: Conway to Sir George Rawdon, 17 Apr 1674; 15264: Humphrey Maunsell to Lady Rawdon, 6 Jul 1674.

elite.[116] Other patrons praised his horses, his masters and his reputation for 'sober students'.[117] Just as importantly for the Rawdons, the cost of sending a son to Foubert, around £170 a year, would be not much more than Cambridge.[118] Foubert later recalled that he had been 'intrusted with the Education of severall of the Nobility and Gentry of England as well as of other Nations', and evidence of British and Irish students extends from the sons of dukes, marquises and earls to those of London merchants and provincial gentlemen.[119] A combination of Protestantism, linguistic prestige and educational expertise drew travellers towards academies in Paris and along the Loire in the two decades after the Restoration.

It can be challenging to work out exactly what young men did there. At licensed but uncontrolled establishments that relied on oral instruction, great variation was possible. Edward Rawdon spent his mornings riding, fencing and dancing at Foubert's academy and dedicated afternoons to French, with a similar pattern holding in other institutions.[120] Perceval lived out of the academy at Angers, going to a professor at the town's university for civil law every morning for several months, receiving masters for French, writing and music in his lodgings and taking tuition—probably in the academy—for fencing, dancing and eventually riding. He and Gailhard spent their evenings reading French, Latin, travel books, divinity, history and the newspapers.[121] Opinion was relatively consistent on how long all of this would take: Gailhard advocated nine or 10 months and one English traveller reported French wisdom that a year was 'enough to perfect al that is learnt in accademye', justifying Conway's argument that the Rawdon brothers would need far longer than three months in Paris.[122] Beyond the curriculum, students were notorious for their hazing rituals, gambling and use of prostitutes.[123]

116 *Memoirs of Isaac Dumont de Bostaquet, a Gentleman of Normandy Before and After the Revocation of the Edict of Nantes*, ed. and trans. D. W. Ressinger (London: Huguenot Society of Great Britain and Ireland, 2005), pp. 62–4, 304–5.

117 Ansell, 'Foubert's Academy', p. 48.

118 HL, mssHA 14526: Conway to Sir George Rawdon, 17 Apr 1674.

119 BL, Harley MS 7614, fo. 101: Humble Petition of Solomon Foubert, [after 1685]; Ansell, 'Foubert's Academy', p. 48.

120 HL, mssHA 15642: Edward Rawdon to Conway, 2 Sep 1676; Motley, *Becoming a French Aristocrat*, p. 142.

121 BL, Add MS 46953, fo. 189: Jean Gailhard to Sir Robert Southwell, 1 Nov 1676; fo. 218: same to same, 3 Jan 1676/7; 46954A, fo. 30: same to same, 14 Feb 1676/7; fo. 83: Helena Perceval and Lady Perceval to Sir Philip Perceval, 30 Mar 1677; fo. 206: Southwell to same, accounts [Jun 1677]; Boutier, 'Grand Tour des gentilhommes', pp. 248–9.

122 Gailhard, *Compleat Gentleman*, ii, pp. 45, 137; BL, Add MS 70013, fo. 116: [Abigail, Lady Harley] to Sir Edward Harley, 12 Jul 1682.

123 Motley, *Becoming a French Aristocrat*, pp. 162–3.

John Rawdon indulged in drinking and 'other dishonest actions' while boarding with Foubert, despite the Frenchman's serious reputation, and Orrery was consumed after his time at Blois and Paris by an 'extraordinary passion … for gaming which overrules him quite'.[124] Families who sent sons to academies tolerated these risks as part of immersion in French society, well known for its gambling, but they probably explain the common decision to place young men with local families.[125]

Foreign travellers, including those from Britain and Ireland, began to abandon French academies in the late 1670s and 1680s. They had come under the king's authority in 1669, as part of a late seventeenth-century trend in which rulers opened or supported such institutions.[126] In France, this enabled the suppression of Protestant academies, with royal edicts forbidding Huguenots from teaching exercises. Foubert's establishment was closed in 1679 and he fled to London, just as the situation deteriorated in Protestant academies along the Loire.[127] The 1685 Revocation of the Edict of Nantes had an immediate impact, with royal licences thereafter given only to Catholics.[128] Orrery left Blois that October, fleeing 'a Country where so little humanity, and so much severity is shewed to persons of a different perswasion to theirs'. Amid fears that his Protestant tutor would be stopped at the border, the earl continued his learning at Utrecht, Brussels and Geneva. To French, writing, mathematics and tennis, he progressively added fencing, dancing and riding. Orrery had not been allowed into the royal academy at Turin as a day pupil and was disappointed with his masters in Geneva, but there was 'no going into France, since all that are there doe flock hither'.[129] Maximilien Misson, author of *A New Voyage to Italy*, had similar experiences with one of Ormond's grandsons, Charles Butler. Misson and his pupil boarded over the winter of 1685–6 at a Parisian academy, intending to stay for a year or 18 months before leaving for Italy,

124 HL, mssHA 14538: Conway to Sir George Rawdon, 10 Apr 1675; NLI, MS 36, fo. 597: Germain Colladon to Dowager Countess of Orrery, 17 Oct 1686.

125 Ormond did 'not intend' that one grandson should 'Lodge or dyet in an Academy': NAL, Forster MS 47.A.40, fo. 11: Ormond to Sir Robert Southwell, 11 Sep 1678. On gambling, see Black, *France and the Grand Tour*, pp. 134–6.

126 Boutier, 'Grand Tour des gentilhommes', p. 241.

127 Ansell, 'Foubert's Academy', p. 49; Jean-Paul Pittion, 'Saumur et les voyageurs britanniques au dix-septième siècle', *Annales de Bretagne et des pays de l'Ouest*, 121:3 (2014), 97–117, at 114.

128 Corinne Doucet, 'Les académies équestres et l'éducation de la noblesse (XVIe–XVIIIe siècle)', *Revue historique*, 628 (2003/4), 817–36, at 834.

129 NLI, MS 36, fo. 577: Earl of Burlington to Dowager Countess of Orrery, 22 Jan 1685[/6]; fo. 661: Germain Colladon to same, 29 Nov 1687; fo. 726: same to same, 15 Feb 1689; 13215, fos 8–11, 25–7, 33–7, 51–7: Orrery's travelling expenses, 1685–9.

but they relocated to Brussels and were soon in England. They avoided France on a second attempt in 1687–8, meaning that Misson's book ignores an important destination.[130] By that time, war had exacerbated the effects of the Revocation of the Edict of Nantes, frustrating access to remaining institutions.[131] Considering the options available to Edward Southwell in 1686, Sir John Perceval observed that 'France, where the recourse out of England was most usual', was 'more impracticable and unsafe than formerly' because of the actions of Louis XIV.[132] Southwell would only travel to the Netherlands in official retinues.

The closure or inaccessibility of French academies opened a gap in the market. Misson singled out Geneva, which drew foreigners after 1685 'by establishing Masters of all Exercises that young Gentlemen are us'd to perform'. Its institutions attracted 'many of the Protestant Countries, as, Germany, Swedeland, Denmark, &c. who for Religion's Sake chuse rather to go to Geneva (where they may speak French) than to France'.[133] Towns that could offer an immersive Francophone experience were well placed to meet ongoing demand. Alongside Geneva, travellers looked to bilingual Turin. Sir Philip Perceval, passing through between the foundation (1677–8) and opening (1680) of the royal academy, noted that the court there lived 'after the french way as much as possibly they can'.[134] The institution, which looked to Parisian precedents, was still welcoming British students a century later.[135] The Duke of Lorraine's academy, which moved between Nancy and Lunéville from 1699, similarly made the most of its proximity to French culture.[136] Even the Dutch, with their French-speaking elite and growing Huguenot community, added facilities for languages and exercises to established universities. The Ingrams favoured Leiden, while another Sir John Perceval and his companion, Francis Clerke, spent the winter of 1705–6 at Utrecht, devoted to 'studies that were of use in their Travels such as fortification, drawing &c not forgetting musick and fencing'.[137] The British

[130] Ansell, 'Reading and Writing Travels', 1459–60.
[131] De Ridder-Symoens, 'Mobility', p. 437.
[132] BL, Add MS 46962, fos 196–7: Sir John Perceval to Sir Robert Southwell, 2 Feb 1685/6.
[133] Maximilien Misson, *A New Voyage to Italy*, fourth English edn, 4 vols (London: 1714), iv, pp. 421–2.
[134] BL, Add MS 46955A, fo. 25: Sir Philip Perceval to Sir Robert Southwell, 30 Jan 1677/8.
[135] Paola Bianchi, 'In cerca del moderno: studenti e viaggiatori inglesi a Torino nel Settecento', *Rivista storica italiana*, 115:3 (2003), 1021–51; Bianchi, 'Appendix III. British Attendees at the Turin Royal Academy', in Paola Bianchi and Karin Wolfe, eds, *Turin and the British in the Age of the Grand Tour* (Cambridge: Cambridge University Press, 2017), pp. 399–410.
[136] Boutier, 'Grand Tour des gentilhommes', pp. 241–3, 250.
[137] BL, Add MS 47072: Perceval, Autobiography, p. 21.

eventually returned to French academies and would dominate, especially along the Loire, as Dutch and German interest waned. They too would leave for good in the late eighteenth century, as educational travel faded and more focused military academies emerged at home and abroad.[138] Before then, with Britain regularly at war with France, establishments in neighbouring territories remained popular options.

Some late seventeenth-century foundations were attractive because, with close links to absolutist projects, they institutionalised access to court. In Turin, the royal academy was at the heart of life in the Piazza Castello, which housed the ducal palace and government buildings, and students gained introductions to noble society.[139] These opportunities ensured a long-lasting reputation, from Orrery's hopes to study there in the 1680s to those of Coote Molesworth in the 1720s. In the 1750s, Katherine Southwell 'heard from every body' that, after some French and civil law in Geneva, boarding in Turin was 'much the best way, attended with many advantages, particularly being well received at Court'.[140] The academy entered Huntingdon's thoughts for his nephews as late as the 1770s, offering a sure route to the best company, though he found it in too 'debauched & low' a state.[141] Lunéville similarly institutionalised access, especially from 1721, when one wing of the ducal chateau housed foreign students and the other courtiers.[142] The younger Edward Southwell visited with his father two years later. Suitably impressed, the elder Southwell sent his son back as a student in 1725. Edward was presented at court and dined with the duke, greeting royalty and dignitaries over four months of dinners and balls. His list of fellow 'academists' contains English, Scottish and Irish names alongside French and German noblemen.[143] Turin and Lunéville offered a residential package of courtly life, previously accessible only with letters of introduction or a well-connected tutor. They encouraged travellers to stay in towns through which they would otherwise have passed.[144]

138 Boutier, 'Grand Tour des gentilhommes', pp. 245; Verhoeven, *Europe within Reach*, pp. 68–9; Matthew McCormack, 'Dance and Drill', *Cultural and Social History*, 8:3 (2011), 315–30, at 326.

139 Bianchi, 'In cerca del moderno', 1028, 1037–8, 1040.

140 Bristol Archives, 45317/2/5/2/23: Katherine Southwell to Edward Southwell III, 27 Jul 1758; 45317/2/5/2/27: same to same, 10 Aug 1758.

141 PRONI, T3765/M/2/14/92: Huntingdon to Lady Rawdon, 21 May 1773; T3765/M/2/14/93: same to same, 1 Jul 1773; T3765/M/2/14/98: same to same, 8 Aug 1774.

142 Edmond Delorme, *Lunéville et son arrondissement*, 2 vols (Marseille: Laffitte Reprints, 1977), i, pp. 18, 22, 23.

143 LLRRO, DG7, Bundle 28/75: Edward Southwell to [Earl of Nottingham], 26 Nov 1723; BNF, MS Anglais 194: Travels of the Hon. Edw[ard] Southwell [II], 1725, fos 22–34v.

144 Boutier, 'Grand Tour des gentilhommes', p. 252.

Universities and Academies in Family Strategy

Just as families assembled domestic education from a mixed economy of provision, so they combined several kinds of foreign institution. The Brydges family included universities and academies in their plans across three generations. James, the future first Duke of Chandos, spent most of 1692–4 in the academy at Wolfenbüttel in Lower Saxony. The institution opened in 1687, at a good moment to take advantage of the flight from France, stood next to the ducal castle and offered access to neighbouring courts. Brydges studied French with a native speaker and used the language with locals, forming romantic attachments with several German women. He spent plenty of time at court and built links with nearby Hanover, including to the polymath Gottfried Wilhelm Leibnitz.[145] In later years, Wolfenbüttel remained important to the family. William Leigh visited the library there in 1710 and saw the room that his father's neighbour and fellow Tory, Sir John Walter, had used at the academy.[146] The Chamberlaynes considered sending their son there in 1711–12, though the plan came to nothing, and family members remembered the establishment even after its closure in 1715.[147] Brydges made his son, John, break his 1723 tour of Germany to visit Wolfenbüttel, 'at which Academy (now laid down) I was for two years'.[148] As late as 1751, his grandson promised a cousin, George Brydges, that he would visit the town, 'where I think I have heard you studied'. Given that George had been in the Netherlands at 18 in 1696, he was probably right.[149]

The short lifespan of the Wolfenbüttel academy forced later generations elsewhere. James Brydges sent his sons, John and Henry, to Leiden in the 1720s, by which point the university catered well for those in search of a gentlemanly education and opportunities for sightseeing. The fact that universities and academies both offered such provision by the early eighteenth century made them interchangeable, and Brydges hesitated over whether to send John instead to Angers. He had heard 'no good spoken of'

145 HL, mssST 57, vol. 1, fos 9–21: James Brydges, letters to contacts in Germany, 1693–5; Boutier, 'Grand Tour des gentilhommes', p. 241. The professor of foreign languages was Gui Leremite, alias Candor.

146 SBT, DR18/20/18b: William Leigh to Theophilus Leigh, 6 Sep 1710; Eveline Cruickshanks, 'Walter, Sir John, 3rd Bt. (*c.* 1674–1722), of Sarsden, Oxon.', in Romney Sedgwick, ed., *The History of Parliament: The House of Commons 1715–1754* (London: HMSO, 1970).

147 SBT, DR18/20/18b: Theophilus Leigh to James Leigh, 16 Apr 1711.

148 HL, mssST57, vol. 22, fos 363–4: James Brydges to Charles Stuart, 4 Oct 1723.

149 HL, mssSTB, Box 12, Folder 53: James Brydges II to George Brydges, 18 Feb 1751; D. W. Hayton, 'Brydges (Bridges), George (1678–1751), of Avington, Hants.', in Hayton, Cruickshanks and Handley, eds, *House of Commons 1690–1715*.

Leiden and knew that John would never 'get a good pronunciation of the French by learning it in Holland'.[150] Stuart, the tutor, evidently rehabilitated Leiden because John settled there after a tour of the Low Countries. The plan was for him to spend two or three years at the university, developing a 'figure' appropriate to his 'Rank' and preparing to appear with 'any tolerable advantage in Conversation'. John studied civil law, French, riding, fencing and dancing, the last three for 'his health, and Gracefullness of carriage'.[151] Leiden also offered access to a French-speaking elite, and he soon began visiting assemblies at The Hague. Brydges hoped that 'publick Company' there would help John to 'wear off the Boy & put on the Man', improving his French and shedding his 'awkward bashfullness', but he believed that his son would only make true progress in France.[152] Stuart found Angers in decline, so John settled in Paris. Geneva was too far for a short stay, but Brydges also worried that it might expose his son to 'ill naturd reflections', presumably on his Anglicanism. John took tuition in French and exercises over six months at an academy in Paris, where the best company would perfect his speech and acquaint him 'with the world & Men'. He returned home via a lengthy tour of Germany.[153]

Brydges planned for his second son, Henry, to spend one or two years at Leiden and another at a Parisian academy. He studied civil law, riding, fencing, dancing, French and Italian at the Dutch university and regularly visited The Hague. He also made other trips, culminating in a three-month tour of the Low Countries and Hanover.[154] Brydges sent Henry to Angers after a year, however, thinking it impossible 'for him to learn to speak the French language fluently without his being in France'. Henry's companion, Captain Robinson, agreed, unlike Stuart, that Angers was the best place for language and exercises. Robinson estimated an enormous £900 a year—perhaps an effort to fleece his patron—but in the end met Brydges's demands to halve the sum. Henry went to the academy for French, riding, dancing and fencing, though poor food and accommodation drove them to lodgings in

[150] HL, mssST57, vol. 19, fos 110–11: James Brydges to Charles Stuart, 11 Jul 1721.

[151] HL, mssST57, vol. 18, fos 249–50: James Brydges to Charles Stuart, 31 Jul 1721; fos 277–8: same to same, 18 Sep 1721; fo. 345: same to John Brydges, 3 Jan 1721/2.

[152] HL, mssST57, vol. 20, fos 15–16: James Brydges to Charles Stuart, 19 Nov 1721; fo. 43: same to same, 10 Dec 1721.

[153] HL, mssST57, vol. 20, fos 74–5: James Brydges to Charles Stuart, 22 Sep 1722; vol. 21, fos 87–8: same to same, 15 Oct 1722; fos 254–5: same to same, 3 May 1723.

[154] HL, mssST12, vol. 3, fos 287–96: accounts; mssST57, vol. 25, fo. 28: James Brydges to Andrew Robinson, 20 Nov 1724; fos 55–6: same to same, 8 Dec 1724.

town.[155] After 10 months at Angers, however, reports reached England that Henry could 'hardly speak three sentences together of French'. Brydges was furious, blaming his son's tendency to waste the day in English company. His hope that Henry would befriend the best French students, forming 'Partys of pleasure' to see local curiosities, were evidently misplaced. French had been 'the principal Cause' of the journey to France, which was probably the only opportunity for Henry to 'make himself Master' of something 'so necessary to him in the future Course of his life'. Brydges searched for another town where his son could seclude himself for two or three years, 'till he can speak it as well as English'. He was favouring Poitiers when John's death forced Henry home.[156] Leiden and The Hague had allowed both young men to learn and test the skills required of a gentleman, but France would ideally have brought them to perfection.

In the 1750s, Henry's son, the younger James Brydges, pursued a similarly mixed education abroad. He immediately settled at Göttingen in the Electorate of Hanover, where he stayed between December 1750 and July 1751. The university had tried to attract wealthy foreigners since its foundation in 1737, offering survey courses alongside extracurricular dancing, fencing, riding, drawing, music and modern languages. Göttingen achieved a higher proportion of noble students than other German universities but, despite the personal union with Hanover, it attracted few from Britain and Ireland before George II's visit in 1748. Brydges contributed to the first peak a few years later, drawn by royal patronage, good teaching and a 'pro-British milieu', and a more sustained rise began in the 1760s.[157] Brydges refused to wait until after Christmas 1750, suspecting that his professors wanted 'as much of my money as possible', and soon started French, German, riding, civil law and history.[158] He took lodgings with Edwyn Francis Stanhope, a gentleman's son from Kent, and received some one-on-one tuition during long days of study. There were 'no Amusements'

155 HL, mssST12, vol. 3, fos 287–96: accounts; mssST57, vol. 26, fo. 295: James Brydges to Andrew Robinson, 6 Sep 1725; vol. 27, fo. 102: same to same, 15 Nov 1725; fo. 235: same to same, 24 Jan 1725/6.

156 HL, mssST57, vol. 26, fo. 356: James Brydges to Andrew Robinson, 27 Sep 1725; vol. 28, fo. 187: same to same, 2 Jul 1726; fo. 339: same to same, 17 Sep 1726; vol. 29, fo. 43: same to same, 23 Oct 1726.

157 Charles E. McClelland, *State, Society, and University in Germany 1700–1914* (Cambridge: Cambridge University Press, 1980), pp. 38–48; Stewart, 'British Students at Göttingen'; Conway, *Britain, Ireland and Continental Europe*, pp. 140–1.

158 HL, mssSTB, Box 12, Folder 53: James Brydges II to Henry Brydges, second Duke of Chandos, 3 Dec 1750.

at Göttingen, but students could easily travel to nearby Hanover.[159] Brydges and Stanhope saved such trips until they had finished their studies, however, touring the region between April and July 1751. They soon left for Orleans on a three-week route through Frankfurt, Strasbourg and Paris. The pair found a house there, arranged for meals, hired a French tutor and made a final tour along the Loire before settling for the winter.[160] They remained outside the academy, having 'no Body to controul us', but dedicated their mornings to study in a well-furnished town. Brydges was back in England the following April.[161]

If universities and academies became interchangeable, how did families choose specific institutions? From the 1680s onwards, war regularly frustrated attendance at French academies while raising the profile of establishments outside France. Beyond that, options rose and fell according to several factors. Some depended on regional context, such as the linguistic purity and Protestant heritage of the Loire Valley. Indeed, the flight from France in 1685, when Catholic-run academies continued unmolested, shows the importance of confession. Other reasons related to individual establishments, from curricula and teaching quality to costs and the necessity of boarding. Although the elder Edward Southwell was unusual in checking on Lunéville himself, many families researched options. Most trusted tutors, but they also found information elsewhere. One Arbuthnot, a Scottish merchant visiting England from Rouen, estimated for James Brydges that a Parisian institution would cost a quarter of what Robinson had proposed for Angers.[162] Sir John Perceval took notes on the academies of Geneva from returning travellers, though his eldest son was only 10, and Katherine Southwell gathered opinion on Geneva and Turin while taking the waters at Tunbridge.[163] Since travel was an opportunity to perform status, fashion was a significant factor. Wolfenbüttel, for instance, was popular in the Tory circles of Oxfordshire and Gloucestershire, to judge by the Brydges family and their associates.

The desire to send sons to fashionable institutions clashed with warnings that educational travellers should avoid countrymen, a rule that endured from

[159] SBT, DR18/20/18: James Brydges II to Barbara Blackley, 8 Feb 1751; HL, mssSTB, Box 12, Folder 53: same to Earl of Oxford, 9 Dec 1750; same to Colonel Robinson, 4 Jan 1751; same to George Brydges, 18 Feb 1751; Stewart, 'British Students at Göttingen', 31.

[160] HL, mssST109, vol. 1: James Brydges II, Travel diary.

[161] HL, mssSTB, Box 12, Folder 53: James Brydges II, marginal note, April 1752; SBT, DR18/20/18: Brydges to Barbara Blackley, 6 Dec 1751.

[162] HL, mssST57, vol. 27, fo. 236: James Brydges to Andrew Robinson, 24 Jan 1725/6.

[163] BL, Add MS 47127: Sir John Perceval, Commonplace book, 1708–22, fo. 142: 'Extract of a letter from Geneva', 1721; Bristol Archives, 45317/2/5/2/23: Katherine Southwell to Edward Southwell III, 27 Jul 1758.

the mid-sixteenth to the mid-eighteenth century.[164] Derby and Sir Philip Perceval both received the message in the 1670s, aimed at immersing them in French, and they deliberately settled among few Englishmen. Sir Robert Southwell had 'very often tould' his nephew Perceval 'how unnescessary is the acquaintance of our owne countrymen while wee are abroad'.[165] James Brydges used the same logic decades later, advising John's tutor that 'too many English men in a Body together dont improve much by travelling'. Henry went to France, by contrast, so that he 'would have no opportunity to speak English, but only hear & speak French'.[166] The longevity of these warnings reflects the fact that they were impossible to heed. After all, if families followed fashions, running into other English-speakers was inevitable. As Chapter 4 will explore, it was in fact necessary: most parents hoped that travellers would not only learn abroad, but also make connections with fellow countrymen.

Could young men not have acquired the same skills at home? Chapter 1 found extensive domestic provision for French, which most travellers used before crossing the Channel. Boys and girls could even learn fashionable pronunciation in England, as French teachers who had left Paris or the Loire Valley advertised their prestige varieties.[167] After a century of failed schemes, masters of exercises also gained a foothold in London after the Restoration, offering their services in preparation or substitution for foreign travel. Even civil law featured in gentlemanly curricula at Oxford by the late seventeenth century.[168] So why invest in acquiring these skills abroad? Foreign travel developed adult male independence in confrontation with financial, moral and physical risk, but it was not enough for young men to leave their families.[169] Indeed, most had been doing so by increments since the age of 6. It mattered where they went and what they learned, as sons went abroad to acquire skills with which they could assert their status at home. To begin with, while teachers claimed that it was possible to learn prestigious French

[164] Gallagher, *Learning Languages*, p. 189; Conway, *Britain, Ireland, and Continental Europe*, p. 204.

[165] Bodleian, MS Carte 50, fo. 103: Ormond to James Forbes, 7 Aug 1673; 243, fo. 87: Forbes to Ormond, 7 Aug 1673; BL, Add MS 46954A, fo. 150: Jean Gailhard to Sir Robert Southwell, 19 May 1677; 46954B, fo. 5v: Sir Philip Perceval to same, 20 Jul 1677.

[166] HL, mssST57, vol. 22, fo. 363–4: James Brydges to Charles Stuart, 4 Oct 1723; *Cassandra Brydges*, p. 205: Cassandra Brydges to Henry Brydges, Jun 1726.

[167] Michèle Cohen, 'The Grand Tour: Language, National Identity and Masculinity', *Changing English*, 8:2 (2001), 129–41, at 131; Gallagher, *Learning Languages*, pp. 46–8.

[168] Levack, 'Law'.

[169] Cohen, 'Grand Tour: Constructing the English Gentleman'; French and Rothery, *Man's Estate*, pp. 137–83; Goldsmith, *Masculinity and Danger*.

outside France, families did not always agree. After all, James Brydges sent his sons from Leiden to France in search of fluency and the best pronunciation. Ultimately, France offered greater opportunities for immersive learning than even French-staffed residential options in London, where avoiding countrymen was hardly possible. The fact that there were never many institutions for learning exercises at home, meanwhile, despite the power of elite patronage, suggests that wealthy families were not overly interested in domestic options. The history of Foubert's academy, transplanted to London in 1679 with royal and noble backing, is a precarious one.[170]

Families preferred sons to learn skills abroad because they thought of them as 'parts' that built a 'compleat gentleman'. Just as some bought 'the Rarities of Towns and Countries ... to shew and keep them when they are come home', Gailhard observed, 'others learn what fine things are taught in such places'.[171] It mattered not only that one could speak French or dance well, but also where one had learned to do so. Although neither the elder Robert Southwell nor John Roades had travelled beyond Britain and Ireland, they recognised the value of civil law from Angers or Italian from Siena. The reputation of Paris for exercises and the Loire for French was similarly widespread, and travellers did well to learn at the most prestigious places. This was also a question of conspicuous consumption, as lengthy settlements required time and money. Orrery's financial accounts list books, writing materials, tennis kit, fencing shoes, foils and additional servants, needed for his exercises, alongside transport and accommodation.[172] As with material goods, distant or difficult provenance added value.[173] While young men might have learned French, riding, fencing or dancing at home, doing so on the Continent allowed them to pursue the most prestigious forms of knowledge, to attend fashionable institutions and to demonstrate family spending power.

Settlements and Tours

This chapter has addressed an imbalance by focusing on settled learning as an integral component of educational travel. To do so, it has had to

170 Ansell, 'Foubert's Academy'.

171 Gailhard, *Compleat Gentleman*, ii, p. 142.

172 NLI, MS 13215, fos. 11, 34, 36, 43, 53: Orrery's travelling expenses, 1685–9.

173 Arjun Appadurai, 'Introduction: Commodities and the Politics of Value', in Appadurai, ed., *The Social Life of Things: Commodities in Cultural Perspective* (Cambridge: Cambridge University Press, 1986), pp. 3–63, at p. 44.

look beyond journals and diaries. The very existence of these sources is a reminder not to tip the scales too far, however, as many believed that mobile tours deserved a special record. Some had prepared for tours by acquiring frames of reference through domestic travel. Others even undertook them without a dedicated period of settlement beforehand, as in William Leigh's travels through Germany and Italy in 1710–12 and Alan Brodrick's similar journey in the following decade, though Brodrick found that constant movement made his language-learning far more difficult.[174] Where travellers planned both institutional learning and sightseeing, moreover, a good place of settlement would provide a base for regional tours. By the second half of the seventeenth century, improved roads and a network of water transport joined Paris and the cities of Holland and Flanders, with an easy ride from Paris to Orleans giving river access to towns along the Loire.[175] The main purposes of tours—political and economic observations, socialising with elites—were inherently urban, and this north-western European cluster offered variety within easy reach of the best academies and universities. Institutional learning should receive more attention than it does, but touring was still an important aspect of elite British travel culture.

If settlement provided opportunities to learn languages and behaviour, tours offered useful observations. Sir Robert Southwell, a future president of the Royal Society, told Sir Philip Perceval that travel 'at this your age, and capacity of observing things in the World, must bee the great Tryall and experiment of your life'.[176] Southwell was well versed in the theory of travel, as is clear from a manuscript compiled before his own voyage, and he continued to engage with the subject in later decades. He was a particularly keen reader of Francis Bacon, 'which next to the Scripture he read most of all other'.[177] Bacon's list of courts, infrastructure, seats of learning, fortifications and cultural events informed the priorities of many travellers, who added art

[174] SHC, 1248/6/303: Alan Brodrick II to Sir Alan Brodrick, 11 Sep 1725.

[175] Gerrit Verhoeven, 'Foreshadowing Tourism: Looking for Modern and Obsolete Features – or Some Missing Link – in Early Modern Travel Behaviour (1675–1750)', *Annals of Tourism Research*, 42 (2013), 262–83, at 270; John Lough, *France Observed in the Seventeenth Century by British Travellers* (Stocksfield: Oriel Press, 1985), pp. 21–2.

[176] BL, Add MS 46953, fo. 102: Sir Robert Southwell to Sir Philip Perceval, 5 Jun 1676. On the Royal Society, see Daniel Carey, 'Compiling Nature's History: Travellers and Travel Narratives in the Early Royal Society', *Annals of Science*, 54:3 (1997), 269–92; Robert Iliffe, 'Foreign Bodies: Travel, Empire and the Early Royal Society of London. Part 1. Englishmen on Tour', *Canadian Journal of History*, 33:3 (1998), 358–85.

[177] King's, Keynes MS 152: Southwell, 'Manuscript concerning travelling'; BL, Egerton MS 1627: Southwell, Memoranda and diary, 1674–85; 1634: Le Grand, 'Life of Southwell', fos 13–13v.

and natural philosophy over the seventeenth century.[178] Southwell was happy with Perceval's account of Turin, which revealed 'the Improvement of your knowledge' as well as 'the pleasing of the Eye'. In Italy, his nephew should keep in mind 'the Modell of our owne Government here in England', to be held 'above all others'. If Perceval kept 'Home in his thoughts', he could 'from everything observable' frame 'little Channels, to drain and draw down the benfit thereof, unto his own particular affairs'.[179] Archbishop William King of Dublin, who owned plenty of educational literature, reminded another Southwell, probably of the Limerick branch, that 'the true Design of Travelling' was 'not meerly to satisfy curiosity of seeing variety, or Learning the exercises that are necessary for a Compleat Gent'. Travel should also 'enlarge a mans Mind, and furnish him with observations, which he may afterwards turn to the advantages of life and by which he many Improve himself[,] his fortune and his Countrey'. If Southwell considered 'Fashions[,] manners and Customs' abroad, as well as their climatic and legal contexts, he could decide 'which of these are to be imitated or introduced into your own Countrey and which are to be rejected'.[180]

The two journals kept in France and Italy by Sir Robert Southwell's grandson, Edward, in 1725–6 offer an example of the results. Edward had practised keeping journals on a foreign voyage of 1723, undertaken with his father, and during a domestic tour the following year.[181] On his return to the Continent, he recorded visits to churches, monasteries, convents, libraries, palaces, homes and fortifications throughout France and Italy, describing the situation, history and economy of each major town. He was particularly interested in interiors and artworks. Edward had a propensity to listing, from envoys at the Congress of Cambrai to the contents of the grand ducal galleries in Florence. Where he observed the Queen of France, the Polish monarchs or the pope, he assessed their characters. If travellers were supposed to compare the foreign with home, however, Southwell's efforts scarcely justified his domestic tour, framed as preparation for his second continental voyage. In sum: Paris, seen from Notre Dame, looked less than

[178] Bacon, 'Of Travel'; Parks, 'Travel as Education'; Chaney, *Evolution of the Grand Tour*, pp. 203–14.

[179] BL, Add MS 46954B, fo. 3: Sir Robert Southwell to Alexandre de Rasigade, 9 Jul 1677; fo. 213: same to same, 14 Nov 1677; 46955A, fo. 59: same to Perceval, 11 Feb 1677/8.

[180] TCD, Lyons MS 2535, fo. 64: William King to Mr Southwell, 21 Jan 1717; Robert S. Matteson, *A Large Private Park: The Collection of Archbishop William King, 1650–1729*, 2 vols (Cambridge: LP, 2003). Perhaps Robert Southwell (d. 1724), Sir Thomas's son.

[181] Clark Library, MS.1975.002: Edward Southwell II, Journal of Our Journey, 1723; BL, Add MS 34753: Southwell, Journal of travels in Lorraine and France, 1723; Brotherton Library, YAS/MS328: Southwell, Account of my Journey.

half the size of London; French villages paid tithes to the local monastery, as 'was the case in England before the Reformation'; the air was better in Genoa than in England; and outside Florence, the homes of noblemen were 'no better then farmers Houses in England, & only remarkable for their agreable Situation'. The fact that Southwell only invoked print in Italy—taking accounts of Genoa and the Tuscan grand dukes from Addison and citing Misson on the Guelphs and Ghibellines—suggests that his time in Italy was more structured by textual precedent than his tour of France.[182]

Like French and exercises, this kind of tour required an investment of time. When Sir Robert Southwell considered keeping Perceval in France for another year after Italy, he acknowledged that 'the Longer he continues abroad, the more he will see and observe'. Perceval came home as planned, however, as his uncle compared travel to arithmetic: 'Although he that goes deeper in may know more', only a finite amount of learning was necessary.[183] James Brydges, on the other hand, worried that his eldest son was moving too quickly, as it was not his 'Desire that you shou'd be so expeditious in Travelling as not to obtain the end of it'. He twice warned that John's route through France and Germany 'was rather riding Post than travelling' and insisted that he 'take a full view of what is worth his seeing' at Paris, Vienna and Prague. John's tours were perfectly timed to see the coronations of Louis XV at Reims and of the Holy Roman Emperor, Charles VI, as King of Bohemia, alongside the usual 'variety of men and manners'.[184] His father's wealth allowed him to reap the benefits of slow travel.

Observation came into its own for those, unlike the Brydges brothers, who went as far as Italy. British travellers rarely settled at Italian universities or academies after the mid-seventeenth century, with Irwin coming to Siena in unusual circumstances. They attended in far lower numbers than the Germans, even after new academies opened in Florence and elsewhere in the late seventeenth century.[185] Turin was the exception, but educational travellers treated the town as a bilingual frontier between French and Italian culture. For a century and a half after 1630, standard routes through Italy afforded several months in Rome around Christmas and Easter, often

[182] BNF, MS Anglais 194: Edward Southwell II, Travels, fo. 55v; BL, Egerton MS 3805: Southwell, Journal of travels in France and Italy, 1725–6, fos 3, 22v–23v, 29, 37v, 38, 58v.

[183] BL, Add MS 46954B, fo. 213: Sir Robert Southwell to Alexandre de Rasigade, 14 Nov 1677.

[184] HL, mssST57, vol. 18, fo. 345: James Brydges to John Brydges, 3 Jan 1721/2; fo. 424: same to Charles Stuart, 21 Apr 1722; vol. 22, fo. 208: same to same, 21 Mar 1722/3; fos 255–6: same to same, 3 May 1723; fo. 256: same to Captain Lancaster, 3 May 1723.

[185] Boutier, 'Grand Tour des gentilhommes', p. 250.

either side of a trip to Naples.[186] As Chapter 4 will show, these lengthy stays fostered a different kind of transient Anglophone community, outside the institutional contexts of France, the Netherlands and Germany. The main influence on encounters with Rome between the mid-seventeenth and mid-eighteenth centuries was the classical education that these men shared, and they structured their time with busy periods of antiquarian sightseeing. For British visitors before the 1760s, Rome exemplified existing knowledge, the city's antiquities an 'intellectual prompt' to classical learning from home. Travellers toured churches and palazzi with the same paid guides, checking off paintings and sculpture that would form and advertise personal taste.[187] Some were enthusiastic: 'no place under Heaven' instructed and entertained like Rome, thought William Leigh, and Metcalfe Robinson had longed to visit 'the famousest place in the world' since he had 'got an insight into its greatness' at school.[188] Francis Rawdon was less enraptured by 1774, visiting the ruins only 'for the sake of having it to say that one has seen them'. Belief in educational travel to the Continent was crumbling, but Rawdon still felt that most would think him a 'barbarian'.[189]

As a result, although Rome was the high point of many tours, the city represents the nadir of travel journals as entertaining or enlightening sources, especially by the eighteenth century. Robert Southwell mixed the fruits of sightseeing in 1660–1 with snippets of conversation with cardinals, expatriates and fellow travellers in a rough commonplace book, but his grandson followed established trends by listing and counting his way around Rome in 1725–6.[190] Where his journals had hitherto taken a narrative approach, Edward presented the city under thematic headings for churches, palaces and antiquities, treating it as a depopulated cabinet of curiosities.[191] He filled '140 Pages in my Journal' with 'the Antient and Modern Curiosities of this famous City', finding 'a sensible pleasure' in seeing the places where his schoolboy history had unfolded.[192] Southwell's approach was typical. When William Leigh determined 'to begin regularly to see what is to be seen', he 'divided the Curiosities of Rome in 4 several

186 Stoye, *English Travellers Abroad*, pp. 117–33; Towner, 'Grand Tour', 318.
187 Sweet, *Cities and the Grand Tour*, pp. 99–163.
188 SBT, DR18/20/18b: William Leigh to Theophilus Leigh, 11 Jul 1711; WYAS, WYL150/6005/6: Metcalfe Robinson to Sir William Robinson, 15 Dec [1704].
189 PRONI, D2924/1/180: Francis Rawdon to Moira, 8 Dec 1774.
190 BL, Egerton MS 1632: Robert Southwell, Notebook during his stay in Italy, 1660–1.
191 Sweet, *Cities and the Grand Tour*, pp. 18, 107, 111.
192 BL, Egerton MS 3805: Edward Southwell II, Journal of travels in France and Italy, fos 71v–129; Add MS 47031, fo. 139: same to Sir John Perceval, 9 Apr 1726.

Branches'—antiquities, churches, palaces and villas—'that there may be no interfereing with things of a different nature'.[193] Alan Brodrick spent seven weeks 'attempting to fix' the same sights in his memory, the '2's and '3's on his lists representing multiple visits.[194] Travellers followed antiquarian conventions by aiming for disinterest, with personal impressions gaining legitimacy only as part of wider changes in travel writing over the late eighteenth century.[195] Between the 1650s and 1750s, if travellers settled in French, Dutch or German towns to acquire new skills, they did so in Rome primarily to confirm existing knowledge.

Plenty of work already covers British engagement with the art and antiquities of Italy, but the point here is that institutional and observational learning were different aspects of elite travel culture. Families might combine or separate them according to their circumstances. Diaries and journals, so often the focus of historical writing, were the products of only one kind of educational travel. In fact, they were integral to its aims, as writing would solidify the benefits of observation. When Southwell called on Perceval to keep a journal in Italy, he explained that 'by the paines of writing all will stick with you soe particularly and fresh in memory'. Perceval's brother, Robert, had written of the diary of his domestic travels 'imprinting' things on his mind.[196] This function could also be fulfilled by letters—Southwell told his son, on his way to the Netherlands in 1691, that 'If you write to me all you Observe it will fix it in your mind'—but it is one of the main reasons why travel journals exist.[197] The other, that they could be shared and reread as a monument to successful travel, will feature in Chapter 5. Diaries and journals are good evidence for how travellers achieved and presented their observational learning, often on mobile tours, but not for the whole of their time abroad.

The party that travelled with Irwin to the Netherlands in 1704 encapsulates the diversity of experience in elite travel culture. The viscount himself used Leiden as a springboard to Italy, a destination only brought forward by his early departure, but his duels at The Hague disrupted the

193 SBT, DR18/20/18b: William Leigh to Theophilus Leigh, 21 Feb 1710/11; DR671/59a: William Leigh, Travel journal, pp. 150–81.

194 SHC, 1248/6/113–114v: Alan Brodrick II, Lists of Roman sights; fo. 209: same to Sir Alan Brodrick, 25 Apr 1725.

195 Sweet, *Cities and the Grand Tour*, p. 6; Katherine Turner, *British Travel Writers in Europe, 1750–1800: Authorship, Gender and National Identity* (Aldershot: Ashgate, 2001).

196 BL, Add MS 46955A, fo. 59: Sir Robert Southwell to Sir Philip Perceval, 11 Feb 1678; 46953, fo. 138: Robert Perceval to Southwell, [22 Jul 1676]. On journals as memory aids, see Sweet, *Cities and the Grand Tour*, pp. 16–18.

197 BL, Add 38015, fo. 388v: Sir Robert Southwell to Edward Southwell, 5 Jan 1690/1.

plans of less privileged companions. His brother, Rich, went straight home from the Netherlands, meaning that his only experience of the Continent, before he returned with the army, was through settled learning in Delft and Leiden. Worsley had settled for a long period at Leiden, though prematurely ended, and there was no suggestion that he would accompany the viscount to Italy. Other gentry families also let institutional learning frame their ambitions, perhaps supplemented by a tour of the surrounding region. The Rawdon brothers went no further than Foubert's Parisian academy in the 1670s and the Chamberlaynes cast John's travels in the 1710s in settled terms, planning for Wolfenbüttel before deciding on Paris. He was about to extend his stay into a tour of France when he died.

Archival research gives the strong impression that younger brothers and the sons of more modest families attended French, Dutch or German institutions rather than undertaking ambitious Italian tours, but the first kind of travel is doomed to under-representation. Without institutional records for most academies, where would one start looking for students? Even where leads can be found, institutional learning often generated little correspondence. The only trace of John Norborne, a Wiltshire gentleman, having spent seven months at Foubert's academy as part of a year in France exists because his tutor, an Anglican cleric, felt the need assert his faith in print.[198] The financial and logistical arrangements necessitated by constant movement mean that tours of Italy have left far more documentation, to set alongside journals individually catalogued in family archives. But shorter-distance educational travel was in fact spread widely among the gentry, who therefore came closer to the Dutch experience than historians have acknowledged. Only one in 10 Dutchmen at Angers appear later in Italian records, with half as many ending up at German and Swiss universities, and the academies of France and Geneva were popular with the sons of burghers as well as noblemen.[199] After all, improving connections meant that it was not much more difficult to get to Paris and the Loire from southern England than from the Netherlands. Gentlemen from Britain and Ireland made the trip in high numbers from the mid-seventeenth century. By overlooking their experiences, British and Irish historians—the originators of the Grand Tour model—have underestimated the gentry's familiarity with the Continent.

[198] *The Visor Pluckt Off from Richard Thompson of Bristol, Clerk in a Plain and True Character of Him* (London: 1681).

[199] Willem Frijhoff, 'Éducation, savoir, compétence: les transformations du Grand Tour dans les Provinces-Unies à l'époque moderne', in Babel and Paravicini, eds, *Grand Tour*, pp. 609–35, at pp. 631–2; Verhoeven, *Europe within Reach*, pp. 63–4.

Settlement was more appealing than a tour to those with limited means. Continual travel was more expensive than staying put, as transport and accommodation were significant elements of any budget, but food, lodgings and other essentials were cheaper if arranged for a long period.[200] Travellers such as William Leigh might blame excessive spending on 'The perpetual motion we have been in for some months last past', promising that a period of settlement would be one of retrenchment.[201] The costs of an academy were not insurmountable, as the Rawdons discovered in comparison with Cambridge, and part of the popularity of these establishments relied on the fact that they were cheaper than employing a range of masters oneself. Neither was it expected, as one tutor put it in the 1670s, 'that anie person, of what qualité soever should live high, so long as he is in an Academy'.[202] A century later, Huntingdon thought his Rawdon nephews would need more money if they did not attend the royal academy in Turin, 'where every thing is (or at least was formerly) on a very oeconomical plan'.[203] The prominence of Wolfenbüttel, Leiden, Angers and Göttingen over three generations of Brydges travel makes clear, however, that the preference for settled learning was not simply a question of money. These institutions were good value because they delivered the main outcomes of educational travel in a structured setting. Otherwise, domestic academies could not have pitched themselves as substitutes or preparations for foreign experience.

A period of settled, institutional learning was the baseline of continental education, to which travellers might add a tour depending on their means, ambitions and obligations. Sir Philip Perceval travelled according to a 'scheme' drawn up by his uncle, such that Southwell could accurately predict his return a year and a half in advance, but others planned itineraries on the hoof.[204] Some way through a period of settled learning, many began to think about what they might do next. John Rawdon went straight from Foubert's Parisian academy into the army, but his uncle Conway considered sending his brothers, Edward and Arthur, on a three-year tour of France and Italy once Edward had also spent time with Foubert. Edward's death after a duel ended the idea and Arthur returned to Ireland. Chapter 2 showed that

200 Black, *British Abroad*, p. 91.

201 SBT, DR18/20/18b: William Leigh to Theophilus Leigh, 8 Feb 1712.

202 National Records of Scotland, GD406/1/5981: James Forbes to Duke of Hamilton, 15 Aug 1676; Ansell, 'Foubert's Academy', p. 54.

203 PRONI, T3765/M/2/14/96: Huntingdon to Lady Rawdon, 22 Feb 1774; T3765/M/2/14/98: same to same, 8 Aug 1774.

204 BL, Add MS 46954B, fo. 213: Sir Robert Southwell to Alexandre de Rasigade, 14 Nov 1677.

these decisions were worked out between their parents and their uncle, but it also included young men who successfully requested further travel after settled learning. Thomas Worsley, for instance, was allowed to spend several months in Germany after his studies at Leiden, though not to follow Irwin to Rome. Coote Molesworth had moved between institutions, from Leiden to Geneva and back, before accompanying his brother to Turin. He argued in vain that he should be allowed to study at the academy there, but he did get permission for a tour of Italy on his way home. Social standing and birth order sketched the outlines of who went on lengthy tours and who did not, but families left room for negotiation.

Conclusion

If settled learning was the baseline for many young men's travels, historians need to rethink the balance between destinations in elite British travel culture. Italy has dominated scholarly accounts, but how important can it have been if many crossed the Channel without knowing whether they were going there? The traditional emphasis on movement over settlement is yet another way in which Grand Tour historiography has privileged Italy over France, the Netherlands and Germany, rushing over Angers, Leiden and Göttingen in order to get to Venice, Florence and Rome. Considering family strategy reveals that an Italian tour was often the optional extra, once the tangible benefits of travel—the French language and exercises—had been secured in northern institutions. French and Rothery argue that, 'although French language and culture was a tremendously powerful force upon elites in Britain and Europe in the century up to 1789, France seems never to have become the primary focus of the Grand Tour'.[205] But is the problem not an exclusive—indeed anachronistic—focus on the Grand Tour, to the exclusion of other forms of educational travel? French and Rothery use travel diaries to place Italy at its centre, but these are partial sources too often taken out of context. In fact, some travellers started journals precisely because they were entering Italy.

Of course, many went to Italy in the seventeenth and eighteenth centuries—enough for Ingamells's 1,000-page dictionary between 1701 and 1800—and they stayed for long periods in Rome, Venice and increasingly Florence and Naples. What is required is a more equitable balance between elements of elite travel culture. James Brydges was furious that Henry

[205] French and Rothery, *Man's Estate*, p. 140.

might come home without fluent French and Orrery was embarrassed to return without his exercises, but discussion of Italy did not elicit such strong responses. If writers and tutors sold travel as a package of northern settlement and Italian tour, many families prioritised the first component. Even for those who did go to Italy, settled learning was usually a crucial part of their travels. Paying greater attention to strategies, decision-making and practice means giving greater weight to France, the Netherlands and increasingly Germany. Between 1650 and 1750, educational travel was primarily about immersion in French culture, in France itself or in neighbouring territories. There is no French equivalent of Ingamells, not least because the book emerged from Brinsley Ford's art-historical preoccupation with Italy.[206] Above all, though, the compilation of a dictionary of British travellers to France would be too large a project.

Scholars writing before the late twentieth-century triumph of the Grand Tour were at greater liberty to emphasise northern settlements and itineraries. John Stoye recognised in the 1950s that, by the late seventeenth century, 'the Italian tour was becoming subordinate to the more indispensable element in a gentleman's time abroad—as it came to be considered—his education in France'.[207] Cohen has more recently agreed, a dissenting voice in the past few decades.[208] As Anglophone historians chip away at the monolithic Grand Tour, in collaboration with continental scholars who have always emphasised institutional options, others should follow. The evidence of family strategy shows that settled learning was not only an overlooked aspect of what most would today call the 'Grand Tour', but it also offered an alternative to those who could not make such a lengthy voyage. Educational courses were complex and families assembled them from a range of options, according to their hopes and capabilities. In turn, educational priorities informed routes, destinations, the length of stays and the age at which young men travelled, shaping encounters with the Continent. There is no sense that the gentlemen of lesser means examined here were unusual; in fact, the use of universities and academies was too routine to leave much manageable evidence. The gentry were far better acquainted with the Continent than historians have appreciated, and the main outcomes of travel—fluency in French and a graceful bearing—far more important aspects of landed culture.

206 Jason M. Kelly, 'Reading the Grand Tour at a Distance: Archives and Datasets in Digital History', *American Historical Review*, 122:2 (2017), 451–63, at 456.

207 Stoye, *English Travellers Abroad*, p. 118, first published in 1952; Clare Howard, *English Travellers of the Renaissance* (London and New York: J. Lane, 1914), pp. 102–4, 121–3.

208 Cohen, 'Grand Tour: Constructing the English Gentleman'.

4

Networking Abroad

At all social levels, though most visibly among the nobility and gentry, a healthy network of family and friends offered mutual support and the prospect of advancement; its absence courted disaster.[1] Unsurprisingly, then, networking was an important aspect of elite education. Historians recognise the bonds that schools and universities created among British and Irish landowners, to the point that educational institutions 'may have been more important socially than intellectually' from the second half of the seventeenth century.[2] By comparison, the social aspects of travel, often the next step in elite education, are usually overlooked in favour of the art and architecture that dominate so many studies of the Grand Tour. The balance is changing with the rise of social network analysis, as quantitative approaches to travel recognise 'a series of intersecting journeys whose details are best understood together'.[3] This chapter takes a traditional, metaphorical approach to networking, but places interpersonal connections—with fellow travellers, expatriates and foreigners—at the heart of educational travel.

The idea that educational journeys were rites of passage summons an image of travellers sent into an unknown continent and plunged, as French and Rothery put it, 'into a social setting in which their previous familial or educational contacts counted for very little'.[4] While this understanding allows the Grand Tour to fit the anthropological template of a rite of passage, it cannot capture the realities of educational travel. Previous chapters have made clear that travel was an extension of prior education and a step for

[1] Thomas, *Ends of Life*, p. 192; Michael Mascuch, 'Continuity and Change in a Patronage Society: The Social Mobility of British Autobiographers, 1600–1750', *Journal of Historical Sociology*, 7:2 (1994), 177–97, at 185–9.

[2] French, 'Gentlemen', pp. 275–6.

[3] Giovanna Ceserani, Giorgio Caviglia, Nicole Coleman, Thea de Armond, Sarah Murray and Molly Taylor-Poleskey, 'British Travelers in Eighteenth-Century Italy: The Grand Tour and the Profession of Architecture', *American Historical Review*, 122:2 (2017), 425–50, at 449.

[4] French and Rothery, *Man's Estate*, p. 175. For a weaker but more plausible interpretation, see Redford, *Venice and the Grand Tour*, pp. 15–16.

which families pooled knowledge and connections. After all, much of the gentry and nobility was on intimate terms with their European neighbours by the late seventeenth century. This chapter finds, moreover, that young men moved through the Continent in what Orrery's tutor called 'a world of English'.[5] They travelled in groups and among a cohort of those abroad for their education, frequenting familiar circles and subject to familiar pressures. They took the help and hospitality of expatriates, whether diplomats, merchants or religious exiles. Increasingly, too, they bumped into others who had crossed the Channel for health or leisure. Educational travel was no removal from British society.

Gaggles of Englishmen are an established feature of historiography, but usually as a negative aspect of travel. Insularity, it is claimed, encouraged criticism of foreigners and ensured that young men misspent their time abroad. Shared activities, such as a hopping race down the grand ducal galleries in Florence, appear as comic relief from engagement with art and architecture.[6] This chapter recognises that British travellers and expatriates gathered abroad, and that many wasted time, but it argues that Anglophone sociability was an integral and deliberate part of educational travel. This priority clashed with advice to avoid countrymen, of course. Learning and networking were in tension, as linguistic progress was inversely proportional to time spent with fellow English-speakers. This tension persisted, however, because it was neither possible nor desirable for young men to seclude themselves from each other on the Continent. Historians reflexively see relationships with countrymen as impediments to successful travel, which depended on diligent observations and interactions with foreign hosts, but time abroad equally forged and strengthened connections between travellers. Not only was travel a common experience among the elite, but it was one that they had undertaken together on the threshold of adulthood.

Travelling Groups

Most young men crossed the Channel in groups. Brothers often travelled together to save money, with James Brydges's decision to send John and Henry separately an indication of his unparalleled means. Henry and Arthur Ingram travelled in the late 1650s and early 1660s, as did Edward and

[5] NLI, MS 36, fo. 726: Germain Colladon to Dowager Countess of Orrery, 15 Feb 1689.

[6] The race appears regularly: Black, *Italy and the Grand Tour*, p. 138; Redford, *Venice and the Grand Tour*, p. 16; Sweet, *Cities and the Grand Tour*, p. 30.

another Arthur 20 years later; John and Richard Molesworth were abroad in the late 1690s. Brothers did not always stay together, though, as birth order often came into play. Irwin and Rich Ingram went to the Netherlands in 1704, but Rich came back to England when his elder brother headed south. Arthur Rawdon only joined Sir John in Italy, arriving by sea in 1739, and the pair returned home together. Young men also travelled with neighbours, schoolfriends, university contemporaries or members of the same political circles. Metcalfe Robinson's father was glad that he would 'contract so early freindship' with his fellow traveller, Thomas Frankland, since they were 'like to be so near neighbours'.[7] Such plans did not always work out, though. The younger Edward Southwell crossed the Channel with Charles Hamilton, a Westminster schoolfellow from another Irish landowning family, and they travelled together as far as Rome.[8] They parted ways there, however, after a violent quarrel. Southwell's writings contain no explanation, but a spy report claims that the pair had come to blows during a political argument, smashing the windows of their coach.[9] They moved out of shared lodgings and never travelled together again. Sir John Perceval recognised that there was 'no travelling with profit or pleasure where humours dont agree', but he pitied Hamilton's 'unaccountable behaviour, he knows not what he has lost by it'.[10]

Travelling parties offered the hope of social as well as geographical mobility. Historians recognise that educational travel allowed middling-sort tutors to advance through aristocratic patronage, affording regular travel, foreign institutional learning and international networks that lay beyond their means.[11] Successful tutors also gained references and even annuities, though such incentives made it more difficult for them to control wayward charges.[12] Jean Gailhard would 'establish a firm and lasting freindship' with Sir Robert Southwell if Sir Philip Perceval's travels went well, while George Townshend received £100 a year from the Ingrams between the 1650s and his death in 1704.[13] Some, such as Gailhard and Maximilien Misson, solidified links

[7] WYAS, WYL150/6002/51: Sir William Robinson to Metcalfe Robinson, 20 Mar [1704].

[8] The future landscape gardener: Ingamells, *Dictionary*, pp. 445–6, 880; Barker and Stenning, *Record of Old Westminsters*, i, p. 416; ii, p. 870.

[9] TNA, SP85/15, fo. 520: report from Baron Philippe von Stosch, alias 'John Walton', 29 Dec 1725.

[10] BL, Add MS 47031, fos 101v–102: Edward Southwell II to Sir John Perceval, 15 Feb 1725/6; fo. 118: Perceval to Southwell, 12 Mar 1725/6.

[11] Klima, 'Introduction', in *Spence: Letters*, pp. 1–22, at p. 2; Boutier, 'Compétence internationale', pp. 173–8; Sánchez-Jáuregui, 'Educating the Traveler', p. 92.

[12] Black, *Italy and the Grand Tour*, p. 11.

[13] NLI, MS 7863, fo. 34: Sir Robert Southwell to Jean Gailhard, 5 Jun 1676; WYAS, WYL100/C/9A, George Townshend to John Roades, 4 Jul 1702.

by dedicating books to former pupils.[14] Artists and architects, too, travelled to Italy in order to attract business, profiting from the flexibility of social boundaries among countrymen abroad.[15] Studies rarely comment, however, on the fact that noblemen travelled with gentlemen of lesser standing. European journeys by those of the highest rank were 'tiny mission[s] to the continent', allowing gentlemen to connect with superiors and travel beyond their means.[16]

Samuel Waring offers one example, as the 27-year-old son of a comparatively modest landowner from County Down. He travelled to Italy in 1687–8 with Ormond's 15-year-old grandson, Charles Butler, and his tutor, Misson. Ormond probably found Waring through Trinity College Dublin, where the young man had taken a degree, and the duke seems to have funded his travels. Waring enjoyed experiences beyond his means, but the journey also demonstrated enduring distinctions. On one Roman outing, Butler rode in the first coach with senior Catholic clergymen and Protestant nobles from home, while Waring travelled in the second with Misson and the servants. When Butler attended gatherings, he sometimes left Waring to explore Rome by himself. At home, however, and elevated to the earldom of Arran, Butler 'often rememberd' Waring, who retained strong ties to his family.[17] Travel could not ennoble gentlemen, but its more relaxed sociability promised memorable experiences and useful connections.

William Leigh's travels with Cornbury similarly offered a 'favorable opportunity' for a gentleman to travel beyond his station, though in his case at inordinate cost.[18] Noble company brought 'many advantages': Leigh attended audiences with the rulers of Hanover, Turin, Florence and Naples, waited on Prince Eugene of Savoy, saw the coronation of the Holy Roman Emperor, stayed on the 'airy and pleasant' Quirinal Hill in Rome and joined high-ranking diplomats on a Dutch state yacht.[19] Surpassing his origins, Leigh told his father that he had been one of 'the Lords that came to Bern in

14 Gailhard, *Compleat Gentleman*, ii, sig. A2r; Maximilien Misson, *A New Voyage to Italy*, first English edn, 2 vols (London: 1695), ii, sig. A2r.

15 Sweet, *Cities and the Grand Tour*, pp. 7–8; Ceserani *et al.*, 'British Travelers in Eighteenth-Century Italy'.

16 Stoye, *English Travellers Abroad*, p. 10.

17 Ansell, 'Reading and Writing Travels', 1457–8, 1461, 1466–7, 1473.

18 HL, mssST58, vol. 5, fos 272–4: Theophilus Leigh to James Brydges, 14 May 1710.

19 SBT, DR18/20/18b: William Leigh to Theophilus Leigh, 14 Jul 1710; same to same, 21 Feb 1710/11; same to same, 14 May 1711; same to same, 22 Aug 1711; same to same, 17 Oct 1711; same to same, 10 Dec 1711; same to same, 19 Jan 1712; HL, mssST58, vol. 10, fo. 266: same to James Brydges, 22 Mar 1712.

Company with my Lord Cornbury'.[20] When his noble companion returned home, however, Leigh reverted to his standing as a gentleman's son. He travelled and lodged with others of his rank and waited on—rather than dined with—English diplomats and noblemen.[21] Leigh's prospects suffered from Cornbury's death soon after his return, though his father, tired by spiralling costs, concluded that the connection 'would have proved of little advantage'.[22] Cornbury nevertheless raised the profile of Leigh's travels as an opportunity for learning and networking.

Viscount Irwin, one of the most prominent noblemen in Yorkshire, offers the opposite perspective to Waring and Leigh. Members of the gentry saw his travels as an opportunity to cement relationships and live beyond their means. The Stricklands of Boynton, in the east of the county, had wanted their son to travel with Irwin. Their plans failed, but Lady Strickland still hoped that William, who instead went to Wolfenbüttel academy, might wait on Irwin abroad.[23] Thomas Worsley of Hovingham, north of York, had greater success as a trustee of the third viscount's will. His only son, also Thomas, had been at Christ's College with Irwin, where he had 'sett his Heart upon Travell' with the young viscount, though 'for some small Time only'.[24] The younger Worsley went to Delft and Leiden, making what Irwin patronisingly called a 'litle voyage into Holland', but he did not follow the viscount to Italy.[25] Detached from his noble patron, Worsley fell in with other Yorkshire gentlemen. He saw Hanover and the Netherlands with Strickland and frequently banded with Metcalfe Robinson, whom Irwin had met in Italy.[26] Worsley was on the Continent for a year and a half, during which his father sent him between £30 and £60 a month, to a total of £654. These were large sums, but they were two or three times smaller than those sent to Irwin in the same remittances.[27] That Worsley did not stay at Leiden after the viscount's unexpected departure suggests that his studies had only been feasible in his retinue.

[20] SBT, DR18/20/18b: William Leigh to Theophilus Leigh, 20 Dec 1711.

[21] SBT, DR18/20/18b: William Leigh to Theophilus Leigh, 30 Jul 1712; same to same, 21 Sep 1712; same to same, 18 Jan 1713.

[22] SBT, DR18/20/18b: Theophilus Leigh to James Leigh, 11 Mar 1712/13.

[23] WYAS, WYL100/C/20: fo. 194: Elizabeth, Lady Strickland to Dowager Viscountess Irwin, 27 Jun [1702]; unfoliated: same to same, 13 Jun [1704].

[24] WYAS, WYL100/C/9A: Thomas Worsley to John Roades, 2 Apr 1704.

[25] WYAS, WYL100/C/9A: Irwin to John Roades, 30 May [1705].

[26] WYAS, WYL150/6005/22: Metcalfe Robinson to Sir William Robinson, 22 Sep [1705]; 35: same to same, 31 Jul [1705].

[27] WYAS, WYL100/C/10: Richard Hey to Thomas Kitchingman, 9 Jan 1705; NYCRO, ZON13/1/278: same to Thomas Worsley, 20 Nov 1705.

The chance to travel with a noble patron could trigger fierce competition, explaining the urgency with which the Leighs secured their place with Cornbury. It might even lead to bribery and violence, of which another of Irwin's Christ's contemporaries, Thomas Maynard, provides a striking example. Maynard's father allegedly offered Irwin's tutor, John Haccius, 'a good Purse of Money' to add his son to the travelling party and, though it was apparently refused, Thomas followed them to Delft.[28] There he proved a 'Troublesome Gentleman', seeking 'to displace evry body who are now with mylord, that he might have all the advantage of Mylords Travells and Acquaintances'. Irwin himself noted that Maynard 'had made severall endeavours to putt us together'.[29] The party hoped to escape him in Leiden, but the situation only worsened. Haccius seems to have punched Maynard, who then spread insults about Irwin among English army officers at The Hague during the War of the Spanish Succession. The viscount in turn struck Maynard in a coffee house, pulling in several noblemen who encouraged a duel. Irwin had the better of it but, with neither side drawing blood, Maynard attacked him again in a fencing school and was almost killed.[30]

Irwin and Maynard received flesh wounds and letters from the local diplomat, but the lasting damage was to Haccius's reputation.[31] The English at The Hague were 'enraged' with him, despite his protests that 'Quarrels will happen in the world', and 'Officers and Men of Quality' brought home the message that he was unfit to accompany Irwin abroad.[32] This became 'the Generall opinion of the Nobility & Gentry of England', such that Roades, the estate steward, could 'goe into what part of the Countrey I will[,] that is throwne upon me'.[33] Irwin's trustees decided to separate him from Haccius but, despite assurances that they had parted in Germany, they suspected that the two men had travelled together to Italy.[34] Maynard ventured south separately and rumours spread in London and York that he had again fought

[28] WYAS, WYL100/C/9A, Irwin to John Roades, [31 Aug 1704].

[29] WYAS, WYL100/C/9A: John Haccius to John Roades, 3 Oct 1704; WYL100/C/20, fo. 98: Irwin to Dowager Viscountess Irwin, 31 Oct [1704].

[30] WYAS, WYL100/C/9B: John Haccius to [John Roades], 16 Dec 1704; anonymous letter, 29 Dec 1704.

[31] WYAS, WYL100/EA/12/17/5: Alexander Stanhope to Irwin, 26 Dec 1704.

[32] WYAS, WYL100/C/9B: anonymous letter, 29 Dec 1704; WYL100/C/10: John Haccius to John Roades, 9 Feb 1705; WYL100/EA/12/17/5: Earl of Halifax to Dowager Viscountess Irwin, 19 Dec 1704.

[33] WYAS, WYL100/C/9A: John Roades to Irwin, 20 Jan 1704[/5].

[34] WYAS, WYL100/C/9B: Thomas Worsley to Dowager Viscountess Irwin, 14 Dec 1704; same to same, 4 Apr 1705.

with the viscount in Rome.[35] Even three years later, when Irwin and Maynard were in the English capital, the authorities avoided trouble by confining them at home.[36] Maynard's attempts to profit by Irwin's travels descended into violence, risking the reputation of both men and destroying Haccius's name, but they show the lengths to which some might go to use travel as a networking opportunity.

If travel deepened connections based on patronage, neighbourhood and domestic education, these concerns also overlapped with political loyalties. Leigh and Cornbury, respectively nephew to James Brydges and son of the Earl of Clarendon, probably travelled together on account of party connections, since both relations were prominent Tories. As will become clear, Leigh's father shared their inclinations.[37] The two men toured Italy and Switzerland with the eldest son of the Tory Viscount Downe and, once Cornbury had returned home, Leigh toured Flanders with three more young Tories. After a period in Brussels, he continued to Paris with the brother of another high-ranking Tory, Viscount Dupplin.[38] The Ingrams, Worsleys, Stricklands and Robinsons were all Yorkshire Whigs, by contrast, and Irwin socialised with the party leadership before his departure from London. Even Maynard, from Suffolk, shared their allegiances.[39] Partisanship inevitably influenced travel between 1702 and 1714, when it suffused elite life.[40] At Brussels, Leigh noted that 'Faction is as violent here as in England' and that the locals were 'full as inveterate one against the other as Whig and Tory'.[41] In these conditions, travel offered the chance to impress future patrons.

[35] WYAS, WYL100/C/10: Thomas Worsley to John Roades, 17 Apr 1706; Brown, ed., *Inglesi e Scozzesi*, pp. 186–7.

[36] WYAS, WYL100/C/10: Thomas Worsley to John Roades, 20 Feb 1708/9.

[37] Hayton and Lancaster, 'Hyde, Edward, Visct. Cornbury (1661–1723)' and Hayton, 'Brydges, Hon. James (1674–1744)', in Hayton, Cruickshanks and Handley, eds, *House of Commons 1690–1715*.

[38] SBT, DR18/20/18b: William Leigh to Theophilus Leigh, 14 May 1711; same to same, 20 Dec 1711; same to same, 21 Sep 1712; same to same, 7 Dec 1712; Eveline Cruickshanks and Ivar McGrath, 'Dawnay, Hon. Henry (1664–1741)'; Cruickshanks, 'Drake, Sir William, 4th Bt. (1658–1716)'; Stuart Handley, 'Hay, George, Visct. Dupplin (aft. 1683–1758)', in Hayton, Cruickshanks and Handley, eds, *House of Commons 1690–1715*; R. S. Lea, 'Longueville, Charles (*c.* 1678–1750)', in Sedgwick, ed., *House of Commons 1715–1754*.

[39] WYAS, WYL100/C/9A: Sir William Strickland to Irwin, 11 Jul 1702; [John Haccius] to John Roades, 13 Apr 1704; R. Hall, 'Whig Party Fortunes in the Yorkshire County Election of 1708', *Northern History*, 32:1 (1996), 111–32; D. W. Hayton, 'Maynard, Thomas (*c.* 1686–1742), of Hoxne Hall, Hoxne, Suff.', in Hayton, Cruickshanks and Handley, eds, *House of Commons 1690–1715*.

[40] Geoffrey Holmes, *British Politics in the Age of Anne* (London: Hambledon, 1987), pp. 20–5.

[41] SBT, DR18/20/18b: William Leigh to Theophilus Leigh, 23 Jul 1712.

Other Educational Travellers

These parties moved among fluctuating communities of educational travellers. Of the 'severall English Gentlemen' whom William Leigh found in Paris in 1713, 'some are travelling to the several provinces; but their places are daily suplied by fresh comers from England'.[42] A decade later, the younger Edward Southwell counted 'now 50 English at Rome more than have been known for ten years past'.[43] Letters often include lists of company and many kept in touch with those they left behind. During onward travel, Robert Southwell, Sir John Perceval and William Leigh continued to correspond with Anglophone circles in Rome or, in Leigh's case, Brussels, receiving lists of arrivals and departures.[44] Like Sir Philip Perceval, they considered themselves part of a cohort of 'all our young Englishmen thatt are abroad'.[45] While Chapter 3 showed travellers invoking this community to justify their actions, parents, guardians and tutors noted its proverbially low standards in calls for sensible spending or studious behaviour.

The largest groups formed in cities, where transport converged and the ends of educational travel—the best tuition, the most exalted company or the most relevant political and social observations—were easily met.[46] The academy and university towns of France, the Netherlands and Germany remain important, but much of the evidence here comes from Italy. Beyond Paris, Leiden and other places of settlement, young men spent their longest periods in Rome.[47] Outside institutions, which structured interactions with fellow students and local culture further north, travellers to Italy built their own communities and sought their own observations. When coupled with the emphasis on keeping a record of Italian tours, travellers there more often leave traces of peers, hosts and impressions of local society. Alongside Paris, Rome and institutional destinations, other centres rose for various reasons. Hanover welcomed travellers before the accession of its electors to the British throne—Thomas Worsley, William Strickland, Metcalfe Robinson and Sir John Perceval visited the court in 1705–6—but it emerged as a

42 SBT, DR18/20/18b: William Leigh to Theophilus Leigh, 19 Jul 1713.

43 BL, Add MS 47031, fo. 141: Edward Southwell II to Sir John Perceval, 9 Apr 1726.

44 BL Add MS 38015, fo. 16: Robert Harpur to Robert Southwell, 2 May 1661; 47025, fo. 81v: James Gibbs to Sir John Perceval, 3 Dec 1707; SBT, DR18/20/18b: Earl of Ailesbury to William Leigh, 27 Jun 1714.

45 BL, Add MS 46956A, fo. 31: Sir Philip Perceval to Sir Robert Southwell, 22 Feb 1678/9.

46 Sweet, *Cities and the Grand Tour*, p. 3.

47 If Towner finds that, on average, travellers made their longest stays in Paris and Rome after 1660 ('Grand Tour', 321), this is not only because he privileges journeys to Italy and mixes travellers of different ages, but also because options for settled learning were diverse.

major stop-off after 1714.[48] John and Henry Brydges, Alan Brodrick and the younger Edward Southwell all recognised the value of 'a handsome appearance' there during the 1720s.[49] By the time that Huntingdon and the Rawdons travelled in the 1770s, Vienna had similarly become a site of extensive Anglophone sociability.[50]

The age and gender make-up of groups in northern destinations such as Paris or Spa was diverse throughout the period, including travellers with various motives. Those who made it to Italy, by contrast, were more uniform before the mid-eighteenth century. A list of 'English Men passing to Rome' in 1675, compiled for the consul in Naples, is dominated by youth, wealth and social standing. Most of its 61 travellers were young, with 23 described as sons, heirs or nephews to better-known figures and seven listed as their governors. There are three pairs of brothers. Of the rest, several—the Earl of Derby, Baron Digby, Sir Giles Estcourt—were in their late teens or early twenties. Older men do feature, but there are no women. The list emphasises prestige and spending power: it includes six titled noblemen, four sons of noblemen and one nephew, four baronets, three sons of baronets and two 'Sons to rich Citysens'. Ten others are recorded as possessing or awaiting fortunes of between £500 and £3,000 a year.[51] In the following decade, when Waring noted his English and Irish company in Venice and Rome, he listed only educational travellers, tutors, merchants and priests.[52] By 1740, Frederick Frankland (45) and his nephew, a third Thomas Worsley (29), were among some older men and repeat travellers in Rome and Naples.[53] Sir John Rawdon, another of their party, had fuelled rumours of marriage with Lady Sophia Fermor, then staying in Siena with her family as part of a three-year foreign residence prompted by her father's reduced finances.[54] By

[48] WYAS, WYL150/6005/32: Metcalfe Robinson to Sir William Robinson, 10 Aug [1705]; BL, Add MS 47025, fo. 74v: James Scott to Sir John Perceval, 20 Jul 1706; Black, *British Abroad*, pp. 54–5.

[49] HL, mssST57, vol. 22, fo. 363: James Brydges to Charles Stuart, 4 Oct 1723; vol. 30, fo. 73: same to Andrew Robinson, 18 Jun 1727; SHC, 1248/5/294: Alan Brodrick II to Sir Alan Brodrick, 17 Aug 1723; BL, Add MS 9714, fo. 164: Edward Southwell to Walter Bowler, 20 Sep 1726.

[50] Goldsmith, 'Social Challenge', pp. 69–70.

[51] TNA, SP93/1, fos 255–6: 'Engl[ish] Men passing to Rome', [early 1675, from known itineraries].

[52] PRONI, D695/228: Samuel Waring, travel notebook, 1687–8, pp. 96–100.

[53] NYCRO, ZON13/2/27: Frederick Frankland to Thomas Worsley III, 19 Mar 1740; ZON13/2/28: same to same, 26 Mar 1740; ZON13/2/32: same to same, 16 Apr 1740.

[54] *Spence: Letters*, p. 284: Spence to Mirabella Spence, 1 Jun 1740; Richard Quaintance, 'Fermor [*née* Jeffreys], Henrietta Louisa, countess of Pomfret (1698–1761)', *ODNB*, online edn, 2015 (https://doi-org/10.1093/ref:odnb/9343).

the late eighteenth century, women were 'commonplace' in Italy and repeat visits much more popular.[55] Before then, though, communities of British and Irish travellers in the country were overwhelmingly young, rich and male. As Chapter 2 showed, these were ideal conditions for status anxiety and competitive spending.

If most educational travellers were united by youth, wealth and standing, several factors drew them even closer together on the Continent. The first was a shared language, unknown to all but a minority of their hosts. They also lived at close quarters. In Rome, for example, the English took accommodation by the Piazza di Spagna as early as the 1630s and continued to rent lodgings there throughout the eighteenth century.[56] Three generations of Southwells and Percevals stayed on or near the piazza between the 1660s and the 1720s, with travellers inheriting apartments and carriages from recent departures.[57] Robert Southwell's notebook reveals debates, in-jokes and book-lending among Englishmen in Rome and Florence.[58] For most, too, private Anglican worship reinforced a sense of community in Italy.[59] The seasonal nature of travel concentrated Anglophone sociability across the Continent, from routes through Italy that fitted around Roman Holy Week and the Venetian carnival to autumn starts at educational institutions and summer trips to France or the Low Countries.

Those who went as far as Rome encountered its sights together. Robert Southwell shared an antiquarian guide with seven Englishmen over 15 days in 1660 and recommended him to others.[60] This was standard practice throughout the period, though courses of six weeks or longer were common by the eighteenth century, as employing an antiquary became another way of displaying wealth.[61] Guides even served several members of the same families: the younger Edward Southwell, for instance, hired the man whom Sir John Perceval had employed 20 years earlier.[62] They saw Catholic as well as classical Rome in groups, with Robert Southwell meeting cardinals

55 Sweet, *Cities and the Grand Tour*, pp. 25–7; John Brewer, 'Whose Grand Tour?', in Sánchez-Jáuregui and Wilcox, eds, *English Prize*, pp. 45–61, at p. 48.

56 Stoye, *English Travellers Abroad*, p. 127; Towner, 'Grand Tour', 322.

57 BNCF, MS Gal. 254, fo. 168v: Robert Southwell to Vincenzio Viviani, 7 Nov 1660; Northants RO, IC 1086: Bruno Talbot to Sir Thomas Isham, 7 May 1678; BL, Egerton MS 3805: Edward Southwell II, Journal of travels in France and Italy, fos 61–2.

58 BL, Egerton MS 1632: Southwell, Italian notebook, fos 5, 25, 29v, 65v.

59 PRONI, D695/228: Waring, travel notebook, p. 47, for Easter prayers at Orrery's lodgings.

60 BL, Add MS 38015, fo. 38v: Robert Southwell to [Edward Howard], [1662].

61 Sweet, *Cities and the Grand Tour*, pp. 101–3.

62 BL, Add MS 47031, fo. 47: Edward Southwell II to Sir John Perceval, 22 Dec 1725; fo. 140: same to same, 9 Apr 1726.

in the same band of Englishmen.[63] His grandson, Edward, recorded that 'We of the English Nation went in a full Body to St Peters' on Christmas Eve of 1725, standing together by the inner portico.[64] This was the context in which young men produced their lists of curiosities. They came to Rome not for a personal encounter, but to extend the collective learning of school and university.

Travellers spent their daily lives together, supporting each other against homesickness but also imposing suffocating obligations.[65] Planning his return to Florence, Robert Southwell worried how fellow travellers would interpret his eagerness to see local academics, sensing that he ought to spend his first evening with Englishmen.[66] Sir Philip Perceval's antisocial behaviour in Rome, 'one of the mallincollyest cittys I have ever ben in', did attract criticism. One countryman suspected that Perceval thought 'alle other men asses, and fools' because he 'sayeth very little' and 'loveth resarvednesse and solitarinesse'.[67] The Ingrams heard similarly 'pitifull' accounts of Irwin from some who had seen him in Venice, which the viscount blamed on his aversion to keeping English company.[68] Metcalfe Robinson, who sent home news of Irwin, also reported on others. At The Hague, the Earl of Hertford was 'mightily improv'd since his leaving England', but Lord Buckhurst and his party lived 'like Hermits' at Utrecht. They refused to gather with countrymen, to the extent that 'some people would scruple to reckon' them among the English.[69] These were damaging sentiments in the goldfish bowl of educational travel, during which young men were supposed to move easily in society. More welcome were Thomas Brodrick's claims, in the context of marrying off his nephew, that the character given of young Alan 'by those whoe come from abroad, strangers as well as englishmen leaves little doubt of his being a great and valluable man, when wee shall bee in our graves'.[70] In return, parents such as Katherine Southwell filled in sons on

[63] BL, Egerton MS 1632: Southwell, Italian notebook, fos 14–15, 21–3, 36, 52v–54, 55v; BNCF, MS Gal. 254, fo. 173: Robert Southwell to Vincenzio Viviani, 23 Dec 1660.

[64] BL, Add MS 47031, fo. 59v: Edward Southwell II to Sir John Perceval, 26 Dec 1725.

[65] Sarah Goldsmith, 'Nostalgia, Homesickness and Emotional Formation on the Eighteenth-Century Grand Tour', *Cultural and Social History*, 15:3 (2018), 333–60, at 349.

[66] BNCF, MS Gal. 161, fo. 304: Robert Southwell to Vincenzio Viviani, 7 Mar 1661.

[67] BL, Add MS 46955B, fo. 64: Sir Philip Perceval to Sir Robert Southwell, 20 Aug 1678; Northants RO, IC 1086: Bruno Talbot to Sir Thomas Isham, 7 May 1678; IC 1099: same to same, 18 Jun 1678.

[68] WYAS, WYL100/C/10: Irwin to Dowager Viscountess Irwin, 22 Jan 1706/7.

[69] WYAS, WYL150/6005/22: Metcalfe Robinson to Sir William Robinson, 22 Sep [1705]; 23: same to same, 24 Apr [1704].

[70] SHC, 1248/6/141: Thomas Brodrick to Sir Alan Brodrick, 9 Feb 1724/5.

the families of fellow travellers.[71] Communities abroad were in such constant communication with those at home that educational travel was no escape from British society.

Peer Group Excess

Historians tend to characterise the Grand Tour as a vector for civility and politeness, concepts that were about self-control. Consequently, they treat drinking, gambling and sex as dangers identified by parents and enjoyed only furtively by sons.[72] Sarah Goldsmith finds several problems with this approach. First, landed families did not necessarily share a moral code among themselves, let alone with middling-sort writers who bemoaned such behaviour in print, and they developed more nuanced understandings of danger. Secondly, with family codes ranging from puritan to libertine, several encouraged traditional assertions of elite male freedom from middling morality. What more often united landed families was a culture of silence, in which the problem was not licentious behaviour but its discussion outside elite company.[73] This understanding of travel fits with findings for Oxford and Cambridge, where the authorities implicitly condoned group-centred excess because they remembered the role that it had played in their own youthful self-assertions.[74] At university and during travel, young men participated in sexual, alcoholic and financial excess as a way of establishing their independence and their membership of an elite in-group. These behaviours stood at the heart of travel as a creator of men and gentlemen.

Such activities leave comparatively little trace in family archives. Time abroad ideally taught civility and politeness, through institutional learning and immersion in a French-speaking elite, but it also looked to verse young men in other forms of masculinity. Throughout the early modern period,

71 Bristol Archives, 45317/2/5/2/23: Katherine Southwell to Edward Southwell III, 16 Aug 1758. Gossip about Edward himself surrounded his facial disfigurement: *The Complete Letters of Lady Mary Wortley Montagu*, ed. Robert Halsband, 3 vols (Oxford: Clarendon, 1965–7), iii, pp. 223–5: Montagu to Lady Bute, 26 Sep 1759.

72 Ian Littlewood, *Sultry Climates: Travel and Sex since the Grand Tour* (London: John Murray, 2001), pp. 11–52; Redford, *Venice and the Grand Tour*, pp. 21–2; Black, *British Abroad*, pp. 202–29.

73 Goldsmith, *Masculinity and Danger*, pp. 18–19, 54–62; Jason M. Kelly, 'Riots, Revelries, and Rumor: Libertinism and Masculine Association in Enlightenment London', *Journal of British Studies*, 45:4 (2006), 759–95, at 774–8, 790.

74 Alexandra Shepard, *Meanings of Manhood in Early Modern England* (Oxford: Oxford University Press, 2003), pp. 93–126.

manhood was also 'learnt by drinking, fighting and sex'.[75] As Goldsmith argues, travel prepared sons for an adult life in which they would need to adopt situational identities, shaping their conduct to different settings.[76] Travellers learned flexible self-presentation in company and rehearsed it in correspondence, to the extent that no single kind of source captures their daily lives abroad. The most common surviving letters are those between a young man and his parents or guardians, whose financial control demanded regular correspondence and whose settled situation offered the best chance for its survival. The need to justify spending to the addressee and a secondary readership of family and friends meant that such sources usually offer few examples of questionable behaviour. Abroad, as at university, such activities were performed for one's peers rather than for those at home.

Older relations could not have been ignorant of continental excesses, not least because several generations shared experience of educational travel. Sir John Perceval was the younger Edward Southwell's second cousin, but Perceval's childhood residence with Southwell's grandfather and correspondence with his father show a closer relationship, as is clear from Edward's letters to him from Italy in 1725–6. The 20-year-old told Perceval that he had 'dined about 16 successive days abroad at Naples, with Irish Officers, with the Factory [merchants] and other English Gentlemen, and the evenings were taken up in assemblies, Balls, Operas, Comedies &ca, and the mornings in seeing Churches and Curiosities'. Southwell asked Perceval, then 42, whether he did not 'Envy me every merry hour I enjoyed after nine at night?' The older man replied that he was jealous of 'every thing you have seen and done except your midnight Assemblys', which were 'partys agreeable to younger Persons'. Perceval had taken part in 1705–7, but had 'disused them so many years' that he now had 'no more notion' of such gatherings 'than the man in the moon'.[77] As at university, authority figures knew all about homosocial bonding during travel but left the details to its participants.

Travellers more often discussed the behaviour of contemporaries. Few letters survive from Sir John Rawdon's Italian travels of 1739–40 but he attracted the attention of others. Joseph Spence knew, before meeting him, that Rawdon was 'pretty much of a rake', 'a combustible gentleman' of 'a very amorous complexion' who fell 'prodigiously in love in every town he makes

[75] Fletcher, *Gender, Sex and Subordination*, p. 92.

[76] Goldsmith, *Masculinity and Danger*, pp. 18–21.

[77] BL, Add MS 47031, fo. 102: Edward Southwell II to Sir John Perceval, 15 Feb 1725/6; fo. 115: Perceval to Southwell, 12 Mar 1725/6.

any stay in'. After their first encounter, Spence confirmed that Rawdon, though 'good-natured', was uncommonly 'amorous' and loved 'company and expense'.[78] Others remembered instances of his idiocy years after they had met him in Italy.[79] Rawdon's promiscuity and foolishness combined in a chapter of *The Adventures of Mr Loveill*, a fictionalised compendium that played with the elite code of silence. The story, set in Rome, revolves around the pursuit of a local woman by a group of Englishmen, who choose one of their number, Mr Dupy, to lure her to their lodgings. When the woman instead draws Dupy into an affair, the others want him out of the way. They leave him alone in public with a Jacobite courtier and convince him that he has committed treason, at which Dupy flees their accommodation and leaves the woman to his friends.[80] Lady Mary Wortley Montagu had no doubt that Dupy was Rawdon, confirming that 'The adventure mention'd at Rome realy happen'd to him.'[81] His letters from Italy may not survive, but Sir John clearly participated in a culture of sexual excess.

Young men readily shared details between themselves, responding to the pressures towards drink, sex and violence by bragging.[82] Letters from Arthur Ingram of Barrowby to his cousin, Henry Ingram, offer one example. Henry travelled through the Netherlands and France in 1716–18, after Arthur had accompanied Henry's elder brother, Viscount Irwin, on his French tour of 1712–13. Arthur trusted that Henry would 'soon become a french man' at The Hague, advising that 'Close conversation with a young wench will soon make you master of the language'. He hoped that there were 'good Whores att Utrecht' and later recommended that Henry stay with 'monsieur Clarett near the Jesuits' at Montpellier because 'his wife will be a good Companion for you'. Madame Clarett was 'a good Christian for the transparent Knight was well diverted there'. Arthur's joke may refer to Irwin, who had died three years previously, as he was soon pleased that his cousin was 'so agreably engaged in an amour with the Knights widow', who was 'as beautifull as Venus' and Henry as 'vigorous as Hercules'. Arthur promised his cousin much 'Cuckoldome & fornication' in London, looking forward to 'tossing

78 *Spence: Letters*, p. 245: Spence to Mirabella Spence, 27 Jan 1740; p. 257: same to same, 9 Mar 1740; p. 284: same to same, 1 Jun 1740.

79 *Montagu Letters*, ii, pp. 457–60: Lady Mary Wortley Montagu to Lady Bute, 22 Jun 1752; *The Yale Edition of Horace Walpole's Correspondence*, ed. W. S. Lewis, 48 vols (New Haven, CT: Yale University Press, 1937–83), xvii, p. 400: Walpole to Horace Mann, 22 Apr 1742; xx, pp. 135–6: same to same, 2 Apr 1750.

80 *The Adventures of Mr. Loveill, Interspers'd with Many Real Amours of the Modern Polite World* (London: 1750), pp. 120–33.

81 *Montagu Letters*, iii, pp. 7–10: Lady Mary Wortley Montagu to Lady Bute, 1 Mar 1752.

82 Shepard, *Meanings of Manhood*, p. 121.

off a bottle of Champaign' and hearing 'an account of your many adventures in ffrance'.[83]

Henry Ingram saw Paris with a group of Englishmen that included William Stapleton, a Yorkshire neighbour who returned home early but kept in touch with those left behind. Stapleton's letter to Ingram represents many sources that promise evidence of sexual activity, in that a later editor has made significant erasures. One passage is now illegible, with 'her acquaintance' visible in several lines on Stapleton's reward for a long crossing to Dover, but other redactions are only partial. At Canterbury, Stapleton told Henry that he 'must remember to [...] the Girl that makes the beds att the Post House; it will require a little force & will be worth your while, for Tho' she is not altogether clean or Handsome she is a try'd sound One'. He similarly called on Piggot, another of their Parisian party, to 'preserve an inch or two of [...] for Old England'.[84] Ian Littlewood presents a singular case of travelling sexuality in James Boswell, but two of his insights are widely applicable: that wealthy young men saw working-class women as perennially available, and that the assumptions of peers suggest a culture of exploitation.[85] Travellers shared formative experiences that included sleeping with, even raping, the same women. While historians usually write of young men drawn into Italian sexuality, such behaviour flourished wherever there was a group of educational travellers.

If drinking and sex are scarce in letters home, so too is violence. There is no trace of duelling in Sir Philip Perceval's correspondence, for instance, but a family history written 65 years later offers some details. Led into 'Excesses' at Angers by the 'ill Example' of another Cork landowner, Lord Kingston, as well as by 'the Fire of his own Temper', Perceval fought 'some dangerous Disputes, which were determined by his Sword'. In one, 'he wounded a Gentleman dangerously, and was himself much hurt'.[86] In the same decade, William Robinson, Metcalfe's father, mortally wounded Edward Rawdon at Foubert's Parisian academy, though the dying man forbade his servant from writing home.[87] Periods of settlement, which gathered communities of young men but offered little excitement, were peak times for duelling. Participants in the culture of civility and politeness bought into what Markku Peltonen

[83] WYAS, WYL100/C/11: [Arthur Ingram of Barrowby] to Henry Ingram, 25 Dec 1716; same to same, 21 Oct 1717; WYL100/C/12: same to same, 31 Mar 1718; [same] to same, 5 Jun 1718; WYL100/C/20, fo. 324: same to same, 25 Sep [1717].

[84] WYAS, WYL100/C/12: William Stapleton to Henry Ingram, 2 Jul 1718.

[85] Littlewood, *Sultry Climates*, pp. 42, 46.

[86] Anderson, *Genealogical History*, ii, p. 380.

[87] *CSPD, Charles II*, xviii, pp. 524–5: B. Magenis to Mr Magenis, 9 Feb [1677].

calls a 'horizontal' understanding of honour, their place in gentlemanly society confirmed by the esteem of its members. Travellers were so annoyed by those who shunned Anglophone sociability because reciprocal civility assured membership of the in-group. In these conditions, contemporaries developed a 'reflexive notion of honour', which demanded a response to the smallest breach as a threat to the status of its victim. If slights were always easy to come by in a competitive social world, spending patterns have shown travel to be a time of heightened sensitivity to personal honour.[88]

Irwin and Maynard reveal this community of honour among Englishmen abroad. Blows and insults were standard justifications for duelling, and those around them reminded them of the fact. Once Maynard had belittled Irwin's courage, the eighth Earl of Huntingdon, a former soldier then travelling on the Continent, encouraged the viscount to give him 'a publick affront' and helped to write the challenge. On the other hand, Lord North and Grey, a serving officer, 'all along push'd Mr Maynard'. After the inconclusive first duel, North would no longer keep Maynard's company until 'he had repaired his honour' by wounding Irwin or being wounded by him. Following their unconventional second clash, when Maynard had continued to attack after the pair had been separated, it was 'very well known' that he was 'no more to be treated so honorably'. Haccius implied that the root of the problem lay in a 'pick' between Huntingdon and North, promising that Irwin would avoid trouble 'if he only Keeps out of so many Lords and Officers Company, who have nothing but Intrigues and fighting in their Heads'.[89] The events were certainly played out for an audience of English soldiers at The Hague.

Patchy detail on duels in family correspondence reflects the same culture of elite silence. The primary audience for duelling, like sexual and alcoholic excess, was one's peers, with the rules of the honour community only binding its members.[90] Haccius assured those at home that they could not have imagined 'how gloriously' Irwin had behaved 'and how favorably ev'ry body talks of him'. The viscount had won because he was 'so good a Fencer & does it with so much nobleness and cold blood, that he has no body in the

[88] Markku Peltonen, *The Duel in Early Modern England: Civility, Politeness and Honour* (Cambridge: Cambridge University Press, 2003), pp. 160–2, 169–71, 174.

[89] WYAS, WYL100/C/9B: John Haccius to [John Roades], 16 Dec 1704; anonymous letter, 29 Dec 1704; WYL100/EA/12/17/5: Irwin to [Alexander Stanhope], [Jan 1705]; WYL150/6005/40: Metcalfe Robinson to Sir William Robinson, 30 Aug [1704]; Lawrence B. Smith, 'North, William, sixth Baron North, second Baron Grey of Rolleston, and Jacobite Earl North (1678–1734)', *ODNB*, online edn, 2008 (https://doi-org/10.1093/ref:odnb/20317). This Huntingdon was uncle to the tenth earl, discussed in Chapter 2.

[90] Peltonen, *The Duel*, p. 176.

world to fear'. Haccius claimed that duels were commonplace, citing another between Englishmen at Utrecht, and argued that the trustees should lock Irwin up if they wanted him not to fight. He advised that 'Courage must be instill'd rather than Cowardice', warning that otherwise 'The family would perhaps be a great looser.'[91] The elder Thomas Worsley also saw the duel as a question of honour and deplored Haccius's lack of involvement, 'when if itt had been Inavoidable, he ought to have stood or ffallen by him'.[92] Irwin and Maynard fought at a moment at which duelling, long recognised as an Italian or French import, was coming to be seen as a marker of British valour. They confirm that participation in Anglophone sociability abroad asserted membership of what Peltonen calls 'the coterie of the honourable—the beau monde'.[93]

Other Kinds of Traveller

Shared experiences acted as a powerful binding agent among young travellers, who spent most of their time together. Particularly in France and the Low Countries, however, they also moved among older friends and family members who had crossed the Channel for other reasons. The younger Edward Southwell's educational travels of 1723 and 1725–6 stood among at least 10 overlapping journeys by his Southwell, Perceval and Dering relations between the mid-1710s and 1730, as they profited from peace and then alliance with France.[94] His father, the elder Edward, had spent more than a month in Flanders and Paris in the summer of 1714, aged 43 and between his first and second marriages. As an MP, he used the parliamentary break to see places 'kept by us & retaken' in the recent war and socialised with army officers and local nobles.[95] Southwell returned to France in the summer of 1718, staying in Paris for six weeks and making a trip along the Loire. He was again alone, after the death of his second wife, and left his work as secretary of state for Ireland for only two months, 'this being the most leisure time with us'. Southwell found the French capital 'a

[91] WYAS, WYL100/C/9B: John Haccius to [John Roades], 16 Dec 1704; WYL100/C/10: same to same, 9 Feb 1705.

[92] WYAS, WYL100/C/9B: Thomas Worsley to Dowager Viscountess Irwin, 14 Dec 1704.

[93] Peltonen, *The Duel*, pp. 162, 178–200.

[94] The 1713 Treaty of Utrecht ended the War of the Spanish Succession, before the Anglo-French alliance of 1716–31.

[95] Bibliothèque Mazarine, MS 4569: [Edward Southwell], Journey from London to Paris, 1714.

more diverting place then Bath or Tunbridge or even Wexford waters', but also believed that 'one ought to have a true Idea of so neighbouring a nation, who may be very great friends and more often very great enemies'.[96] In the same summer, Southwell's first cousin once removed, Sir John Perceval, and his wife, Catherine née Parker, saw France and the Low Countries over several months, leaving their children with the Derings. In his lengthy journal, Perceval explained it as 'a journey of pleasure with my wife to see France'.[97] One of the Derings may also have crossed the Channel that summer, having 'resolved upon Seeing Paris'.[98]

The Southwells, Percevals and Derings all returned to the Continent in 1723. Sir John and his wife took 'a journey of necessity' in June, resolving to 'experience' the waters at Spa after Lady Perceval's two miscarriages. They travelled for three months, again without their children, and included a tour of Flanders. Sir John's journal defers to his previous accounts of towns that he had already seen twice, in 1705–7 and 1718.[99] Perceval's cousin twice over, Daniel Dering, and his wife, Mary née Parker, Lady Perceval's sister, also travelled to Spa that summer with Daniel's infirm uncle, Charles Dering, adding visits to Aix and Hanover.[100] At Hanover they met Daniel's cousin, the elder Edward Southwell, passing through with his son and namesake on a five-month tour of the Netherlands, Flanders, western Germany, France and Lorraine.[101] Though the elder Edward had missed out on a long educational voyage, apart from his visits to the Low Countries in the 1690s, he pieced together continental experience in later life. His superior in Ireland, the Duke of Grafton, thought that the latest journey arose from Southwell's love of 'running about' as much as his son's education, while Archbishop King noted that 'you only want Italy and Spain to compleat your Travells for which I suppose you will take another opportunity'.[102] The Derings met

96 TCD, Lyons MSS 1995–2008, fo. 1865: Edward Southwell to William King, 6 Aug 1718; fo. 1881: same to same, 9 Oct 1718.

97 BL, Add MS 47025, fo. 240: Daniel Dering to [Catherine, Lady Perceval], 19 Jul 1718; 47059, fos 19–162v: Sir John Perceval, 'My Journy to France Holland & Flanders in 1718'; 47072: Perceval, Autobiography, fo. 106.

98 BL, Add MS 47025, fo. 245v: Sir John Perceval to [Daniel Dering?], 9 Aug 1718.

99 BL, Add MS 47059, fos 1–18: Sir John Perceval, 'My Expedition to Spa 1723', at fos 2v, 12; 47072: Perceval, Autobiography, fo. 106.

100 BL, Add MS 47030, fo. 14: Sir John Perceval to Kitty Perceval, 16 Aug 1723; fos 14–14v: Daniel Dering to Sir John Perceval, 20 Aug 1723; fo. 15: same to same, 25 Aug 1723; fo. 16: Sir John Perceval to Philip Perceval, 12 Sep 1723.

101 Clark Library, MS.1975.002: Edward Southwell II, Journal of Our Journey; BL, Add MS 34753: Southwell, Journal of travels in Lorraine and France.

102 TCD, Lyons MS 2537, fo. 30: William King to Edward Southwell, 2 Dec 1723. Southwell did not see Italy or Spain.

the Percevals and returned with them to England, but the Southwells spent seven weeks in Paris before coming home in November.

The families overlapped abroad for a third time two years later, when the younger Edward Southwell travelled through France, Italy and Germany with Charles Hamilton.[103] Sir John and Lady Perceval meanwhile began a journey of 11 months that finally improved the latter's health. They settled in Paris from August 1725 until May the following year with their eldest children, Johnny (14) and Kitty (13).[104] Southwell and Hamilton spent six weeks in the French capital, during which they regularly dined with the Percevals and saw the sights of Paris and Versailles with them.[105] When the young men moved on to Italy, the Percevals continued to explore the French capital, following recommendations from Southwell's father and drawing comparisons with their experiences of 1718.[106] The Percevals joined other educational travellers and families from 'our English Gentry' for sightseeing and sociability, with the organisation of one ball keeping Lady Perceval up until five in the morning.[107] They left Paris for a two-month tour of the Low Countries, hoping to meet the Derings and the Parkers, Lady Perceval's family, who were then considering shorter trips for their health.[108] Such plans were popular among their kin, as the elder Edward Southwell was also at Spa two years later.[109] As an educational traveller, then, his son moved among relations variously abroad for 'leisure', 'pleasure' and health.

Anglophone historians only notice shorter trips after 1763, when the aristocratic 'golden age' of Spa coincided with the growing desire for cheaper Dutch, Flemish and French itineraries among middling-sort travellers.[110] In the same period, the British elite demonstrated 'great familiarity' with Paris during long 'sabbaticals' in the city, where the wealthiest families developed

[103] BNF, MS Anglais 194: Edward Southwell II, Travels; BL, Egerton MS 3805: Southwell, Journal of travels in France and Italy.

[104] BL, Add MS 47030, fo. 154v: Sir John Perceval to Daniel Dering, 26 Aug 1725; 47031, fo. 173: same to same, 19 May [1726]; 47072: Perceval, Autobiography, fo. 106.

[105] BNF, MS Anglais 194: Edward Southwell II, Travels, fos 44, 46, 62; BL, Add MS 47030, fo. 158: Sir John Perceval to Daniel Dering, 31 Aug 1725.

[106] BL, Add MS 47030, fo. 158: Sir John Perceval to Daniel Dering, 31 Aug 1725; 47031, fos 98–98v: same to same, 3 Feb 1725/6.

[107] BL, Add MS 47031, fo. 111: Sir John Perceval to Daniel Dering, 19 Feb 1725/6; fo. 122: same to same, 18 March 1725/6; fo. 145: same to same, 16 April 1726; fo. 163: same to same, 4 May 1726.

[108] BL, Add MS 47031, fo. 124: Daniel Dering to Sir John Perceval, 24 Mar 1725/6; fo. 143v: Perceval to Sir Philip Parker, 16 Apr 1726.

[109] BL, Add MS 47032, fo. 80: Edward Southwell to Sir John Perceval, 3 Aug 1728.

[110] Black, *British Abroad*, pp. 9–10; Richard Bates, 'The Petit Tour to Spa, 1763–87', in Sweet, Verhoeven and Goldsmith, eds, *Beyond the Grand Tour*, pp. 127–46.

an alternative season to London.[111] Although landowners would have greater spending power between the 1760s and the 1790s, when average estate revenues rose by 40 or 50 per cent, the Southwells, Percevals and Derings suggest that each of these trends in travel was under way among the gentry much earlier in the century.[112] Late seventeenth- and early eighteenth-century improvements in transport, as well as the cultural attractions of Paris and London, certainly facilitated shorter pleasure trips among Dutch and Flemish families. If Gerrit Verhoeven suspects that English, French and German research would confirm a European trend, the Southwells, Percevals and Derings suggest that gentry families did travel regularly to the Continent well before the late eighteenth century.[113] Metcalfe Robinson, for instance, made a similar trip after the Peace of Utrecht, hoping to recover his health in France before 'a little tour of Flanders'. Robinson's undated letters from 1714 are mixed in with those from his Italian journey of 1704–5, however, a situation that may hide repeated travel among others of his standing. He was another seasoned traveller, revelling in things 'one don't meet with but at Paris', and returned to the Continent at least once more in the following years.[114] As with institutional learning, British experience was closer to the Dutch than historians have recognised.

There are signs that shorter trips for health or pleasure were of far longer standing. Late sixteenth- and especially early seventeenth-century passes, some mentioning Spa and others foreign travel for health, suggest family journeys in the summer months, in some cases almost annually.[115] Equally, belief in the restorative properties of French air was an important factor in Arthur Rawdon's travels during the 1670s and sent Lady Perceval to the Loire in the same decade. On his way to Italy, her son worried that 'the aer of france' was often a doctor's 'last remedy'.[116] Lady Perceval settled in Saumur—'a towne thatt is selldome without Englesh Ladys'—for over a year and in Paris for four months with her daughters, Kate and Helena, who socialised with locals, saw the sights and received tuition in French, dancing

111 Robin Eagles, *Francophilia in English Society, 1748–1815* (Basingstoke: Macmillan, 2000), pp. 148, 153.
112 Mingay, *English Landed Society*, pp. 51–2.
113 Verhoeven, 'Foreshadowing Tourism'.
114 WYAS, WYL150/6005/1: Metcalfe Robinson to Sir William Robinson, 22 Apr [1714]; 2: same to same, 4 May [1714]; 30: same to same, 9 May [1714]; WYL150/6002/56: Sir William Robinson to Metcalfe Robinson, 14 Oct [*c.* 1716]. 1714 letters dated from discussion of peace with France and death of Duc de Berry.
115 Ghazvinian, 'Certain Tickling Humour', pp. 249–61.
116 BL, Add MS 46954B, fo. 25: Sir Philip Perceval to Sir Robert Southwell, 1 Aug 1677.

and music.[117] The family reunited in Paris over the winter of 1678–9, with Sir Philip returning from Italy, and they moved in an Anglophone circle that shared visits, dances and trips to the opera. The group gathered around the former governor of Jamaica and his wife, comprising a handful of male educational travellers but also several women, some of whom had also come to France with their children for the air.[118] In north-western Europe, young men had long moved among countrymen and countrywomen abroad for other reasons.

Anglophone communities were well established by the early eighteenth century. At Spa in 1723, Sir John Perceval noted that 'the entertainments which Englishmen where-ever they meet make each other runs the weeks account pretty high'. He found that 'the pleasure that persons of the same Country have to meet in foreign Countrys' obliged them to give 'marks of their mutual esteem by dinners and suppers, so much the fashion of some countrys and especially of our own'. Perceval recorded sociability among the 'company of England', listing 20 gentlemen, ladies and soldiers among 'several others whome I have now forgotten'. Two locals, the Count and Countess of Wignacourt, were 'The only foreigners I was acquainted with'.[119] At the same time, in Paris, the elder Edward Southwell reported 'a vast concourse of English Gentlemen & Ladys'.[120] His son's journal gives 81 of the 'English at Paris in Sept 1723', including several Jacobites among the noblemen and noblewomen, gentlemen and ladies. They had ample entertainment amid rounds of dining and sightseeing, as the Southwells went to the theatre 17 times in 20 days.[121]

By the early eighteenth century, families had even begun to habituate their sons gradually to educational travel, before an independent journey in their late teens or early twenties. Between 1723 and 1725, the younger Edward Southwell travelled through the Low Countries, Germany, France and Lorraine with his father, toured England with a cousin and returned to the Continent with Hamilton. In 1725–6, Sir John Perceval's 14-year-old

[117] BL, Add MS 46954B, fo. 199: Sir Philip Perceval to Helena Southwell, 13 Nov 1677; fo. 224v: Lady Perceval to William Taylor, 29 Nov 1677; 46955B, fo. 44v: same to Sir Robert Southwell, 20 Jul 1678.

[118] BL, Add MS 46956A, fo. 15: Sir Philip Perceval to Sir Robert Southwell, 27 Jan 1678[/9]; fo. 17: Lady Perceval to same, 27 Jan 1678/9; fo. 31: Sir Philip Perceval to same, 22 Feb 1678/9; fo. 33: Alexandre de Rasigade to same, 25 Feb 1679; fo. 41: Lady Perceval to same, 9 Mar 1679; TNA, SP44/51, fo. 94: 29 Nov 1678, Pass for Elizabeth, Viscountess Mordaunt.

[119] BL, Add MS 47059: Perceval, 'My Expedition to Spa', fos 4v–6.

[120] BL, Add MS 47030, fo. 18: Edward Southwell to Sir John Perceval, 27 Sep 1723.

[121] BL, Add MS 34753, fos 31–31v: Edward Southwell II, Journal of travels in Lorraine and France.

son, Johnny, began a similar course. Although the family came abroad for Lady Perceval's health, one of their priorities in Paris was finding tuition in French, drawing and music for Johnny and his 13-year-old sister, Kitty. The children were 'great Journalists', presumably recording the visits to churches, convents and palaces that emerge from their father's letters.[122] Only Johnny's diary for the Low Countries survives, relating outings with his father and his tutor, Mr Wolley, for sightseeing, sociability and financial arrangements, occasionally including Kitty and rarely Lady Perceval. The existence of two near-identical texts, one labelled 'Percivale' and the other 'Wolley', suggests that Johnny co-wrote or copied the journal with his tutor.[123] The exercise was supposed to prepare him for independent travel, as was a tour of England two years later, but Johnny returned to the Continent on another family trip, accompanying his ailing uncle, Daniel Dering, to the Netherlands in 1730. Dering settled for treatment by the famous Dr Boerhaave at Leiden, where Sir John hoped that his son would study at the university, but Dering's death ensured that Johnny soon came back to England with the Parkers, who had been on another trip to the Low Countries.[124] Despite their long gestation, the young man's plans for solo travel were overtaken the following year by entry into the Irish parliament.[125] Sir John had boasted that the initial journey of 1725–6 had shown his son to be 'the Eagerest lad for travelling I ever knew' and had hoped that the boy might emulate the younger Edward Southwell's letters 'when it becomes his turn to Voyage as you do', indicating that families still set independent travel apart.[126] Widespread familiarity with the Continent now allowed some to prepare by increments, however. Even once young men crossed the Channel on their own, they continued to move among older friends and relations; again, educational travel was no removal from British society.

[122] BL, Add MS 47031, fo. 24: Sir John Perceval to Edward Southwell II, 30 Oct 1725; fo. 97v: same to Daniel Dering, 3 Feb 1725/6; fo. 135: same to same, 9 April 1726.

[123] Clark Library, MS.1968.006: [John Perceval, future second Earl of Egmont], Journal of a tour in France and the Netherlands, 2 vols, 1726.

[124] BL, Add MS 47032, fo. 212: John Perceval [future second Earl of Egmont] to Sir John Perceval, 28 Aug 1730; fo. 219v: same to same, 15 Sep 1730.

[125] HMC, *Manuscripts of the Earl of Egmont: Diary of Viscount Percival, Afterwards First Earl of Egmont*, 3 vols (London: HMSO, 1920–3), i, pp. 180–93.

[126] BL, Add MS 47031, fo. 39: Sir John Perceval to Edward Southwell II, 1 Dec 1725; fo. 169v: same to same, 7 May 1726.

Expatriates

Alongside those travelling for education, health or pleasure, the 'world of English' included men and women abroad for longer periods. From diplomats and merchants to religious and political exiles, expatriates facilitated Anglophone sociability, engagement with local society and the practicalities of travel. Educational travellers did not cross the Channel to visit them, but they shed light on the national, social and confessional self-understandings of English-speaking communities on the Continent. Above all, they confirm the unifying effect of time abroad on gentlemen from Britain and Ireland. At the same time, though, travellers of Irish residence or Catholic confession tapped into networks that were not so open to the English Protestant majority.

Diplomats allowed relations, dependents and neighbours to venture abroad more economically, as Chapter 2 has shown. Alongside consuls, who oversaw regional British trade, they were also central to the travels of a broader swathe of young men. Representatives in the Netherlands, France and Italy hosted countrymen, offering advice, assistance and a focal point for Anglophone sociability. These expensive and time-consuming duties, which often fell on wives, were a developing feature of the ambassadorial role across the seventeenth and eighteenth centuries, frustrating those officials who lacked substantial independent means.[127] Travellers often recorded 'civilities' from a local official, a term that covered a range of services: some worshipped in the ambassador's chapel, others used embassies and consuls as forwarding addresses and still more found accommodation, which was 'very reasonable' at the consul's house in 1660s Venice.[128] Hosting duties also included resolving disputes, as in the case of Irwin's duel with Maynard. The envoy at The Hague, Alexander Stanhope, showed due deference, stressing that the combatants had behaved 'like men of honour', but his threat to place them under guard hastened Irwin's departure for Italy.[129]

[127] Stoye, *English Travellers Abroad*, pp. 311–12; Mori, *Culture of Diplomacy*, pp. 151, 153, 161, 163.

[128] PRONI, D695/228: Waring, travel notebook, p. 97; NLI, MS 36, fo. 578: Burlington to Dowager Countess of Orrery, 3 Feb 1685[/6]; SBT, DR18/20/18b: Theophilus Leigh to James Leigh, 21 Jul 1712; BL, Add MS 38015, fo. 37: Robert Southwell to [Edward Howard], [1662]; SHC, 1248/6/127: Alan Brodrick II to Thomas Brodrick, 19 Jan 1724/5.

[129] WYAS, WYL100/C/9B: Thomas Worsley to John Roades, 28 Jan 1704[/5]; WYL100/EA/12/17/5: Alexander Stanhope to Irwin, 26 Dec 1704.

Diplomats and consuls offered entries into local society. The representative in Turin, Richard Hill, arranged for the safe arrival of Robinson and Frankland in wartime, entertained them at his home, introduced them 'into the Assemblys & the best acquaintance' and acquired passes for their onward travel.[130] François-Louis de Pesmes de Saint-Saphorin, the Swiss-born British official, similarly furnished John Brydges with a line of credit and introductions to Viennese society, though John embarrassed his father by writing of 'Sensaforen'.[131] These duties clashed, on some level, with representatives' parallel role of structuring Anglophone sociability, so that they might at once facilitate meetings but hamper deeper engagement with Dutch, French or Italian circles.[132] Diplomats and consuls also assisted travellers and those at home in buying art, a role that brought them financial and political capital throughout the period. John Molesworth, for instance, helped his Whig patron, the Earl of Macclesfield, to furnish his home with statuary.[133] Dealings with Italy began to dominate in the eighteenth century, as the growing interest of British collectors and the geopolitical decline of the Italian states offered the incentive and opportunity for officials to work in this way.[134] Across Europe, though, they were points of access to local society and culture.

Diplomats paid more attention to high-ranking travellers, as the good opinion of powerful families promised to reward their efforts abroad.[135] Some even saved family strategies, helping noble parents whose sons had taken sexual or financial excesses too far.[136] Macclesfield's patronage of Molesworth had in fact begun with attempts to extricate the former's son from a relationship with a Venetian woman. The earl suggested confining her to a convent but, whatever Molesworth's solution, it won him political and financial support.[137] He had taken risks during his first posting in 1711, making 'a great noise' by declining to wait on Cornbury in Florence, but had clearly learned his lesson.[138] James Brydges used his Whitehall connections

130 WYAS, WYL150/6005/7: Metcalfe Robinson to Sir William Robinson, 29 Sep [1704]; 11: same to same, 8 Oct [1704]; 40: same to same, 30 Aug [1704].

131 HL, mssST57, vol. 21, fo. 263: James Brydges to John Brydges, 18 Jun 1723; vol. 22, fos 254–5: same to Charles Stuart, 3 May 1723.

132 Mori, *Culture of Diplomacy*, pp. 153, 155, 162.

133 NLI, microfilm P.3752: Earl of Macclesfield to John Molesworth, 18 Jan 1724/5.

134 Jacobsen, *Luxury and Power*, p. 85; Mori, *Culture of Diplomacy*, pp. 158–9.

135 Stoye, *English Travellers Abroad*, p. 2; Mori, *Culture of Diplomacy*, p. 155.

136 Mori, *Culture of Diplomacy*, p. 154; Black, *British Abroad*, p. 219.

137 NLI, microfilm P.3752: Macclesfield to John Molesworth, 29 Mar 1722; same to same, 10 Feb 1722/3.

138 SBT, DR18/20/18b: William Leigh to Theophilus Leigh, 22 Aug 1711.

to ensure similar attention for his sons and nephews.[139] Ever conscious of his status, Brydges wanted to know that his name ensured a good reception in Paris, Hanover and Vienna, demanding reports of his son's treatment by any British official so that he might know 'in what manner I am to carry my self towards him for it'.[140] Personal attention would confirm his high standing.

The English had had no official representation in Rome since the Reformation, apart from a few years under James II, and thus no clear focal point in the city. Between the late 1670s and the early 1690s, the role was filled by Philip Howard, Cardinal Norfolk, who received travellers and introduced them to Roman society. Sir Philip Perceval took advantage of his uncle Southwell's familiarity with Norfolk, whom he had known at court in London, and the Boyles secured the cardinal's 'Civilty' for Orrery by reminding him of their kinship.[141] Perceval, Orrery and Waring all cast Norfolk at the centre of their engagement with Rome. The cardinal took groups to the celebrations of Holy Week, introduced them to the pope and hosted them at his Dominican monastery, where he watched Waring play nine-pin bowling.[142] Strikingly, Norfolk's implication in the Popish Plot did little to damage his convenience as a guide to Rome.[143] Each of these travellers described the cardinal as courteous and obliging, with Waring persuaded that his 'odd' penchant for 'dwelling upon religion & making distinctions' was done 'only for discourse sake & drollerie'.[144] He may have been naïve, as Howard's 'bonhomie' facilitated some effective proselytisation of visiting countrymen.[145]

This quasi-diplomatic role was filled by the Jacobite court from 1719, when it settled permanently in Rome. Travellers listed the Pretender's followers among English communities on the Continent but were often

[139] HL, mssST57, vol. 22, fos 254–5: James Brydges to Charles Stuart, 3 May 1723; fos 363–4: same to same, 4 Oct 1723; mssST58, vol. 7, fo. 16: William Leigh to James Brydges, 1 Nov 1710.

[140] HL, mssST57, vol. 21, fo. 263: James Brydges to John Brydges, 18 Jun 1723.

[141] BL, Add MS 46955A, fo. 190v: Sir Philip Perceval to Sir Robert Southwell, 4 Jun 1678; NLI, MS 36, fo. 667a: Dowager Countess of Orrery to Cardinal Norfolk, 7 Dec 1687.

[142] BL, Add MS 46955A, fo. 100: Alexandre de Rasigade to Sir Robert Southwell, 9 Apr 1678; 46955B, fo. 64v: Sir Philip Perceval to same, 20 Aug 1678; PRONI, D695/226: Samuel Waring, travel journal, 1687–8, pp. 24–8.

[143] Alan White, 'Howard, Philip [*name in religion* Thomas] (1629–1694)', *ODNB*, online edn, 2004 (https://doi-org/10.1093/ref:odnb/13930).

[144] PRONI, D695/226: Waring, travel journal, p. 23; D695/228: Waring, travel notebook, p. 128; BL, Add MS 46955B, fo. 8: Alexandre de Rasigade to Sir Robert Southwell, 18 Jun 1678.

[145] Eoin Devlin, 'English Encounters with Papal Rome in the Late Counter-Reformation, *c.*1685–*c.*1697', Ph.D. thesis (Cambridge, 2010), pp. 43, 60–1, 79–80, 83.

reluctant to engage with them. James Brydges regretted that his son had lodged in the same Parisian house as the Earl of Mar, an estranged Jacobite. Brydges, a Tory, worried that there were 'People Mallicious enough to make unjust remarks thereon', but he decided that moving would 'show to[o] great a regard to such animadversions'.[146] Sir John Perceval, an independent, had the same problem and solution at a Ghent inn.[147] Regardless of their politics, Brydges and Perceval were perhaps too prominent to risk association with Jacobitism. The comparatively humble Leighs had no such concerns. According to family tradition, Theophilus and William both held 'high Tory principles (to call them by their softest name)' and took pride in the loyalty that Theophilus had shown to the Stuarts in 1688.[148] William travelled with other Jacobite sympathisers and regularly met the exiled Earl of Ailesbury, a friend of his father, in Brussels.[149] Leigh benefited from the earl's introductions, met him in Paris and corresponded with him for decades.[150] Although conclusions are hampered by contemporary secrecy and fuzzy divisions between Jacobite and Tory, interactions with the Pretender's followers evidently varied by family.

In Rome, a lack of options may have drawn all travellers to the exiled court. Edward Corp has used the letters of Baron von Stosch, a Prussian antiquarian who spied on the Pretender and his followers between the 1720s and the 1750s, to describe the Stuarts as 'a substitute or surrogate embassy'. Protestant and Catholic travellers of Whig and Tory inclinations sought passports, diplomatic protection, access to portraitists, introductions to the Roman elite and a hub of Anglophone sociability.[151] Of those who went to Rome between the arrival of the Jacobite court and its declining political threat after Culloden, however, the letters of Alan Brodrick in 1725 and Edward Southwell in 1725–6 dwell on artwork and antiquities rather than company. Stosch mentions neither man, aside from Southwell's fight with Hamilton, but the spy does claim that most Englishmen paid their respects

146 HL, mssST57, vol. 18, fo. 423: James Brydges to Charles Stuart, 21 Apr 1722.

147 BL, Add MS 47028, fos 232v–233: Sir John Perceval to Daniel Dering, 21 Jun 1718.

148 SBT, DR18/31/884a: Leigh, 'History of the Leigh Family', pp. 40, 69.

149 Ingamells, *Dictionary*, p. 284. Dupplin's brother, Mr Hay, may be the Jacobite courtier, as he 'designes himself for France': SBT, DR18/20/18b: William Leigh to Theophilus Leigh, 7 Dec 1712; Margaret D. Sankey, 'Hay, John, of Cromlix, Jacobite duke of Inverness (1691–1740)', *ODNB*, online edn, 2011 (https://doi-org/10.1093/ref:odnb/12729).

150 SBT, DR18/20/18: Ailesbury to William Leigh, 17 Jul 1737; DR18/20/18b: William Leigh to Theophilus Leigh, 29 Jun 1712; same to same, 7 Dec 1712; Ailesbury to William Leigh, 27 Jun 1714.

151 Edward T. Corp, *The Stuarts in Italy, 1719–1766: A Royal Court in Permanent Exile* (Cambridge: Cambridge University Press, 2011), pp. 1–8.

to the Pretender's eldest son in this period. Sir John Rawdon does not appear in Stosch's reports for 1739–40, either, by which time some apparently visited the Pretender's sons and others held back.[152] Dupy's story suggests easy interaction between travellers and Jacobite courtiers—unsurprisingly, given that members of Rawdon's group were sympathetic to the cause—but the plot also relies upon fear of association with the exiles.[153] Having identified Rawdon, Lady Mary Wortley Montagu remembered that he had subsequently 'endeavor'd to manifest his Loyalty by railing at [the Pretender] in all Companys with all the warmth imaginable'. Fellow travellers then gulled him into staying at home and eating sparingly for fear of Jacobite assassins.[154] Horace Walpole equally recalled Rawdon's 'boasted loyalty of having been kicked down stairs for not drinking the Pretender's health, though even that was false'.[155] These tales rely on Rawdon's credulity, but they echo the reluctance with which many met Jacobites in other countries. Travellers made use of the Pretender's court in Rome, but some were not as happy about it as Stosch would suggest.

Alongside official and unofficial representatives, expatriate merchants and bankers offered widespread assistance. Families generally had two options for supplying travellers, the bill of exchange or the letter of credit, with the former moving a fixed sum between two named merchants and the latter allowing more flexible itineraries and withdrawals.[156] Irwin's trustees preferred predictable bills, which the family had used in the previous generation, though Irwin argued that a letter of credit would save on commission and suit his independence.[157] He initially relied on a Leeds merchant, Thomas Kitchingman, and his Rotterdam correspondent, Richard Hey, but lamented that Hey was not 'a Merchant of Consequence', lacking the connections needed as he moved southwards. Irwin instead recruited Samuel Williams, 'a man of great reputation and Civility' in Venice who had links with London.[158] Several generations of Southwells and Percevals

[152] TNA, SP85/16, fo. 75v: report from Stosch, 11 May 1726; SP98/41, fo. 343: same, 13 Sep 1739.

[153] TNA, SP98/43, fo. 13: report from Stosch, 7 Feb 1740; fo. 33: same, 17 Apr 1740.

[154] *Montagu Letters*, iii, pp. 7–10: Lady Mary Wortley Montagu to Lady Bute, 1 Mar 1752.

[155] *Walpole's Correspondence*, xx, pp. 135–6: Horace Walpole to Horace Mann, 2 Apr 1750.

[156] John Booker, *Travellers' Money* (Stroud: Sutton Publishing, 1994), pp. 22–4.

[157] WYAS, WYL100/C/9A: John Roades to Irwin, 16 Aug 1704; WYL100/C/20, fo. 242: Irwin to Dowager Viscountess Irwin, 14 Aug [1706].

[158] WYAS, WYL100/C/9A: John Roades to [John Haccius], 8 Jul 1704; Irwin to [Roades], 30 May [1705].

similarly used British commercial networks to supply travellers.[159] Merchants and bankers offered more than financial services, recommending itineraries, offering hospitality, providing introductions, forwarding letters and sending home luggage. A British merchant even gave the two Edward Southwells a tour of Amsterdam in 1723.[160] In Italy, services were easy to find at the 'factory' in Livorno, where Waring 'mett with many English, which with the forrest &c made us almost thinke our selves in England againe'.[161] Similar communities offered assistance in Venice, Rotterdam and other ports on standard itineraries.[162]

Expatriate hosts, whether ambassadors, cardinals or merchants, reinforced a sense of community by hosting meals for travelling countrymen. Sir Philip Perceval dined among Englishmen with the consul in Milan, for instance, and Cornbury and Leigh ate three or four times a week with the ambassador in Berlin.[163] On special occasions, such as a coronation or royal birthday, hosts gathered the most prominent countrymen.[164] They served familiar menus, to which Waring's notebooks pay special attention. In Rome, James II's representative, Sir John Lytcott, provided noblemen, tutors and companions with 'a plentifull English dinner, a bagpuddin to be shure'. At his monastery outside the city, Cardinal Norfolk offered a similar group 'large rosted ribbs of beife' and 'two good large plum puddings'. A Livorno merchant, John Horsey, served soup, veal, plum pudding, quails, artichokes, beans, peas and strawberries, 'the first time we teasted any of the three last this yeare'.[165] Certainly, roast beef and boiled puddings were stereotypically English fare.[166] They marked a notable occasion, since travellers otherwise

159 BL, Add MS 46939, fos 72–3v: Sir John Perceval (1629–65), Accounts, 7 Aug 1660, including payments to Robert Southwell in France; 47025, fos 79–79v: Sir John Perceval (1683–1743), Accounts during travel, 1705–7.

160 Clark Library, MS.1975.002: Edward Southwell II, Journal of Our Journey, 7 Aug 1723.

161 PRONI, D695/226: Waring, travel journal, p. 36; Stefano Villani, '"Una piccola epitome di Inghilterra". La comunità inglese di Livorno negli anni di Ferdinando II: questioni religiose e politiche', *Cromohs*, 8 (2003), 1–23.

162 Stoye, *English Travellers Abroad*, pp. 125, 131; Conway, *Britain, Ireland and Continental Europe*, p. 220.

163 BL, Add MS 46955B, fo. 135: Alexandre de Rasigade to Sir Robert Southwell, 11 Oct 1678; SBT, DR18/20/18b: William Leigh to Theophilus Leigh, 6 Sep 1710.

164 BL, Add MS 38015: Robert Harpur to Robert Southwell, 2 May 1661; SBT, DR18/20/18b: William Leigh to Theophilus Leigh, 16 Feb 1712.

165 PRONI, D695/226: Waring, travel journal, pp. 26–8, 36; D695/228: Waring, travel notebook, p. 48.

166 [Henri] Misson, *M. Misson's Memoirs and Observations in his Travels over England. With Some Account of Scotland and Ireland*, trans. John Ozell (London: 1719), pp. 313–17: 'two Dishes are their Dinners; a pudding, for instance, and a Piece of roast Beef'; 'Give an English Man a Pudding, and he shall think it a noble Treat in any Part of the World.'

aligned their diets with those of their hosts, as had Irwin's party in Delft.[167] Like language, food helped travellers to recognise what connected them in foreign environments, across boundaries of confession and residence. Catholics such as Norfolk and Lytcott entertained groups of Protestant Englishmen that included Irish landowners like Perceval and Waring.

More generally, English Catholics offered Protestant travellers hospitality in Flanders, France, Germany, Bohemia and Italy.[168] Metcalfe Robinson's party were shown the 'relicks & Jewells' of Loretto by 'a little English Jesuit' who came back to their inn to sing for them.[169] Cornbury and Leigh likewise met with 'a great deal of Civility' over three days at the English Benedictine monastery of Lamspringe in Lower Saxony. Though the young men 'only gave the hearing to it', their hosts brought out papers that proved Charles II had died a Catholic.[170] In the Low Countries, too, Protestants regularly visited English convents. Shared national and social backgrounds helped travellers and nuns, usually the daughters of gentry families, to cross the confessional divide.[171] The same was true of monasteries, as Leigh's correspondence attributes their welcome at Lamspringe to the abbot's genteel Yorkshire origins. If travellers received hospitality, religious communities saw reminders of home and opportunities for soft power. In their papers, historians find Scottish Jesuits looking to convert noble countrymen in France and Italy during the Interregnum; a recusant priest asking clerics in Rome to show his MP around the city in 1725, as a 'service hereafter to the Catholicks in this neighbourhood'; and nuns catechising visitors to the Low Countries throughout the seventeenth and early eighteenth centuries.[172] Indeed, in a journal containing several anti-Catholic diatribes, Leigh rethought his visit to Lamspringe: such hospitality was common 'in the popish Countries', but 'not so much out of pure generosity and good breeding as proceeding from

[167] On a longer stay, see Melissa Calaresu, 'Thomas Jones' Neapolitan Kitchen: The Material Cultures of Food on the Grand Tour', *Journal of Early Modern History*, 24:1 (2020), 84–102.

[168] Stoye, *English Travellers Abroad*, pp. 76–7, 121–2, 183; Conway, *Britain, Ireland and Continental Europe*, pp. 198–9.

[169] WYAS, WYL150/6005/46: Metcalfe Robinson to Sir William Robinson, 21 May [1705].

[170] SBT, DR18/20/18b: William Leigh to Theophilus Leigh, 6 Sep 1710.

[171] Liesbeth Corens, 'Catholic Nuns and English Identities. English Protestant Travellers on the English Convents in the Low Countries, 1660–1730', *Recusant History*, 30:3 (2011), 441–59.

[172] Chaney, *Grand Tour and Great Rebellion*, pp. 100–8; Gabriel Glickman, *The English Catholic Community, 1688–1745: Politics, Culture and Ideology* (Woodbridge: Boydell, 2009), p. 78; Claire Walker, *Gender and Politics in Early Modern Europe: English Convents in France and the Low Countries* (Basingstoke: Palgrave Macmillan, 2003), pp. 121–2.

self interest, and intrieage'.[173] Either way, Catholic hopes were raised by a sense of national, social and linguistic community.

One might not expect such community in the Irish case, where Protestant landowners funded their travels from estates expropriated from Catholic neighbours within living memory. But the letters and journals of Robert Southwell and Samuel Waring are in fact full of Irish clergymen on the Continent; they are even more frequent than English Catholics in the papers of English Protestants. Irish clergymen forwarded correspondence for Southwell in Rome and Bologna, and another in Paris wrote introductions for him to Montpellier. His papers contain draft letters of thanks in Latin and Italian for the hospitality of Irish priests and nuns with Irish connections. Later in life, Southwell also remembered the 'great Entertainment' afforded by the Irish Franciscans of Prague. He even discussed theology and church organisation, among many stories and jokes, with Franciscans and Jesuits in Paris and Rome.[174] Similarly, Waring was shown 'curiosities' by 'two of our Irish fathers' in Rome and attended Franciscan disputations, taking away copies of their theses.[175] Southwell and Waring regularly comment on the civility and learning of Irish Catholic clergymen abroad.[176]

These interactions built on shared memories of Ireland, with which Southwell 'opened' one Franciscan in Rome to further conversation.[177] They afforded help and hospitality to Protestant travellers. From the Catholic side, as with English and Scottish counterparts, they offered contact with home and the hope of influencing its rulers. Southwell travelled at the Restoration and Waring in 1687–8, a period in which confessional divides were bridged by a noble 'community of honour' at home and relationships forged abroad during the royalist exile.[178] Domestic dealings with priests, however, whom writers including Edward Southwell and Robert Molesworth identified as

[173] SBT, DR671/59a: Leigh, Travel journal, p. 13.

[174] BL, Egerton MS 1632: Southwell, Italian notebook, fos 11, 31v–33, 75v; 1634: Le Grand, 'Life of Southwell', fo. 17v; Add MS 38015, fos 3v: John Ponce to Antoine Leurin, 3 Oct 1659; fos 25, 25v, 29–30, 32: draft letters, [*c.* 1661]; 58219: Southwell, Diary and commonplace book of travels in France, 1659, fos 23–5; BNCF, MS Gal. 161, fo. 314: Gregory Fallon to Vincenzio Viviani, 16 Apr 1661; 254, fo. 191v: Southwell to same, 1 Aug 1661.

[175] PRONI, D695/226: Waring, travel journal, p. 24; D695/228: Waring, travel notebook, p. 49.

[176] BL, Add MS 38015, fo. 25: Southwell, draft letter, [*c.* 1661]; fos 36v, 37v: same to [Edward Howard], [1662]; PRONI, D695/228: Waring, travel notebook, pp. 98, 128.

[177] BL, Egerton MS 1632: Southwell, Italian notebook, fo. 11.

[178] S. J. Connolly, *Religion, Law, and Power: The Making of Protestant Ireland, 1660–1760* (Oxford: Clarendon, 1992), pp. 126–7; Ohlmeyer, *Making Ireland English*, p. 167; Mark R. F. Williams, *The King's Irishmen: The Irish in the Exiled Court of Charles II, 1649–1660* (Woodbridge: Boydell, 2014).

the instigators of native resistance, only grew with Protestant self-assuredness after military victory in 1688–91.[179] Afterwards, as landowners began to see the Catholic clergy as a tool of control, a foreign education even helped priests to enter polite society at home.[180] It had allowed travellers and clergymen to interact abroad at an earlier date, in a dynamic that endured into the eighteenth century.[181] Protestant landowners travelled abroad as Englishmen, but their Irish residence let them tap into a network that appears less regularly in the papers of English fellow travellers.

Foreigners

If educational travellers were surrounded by countrymen and some countrywomen, how did they get on with the locals? Foreign hosts receive little treatment beyond the anecdotal in most studies of the Grand Tour, not least because much evidence is buried in continental archives, with British diaries and journals prioritising sightseeing over personal interactions. The work of Ilaria Bignamini is one exception, exploring the roles of the Venetian government, papal museums and souvenir manufacturers in the Italian Grand Tour.[182] Others have recently followed, thinking about how the rulers, artists and entrepreneurs of Rome attracted foreigners and tracing two-way exchanges between the British and locals in Turin.[183] Progress has also been made on travel writing, where Melissa Calaresu and Wendy Bracewell explore instances of host communities writing back against the published opinions of visitors.[184] This work has begun to recover the perspectives and

179 BL, Add MS 28724: [Edward Southwell], 'Discourse concerning the securing the Government of the Kingdome of Ireland to the Interest of the English Nation', [*c.* 1695]; Robert Molesworth, *The True Way to Render Ireland Happy and Secure* (Dublin: 1697); Hayton, *Anglo-Irish Experience*, pp. 170–1.

180 David Dickson, 'Jacobitism in Eighteenth-Century Ireland: A Munster Perspective', *Éire-Ireland*, 39 (2004), 38–99, at 85; Barnard, *New Anatomy*, pp. 112–13.

181 Barnard, *Making the Grand Figure*, pp. 312, 318; Connolly, *Religion, Law, and Power*, pp. 127–8.

182 Ilaria Bignamini, 'The Italians as Spectators and Actors: The Grand Tour Reflected', in Hornsby, ed., *Impact of Italy*, pp. 29–47.

183 Jeffrey Collins, 'Sites and Sightseers: Rome through Foreign Eyes', in Pamela M. Jones, Barbara Wisch and Simon Ditchfield, eds, *A Companion to Early Modern Rome, 1492–1692* (Leiden and Boston, MA: Brill, 2019), pp. 564–81; Bianchi and Wolfe, eds, *Turin and the British*.

184 Melissa Calaresu, 'Looking for Virgil's Tomb: The End of the Grand Tour and the Cosmopolitan Ideal in Europe', in Jaś Elsner and Joan-Pau Rubiés, eds, *Voyages and Visions: Towards a Cultural History of Travel* (London: Reaktion, 1999), pp. 138–61; Wendy Bracewell,

activities of hosts, especially Italians, who are too often passive in accounts of travel. Like the discussion of Anglophone communities, this section will consider Italy more frequently than Chapter 3 on settled learning. Though not all travellers ventured so far south, the emphasis placed on written observations in the country and the fact that engagement with host cultures were no longer mediated by educational institutions mean that young men have generally left more evidence of interactions with Italian hosts. They were not straightforward, however, and easier sociability remained a significant draw to France, the Netherlands and Germany.

In Italy and elsewhere, interactions between travellers and hosts were habitually cross-confessional. Although travel books embraced a 'confessional geography' that separated the Continent into Catholic and Protestant lands, a conception evident in the papers of travellers from Britain and Ireland, this understanding coexisted with unifying ideas of Christendom.[185] Shared notions of honour, hospitality and gift exchange, as well as economic, political and religious self-interest, equally promoted interactions across the confessional divide. Later in the eighteenth century, the experiences of Sir Thomas Gascoigne suggest that English Catholics may have received warmer welcomes from their co-religionists in places such as Spain and Naples, but only a wider investigation would confirm this suspicion.[186] Even so, Protestant travellers met with a baseline of good treatment upon which Catholic countrymen could improve.

Across the Continent, hosts singled out English travellers for a special welcome. Cardinal Pallavicino declared 'a particular passion for the English', adding to Robert Southwell's impression, formed during audiences with other cardinals, that 'the Italians hig[h]ly prefer the English man'.[187] Orrery and Edward Southwell both partook of the 'universall Civility' shown to Englishmen in Amsterdam by Alvaro da Fonseca, a diamond merchant who had lived in London. Fonseca, 'for many years the Patron of the English Nation', showed Southwell the sights, fed him, arranged transport and helped with shopping. The duty was reciprocal, though, as Orrery's accounts record

'Arguing from Experience: Travelees versus Travelers in Early Modern Exchanges', *Renaissance Studies*, 3:4 (2019), 548–67.

185 Claydon, *Europe and the Making of England*, pp. 13–66; Conway, *Britain, Ireland and Continental Europe*, pp. 162–88, 210. Sir John Perceval rejoiced at reaching the 'Protestant ground' of the United Provinces: BL, Add MS 47031, fo. 181: Perceval to Daniel Dering, 5 Jun 1726.

186 Alexander Lock, *Catholicism, Identity and Politics in the Age of Enlightenment: The Life and Career of Sir Thomas Gascoigne, 1745–1810* (Woodbridge: Boydell, 2016), pp. 64–71.

187 BL, Egerton MS 1632: Southwell, Italian notebook, fos 5v, 52v–53.

'a dinner which my Lord out of Civillity was forced to give to Mr ffonsecas family' in Brussels.[188] Metcalfe Robinson found that Sophia of Hanover, the daughter of Elizabeth Stuart, had 'all the English about her always, & loves to speak our language'.[189] In Turin, too, the Duchess of Savoy told William Leigh that 'because her mother was an Englishwoman, she always had a great respect and tenderness for the nation, although they had driven her out of the succession'. The viceroy of Naples had spoken 'in terms very obliging of our nation in general and the Queen in particular', and in Brussels Leigh was 'a wittness of the respect & veneration the Nobility & Gentry have for the Queen of England'.[190] These receptions are so rooted in individual circumstance or contemporary context as to offer little insight: Pallavicino may have shared Norfolk's proselytising imperative; Fonseca, the Electress of Hanover and the Duchess of Savoy had personal reasons for their Anglophilia; and Leigh travelled towards the end of the War of the Spanish Succession, when Queen Anne was a powerful friend. Hosts nevertheless welcomed travellers as representatives of England.

Travellers felt particularly welcome in Florence, especially under Cosimo III between 1670 and 1723.[191] On arrival, Charles Butler received '32 noble flasks of the best & severall sorts of florence wines, four severall large dishes of varietie of sweet meats, a dussen cheeses ... a dussen of Bolonian sasages, two dusen of cappens, a creele of live pidgeons, a nother of chickins 6 or 8 dussen of small birds puld, as many of fedd thrush with the feathers on, 2 or 3 basketts of Ice to coole the wine'.[192] Robinson and Frankland similarly received 'a present of Wine, Sweetmeets & Fowl' after an audience with the ruling family.[193] Their welcome was remarkable for its context, at a strongly counter-reforming court, but travellers ascribed Cosimo's hospitality to his own travels of 1668–9, which had included several months in England.[194] Over 40 years later, William Leigh explained

[188] BL, Add MS 58220: Edward Southwell, Journal of a visit to the Netherlands, 2 vols, 1696, i, fos 3, 3v, 4v, 6, 8, 11, 11v; NLI, MS 13215, fo. 33: Orrery's travelling expenses, 1685–9; Edgar Samuel, 'Fonseca, Alvaro da (1657?–1742)', *ODNB*, online edn, 2008 (https://doi-org/10.1093/ref:odnb/49976).

[189] WYAS, WYL150/6005/32: Metcalfe Robinson to Sir William Robinson, 10 Aug [1705].

[190] SBT, DR18/20/18b: William Leigh to Theophilus Leigh, 14 May 1711; same to same, 17 Oct 1711; same to same, 30 Jul 1712.

[191] Sweet, *Cities and the Grand Tour*, p. 68.

[192] PRONI, D695/228: Waring, travel notebook, p. 37.

[193] WYAS, WYL150/6005/8: Metcalfe Robinson to Sir William Robinson, 29 Nov [1704].

[194] Marcello Fantoni, 'The Grand Duchy of Tuscany: The Courts of the Medici 1532–1737', in John Adamson, ed., *The Princely Courts of Europe: Ritual, Politics and Culture under the Ancien Régime 1500–1750* (London: Seven Dials, 2000), pp. 255–73, at p. 272.

that the grand duke had had 'a more then common regard for our nation ever since the time he Honoured it with his presence'. Cosimo even reminisced with Cornbury about civilities received from his family.[195] The debauched Gian Gastone had ruled for a couple of years when Martha Courthope hoped that her great-nephew Brodrick would also see 'the grand D of florence', 'one of the finest old men of his time', so she probably had Cosimo in mind.[196] Travellers continued to see Florence as 'a home from home' after the fall of the Medici in 1737, though, as the British diplomat, Horace Mann, introduced them to local society over the rest of the century. Before then, despite geopolitical and mercantile reasons why Tuscan rulers would have sought friendly relations, it suited British understandings of honour and travel to believe that their welcome was a product of reciprocal civility.[197]

The Medici did develop a reciprocal relationship with the Southwell and Perceval families. Robert Southwell arrived in 1660, hoping to connect one of his patrons, Robert Boyle, and the 'experimenters' of the Accademia del Cimento. Southwell impressed at meetings of 'their virtuosi' and studied with Vincenzio Viviani, the court mathematician and self-styled last disciple of Galileo.[198] Exploring the grand duke's gallery, he also noticed Holbein's portrait of Sir Richard Southwell, the Tudor administrator and client of the Howards, which the Earl of Arundel had given to the Medici in 1621. The picture allowed Southwell to remind Bernard and Edward Howard, brothers of the Duke of Norfolk and fellow travellers in Italy, of their 'past relation'.[199] When he returned to Florence, after passing the winter in Rome, he and Viviani dived into 'an ocean of talking'. A few months later, on the way home, he delivered copies of the mathematician's treatise, *De Maximis et Minimis*, and advertised it to booksellers. Fellow travellers jokingly called Southwell 'a subject of the Grand Duke', Ferdinando II, upon whom the young man

195 SBT, DR18/20/18b: William Leigh to Theophilus Leigh, 22 Aug 1711.

196 SHC, 1248/6/205v: Martha Courthope to Alan Brodrick II, 23 Apr 1725.

197 Sweet, *Cities and the Grand Tour*, pp. 68–70.

198 *The Correspondence of Robert Boyle*, ed. Michael Hunter, Antonio Clericuzio, and Lawrence Principe, 6 vols (London: Pickering & Chatto, 2001), i, pp. 428–31: Robert Southwell to Boyle, 10 Oct 1660; W. E. Knowles Middleton, *The Experimenters: A Study of the Accademia del Cimento* (Baltimore, MD: Johns Hopkins University Press, 1971).

199 BL, Egerton MS 1634: Le Grand, 'Life of Southwell', fo. 15; Add MS 38015, fo. 22: Southwell, draft letter, [*c.* 1661]; fo. 37: same to Edward Howard, [1662]; David Howarth, *Images of Rule: Art and Politics in the English Renaissance, 1485–1649* (Basingstoke: Macmillan, 1997), pp. 244–5.

prevailed for a letter of recommendation to Charles II.[200] Once home, he even dispatched two Irish wolfhounds to the Tuscan court.[201]

The Southwells maintained links with Florence. Sir Robert waited on Cosimo and his courtiers in Lisbon and Paris during the 1660s, his diplomatic missions coinciding with the heir's travels.[202] When Cosimo was blown off his course to England in 1669, he even met Southwell's father and the Percevals at Kinsale. Two courtiers, Lorenzo Magalotti and Giovanni Battista Gornia, produced accounts of their Irish landing, praising the elder Southwell for his wealth, manner and hospitality. Indeed, Sir Robert maintained a friendly correspondence with Magalotti, whom he had met in Florence and Lisbon and with whom he shared natural philosophical interests.[203] Both courtiers changed their tone, however, when discussing local Catholics, who were 'treated as slaves' in 'a conquered land'. They lived in huts, 'begging and ill-treated', and paid 'their meagre earnings' to the Southwells. The church of Kinsale was 'now profanely used in the service of the Anglican sect', which Gornia thought little different to Calvinism, and Catholics avoided 'harassment' by worshipping in secret.[204] The Tuscans recognised civility, but also harsh treatment of their co-religionists.

Reciprocal hospitality outweighed confessional division in the following decades. A few years later, the new grand duke sent 40 bottles of 'excellent Florence wine' to Sir Robert Southwell in London.[205] When Sir Philip Perceval arrived in Tuscany in 1678, Cosimo remembered meeting him as a child, lavished gifts upon him and talked 'a great while' of his uncle.[206] Connections even outlived Southwell and Viviani, who corresponded for

[200] BNCF, MS Gal. 161, fo. 318: Giovanni Battista Riccioli to Vincenzio Viviani, 4 Aug 1661; fo. 331: Robert Southwell to same, Oct 1661; 254, fos 172–172v: same to same, 23 Dec 1660; fo. 181v: same to same, 9 Apr 1661; 258, fo. 40: Ferdinando II to Charles II, 4 Mar 1660[/1]. Translations mine.

[201] BL, Add MS 38015, fo. 28v: Robert Southwell to Giuseppe Francesco Borri, [1662].

[202] BL, Add MS 35100, fo. 16: Francis Parry to Joseph Williamson, 21 Jan 1668/9; TNA, SP78/127, fo. 34: Sir Robert Southwell to [same], 3 Aug 1669.

[203] BL, Add MS 38015, fo. 129: Sir Robert Southwell to Lorenzo Magalotti, 2 Jan 1672.

[204] Bibliothèque Mazarine, MS 4570: Giovanni Battista Gornia, 'Viaggio fatto dal serenissimo Principe Cosimo Terzo di Toscana per la Spagna, Inghilterra, Francia, ed altri luoghi negli anni 1668 e 1669', pp. 6–7 (translations mine); Lorenzo Magalotti, 'An Unexpected Visitor: Cosimo de'Medici's Visit to Cork in 1669', trans. Anne O'Connor, *Journal of the Cork Historical and Archaeological Society*, 111 (2006), 1–6.

[205] Kent Archives, U1713/C1/7: Sir Edward Dering to Sir Robert Southwell, 10 Jun 1671.

[206] BL, Add MS 46955B, fo. 107: Sir Philip Perceval to Lady Perceval, 17 Sep 1678; fo. 109: Alexandre de Rasigade to Sir Robert Southwell, 20 Sep 1678; fo. 110: Sir Philip Perceval to same, 20 Sep 1678.

decades but died in 1702 and 1703.[207] In 1707, Sir John Perceval was well received in Florence because 'the good Old Duke' remembered 'the civilities' offered by 'Old Mr. Southwell Sr. Roberts father' at Kinsale.[208] Afterwards, Perceval obliged Florentine requests for Irish whiskey and sent one courtier 'a little kerry stone to make a seal'. Cut with intaglios of the grand duke, the stone became 'an eternal memory' of Perceval connections to Florence.[209] Strong links died with Cosimo in 1723, however. The younger Edward Southwell visited Florence two years later, dutifully describing the portrait of Sir Richard. Like Sir Philip Perceval before him, he noted its plates bearing the Southwell, Howard and Medici arms. Edward lodged at an inn, however, and never met Gian Gastone.[210] Over 60 years, the Southwells, the Percevals and the Tuscan court demonstrate that shared scholarly interests, codes of civility and reciprocal hospitality permitted cross-confessional interactions. Although Cosimo made all his 'English' visitors feel special, they offer a developed case of how families wove together memories and souvenirs over several generations.

Continental hosts usually treated Robert Southwell, Orrery and other Irish landowners as uncomplicated Englishmen. Travellers of their background were relatively few, with one Irish resident of 'English extraction' claiming that Sir Philip Perceval was 'the first Person of quality' whom he had met 'of our countrey' during four years in Florence.[211] Their English language and company were their most apparent attributes on fleeting visits, while hosts who spoke Romance tongues rarely differentiated between 'English' and 'British'.[212] Only much deeper connections could bring nuanced identifications. Southwell unusually requested, while still abroad, that some Italian correspondents address him as 'Gentilhuomo Irelandesi', but Viviani brought him into particularly close contact with academic networks in the country.[213] As the family's links with Florence endured,

207 Luciano Boschiero, 'Robert Southwell and Vincenzio Viviani: Their Friendship and an Attempt at Italian–English Scientific Collaboration', *Parergon*, 26 (2009), 87–108.

208 BL, Add MS 47072: Perceval, Autobiography, pp. 21–2.

209 BL, Add MS 47025, fo. 81: Lorenzo Magnolfi to Sir John Perceval, 22 Nov 1707; fo. 94: same to same, 19 Aug 1708.

210 BL, Add MS 46955B, fo. 110v: Sir Philip Perceval to Sir Robert Southwell, 20 Sep 1678; 47031, fos 46v–47: Edward Southwell II to Sir John Perceval, 22 Dec 1725; Egerton MS 3805: Edward Southwell II, Journal of travels in France and Italy, fos 39–58v.

211 BL, Add MS 46955B, fo. 122: John St John to Sir Robert Southwell, 1 Oct 1678.

212 Paul Langford, *Englishness Identified: Manners and Character, 1650–1850* (Oxford: Oxford University Press, 2000), p. 12.

213 BNCF, MS Gal. 158, fo. 157: Vincenzio Viviani to Arrigo Broune, 1661; 161, fos 276–276v: Robert Southwell to Viviani, 2 Nov 1660; fo. 322: Carlo Antonio Manzini to same, 6 Sep

Southwell and his great-nephew, Sir John Perceval, sent gifts resonant of their Irish estates. Otherwise, Protestant landowners encouraged the perception that they were Englishmen, with letters from family and continental hosts usually addressed to 'Gentilhomme Anglois'.[214] If a traveller spoke English, arrived in English company and presented himself as an Englishman, hosts saw no need to scruple.

Beyond questions of identity, the Southwells' and Percevals' engagement with Florence was unusual because educational travellers tended to make temporary, even superficial connections with foreign hosts. Several historians cite friendships among the European aristocracy as a useful and long-lasting outcome of travel but they draw their examples almost exclusively from the nobility, whose patterns of continental engagement cannot represent wider practices among the gentry.[215] A fuller account ought to recognise the difficulties that gentlemen found in accessing the best company, their tendency to see foreigners as learning opportunities and, despite the Southwells, Percevals and Medici, the scarcity of evidence for sustained relationships after travel.

The degree to which travellers socialised with foreign elites depended upon several factors. The first, of course, was the host culture. As will become clear, one of the attractions of Paris and The Hague was the ease with which circles there opened for British visitors. In Italy, by contrast, the situation was fragmented. Travellers from Britain and Ireland evidently received a warm welcome in Florence and their good French facilitated engagement with court society in bilingual Turin. If lower competency in Italian was one bar to interactions elsewhere, another could be the disposition of locals. Alan Brodrick found that 'People spend most of their Time' in Venice viewing art and architecture, 'since the Conversation of strangers is what the Venetians dare not give in to'. Government paranoia about betraying state secrets was part of the myth of the city, but—along with the relative absence of court culture—it evidently shaped real interactions. Brodrick had not been long in the country, but others warned him that meeting the elite was 'very little more difficult here, than in other parts of Italy'. In Rome, where travellers regularly complained of a cold reception from local families, he

1661. He still signed a Latin letter as 'Anglus': King's, Keynes MS 152, fo. 10v: Southwell to Ovidio Montalbani, 12 Apr 1661.

[214] BL, Add MS 38015, fo. 12: 'Gentil' to Robert Southwell, 30 Mar 1660; 46954A, fo. 83: Helena Perceval to Sir Philip Perceval, 30 Mar 1677; fo. 112: Richard Mulys to same, 12 Apr 1677.

[215] Conway, *Britain, Ireland and Continental Europe*, pp. 211–12; Ohlmeyer, *Making Ireland English*, p. 443.

found that 'strangers' were 'as it were necessary to One another'.[216] These disappointments help to explain frequent but impersonal commentary on Roman and Venetian hosts.

A second factor affecting interactions was a traveller's status, especially in Italy. Even with some allowance for exaggeration, everything seems to have come easily to Viscount Irwin. He told his mother that 'where ever I have been' during a year and a half in the country, 'I have had the best company in the Town'. In Venice, few 'Strangers' before him could boast of 'being esteemed by persons of sence and breeding, and not keeping continually together as the English do, without knowing anything of the Country but what their Landlord tells them'. In Lucca, he had met 'the first quality' of Genoa, who received him with great 'civility': 'I was as familiar with them, though the greatest Families of all Genova, as I am with my own Countrymen.'[217] Irwin's fellow Yorkshireman, Metcalfe Robinson, encountered the same places differently as the son of a baronet. He thought everyone 'affable & obliging' at Turin apart from 'the noble Genoese being just the contrary'. While he enjoyed Genoa itself, he found that 'the good companys are inaccessible to strangers without recommendations' and hoped that things would not be 'the same in other parts, we being so slenderly provided therewith'. Unfortunately, it did prove 'impracticable' for strangers to converse 'amongst the people of quality & Ladies' in Rome. At Venice, too, where Irwin had reflected on his good company throughout Italy, Robinson found that 'tis nothing but the same thing over again to a stranger'. Travellers only spoke among themselves, since the locals were 'unaccessible'. After three weeks, Robinson regretted that 'all the acquaintance I could make amongst 'em may be reduc'd to a poor noble or two, & as many Virtuosos'.[218] In the same cities, a gentleman had different experiences to his noble neighbour.

Difficulties of access could be pronounced in Italy, but the gentry faced them elsewhere. Robinson had hoped for a change in Vienna but found that the local elite avoided 'a stranger, perticularly if he be without a title'. Despite the efforts of the British envoy 'to make us pass our time more agreably', the Viennese had 'no notion of any bodys being company for 'em but a Mylord'. John Brydges, then Marquis of Carnarvon, certainly had little problem in the city two decades later. Robinson felt that it was 'the same in

216 SHC, 1248/6/90: Alan Brodrick II to Sir Alan Brodrick, 7 Oct 1724; fo. 209: same to same, 25 Apr 1725; Sweet, *Cities and the Grand Tour*, pp. 139, 207.

217 WYAS, WYL100/C/10: Irwin to Dowager Viscountess Irwin, 22 Jan 1706/7; WYL100/C/20, fo. 242: same to same, 14 Aug [1706].

218 WYAS, WYL150/6005/3: Metcalfe Robinson to Sir William Robinson, 21 Nov [1704]; 5: same to same, 8 Nov [1704]; 13: same to same, 24 Jan [1705]; 45: same to same, 3 Jun [1705].

Italy & most places abroad, where they have nothing but Counts, Marquis, & then Citizens, Artisans &c'.[219] Letters of recommendation were essential to successful travel in such circumstances. Sir Robert Southwell's diplomatic contacts made 'all things easy' for Sir Philip Perceval in Italy and the younger Edward Southwell gained 'a free Introduction' to the assemblies ofVienna on account of his 'large Packett' of '12 or 14 letters'. In turn, Edward's widow, Katherine, worked hard at home to secure references for their son in Italy.[220]

Why did educational travellers want to overcome these obstacles, if diplomatic, mercantile and religious networks could facilitate tours with little local assistance? It was partly a question of conspicuously keeping the best company, as had Irwin, but elite conversation also allowed young men to pursue educational objectives. Advisors often described these goals in nebulous terms. Robert Southwell fancied that 'the Sight of the World (I meane some neerer parts, well civiliz'd, and of polite practice) does very much open a man's head, and take him off from a narrow dotage on his owne spott', particularly if travellers extended their 'view and practice to the Conversation of Men'. Archbishop King similarly encouraged interaction 'with men of different sentiments', one of 'the greatest advantages of Travelling'.[221] High-level interactions promised material for comparative observations—Sir John Perceval held that 'men should fix at Courts, and not loose their time in lesser towns where nothing but the outside of Houses and inside of Churches is to be seen'—but in practice families saw conversation with foreign elites as an opportunity to develop linguistic and social skills.[222] In the Netherlands, Orrery pursued 'such companies where he may learn good breeding & good manners ... those things which can make him an accomplished gentleman'.[223]

Foreign company allowed travellers to perfect the linguistic skills on which many had worked in institutional settings, as there was no better environment in which to absorb and imitate prestige varieties of speech.[224]

[219] WYAS, WYL150/6005/37: Metcalfe Robinson to Sir William Robinson, 11 Jul [1705].

[220] BL, Add MS 46955A, fo. 25: Sir Philip Perceval to Sir Robert Southwell, 30 Jan 1677/8; 47031, fo. 192v: Edward Southwell II to Sir John Perceval, 4 Jul 1726; fo. 199v: same to same, 27 Jul 1726; Bristol Archives, 45317/2/5/2/22: Katherine Southwell to Edward Southwell III, 23 Jul 1758; 45317/2/5/2/23: same to same, 27 Jul 1758.

[221] BL, Add MS 38015, fo. 34: Robert Southwell to Gabriel Towerson, [1662]; TCD, Lyons MS 750/4, fo. 101: William King to Samuel Molyneux, 6 Jan 1712.

[222] BL, Add MS 47031, fo. 84: Sir John Perceval to Edward Southwell, 22 Jan 1725/6. On the learning opportunities and dangers of conversation, see Gallagher, *Learning Languages*, pp. 157–207; Ansell, 'Educational Travel', 944–5.

[223] NLI, MS 36, fo. 597: Germain Colladon to Dowager Countess of Orrery, 17 Oct 1686.

[224] Gallagher, *Learning Languages*, p. 188.

Derby, Sir Philip Perceval and the Brydges brothers were sent into as much French-speaking company as possible to speed up their linguistic progress.[225] Travellers certainly felt that this approach was successful. William Leigh thought attendance at nightly gatherings in Brussels had improved his French and granted 'a little assurance'. He had 'learnt to express my self as clearly and correctly as I can', 'the Company being always desirous of informing one whenever he makes a mistake in the Language, and not, as is too common in England to laugh at a Forreigner for not talking English as well as the Inhabitants'.[226] Looking back on his travels, Alan Brodrick was pleased with his progress in French, but regretted that he had only learned 'as much Italian as the small share of Conversation, which strangers can have with the People of the country, wou'd admit of'.[227]

If the company of local elites was vital to linguistic improvement, it also developed social skills. Coote Molesworth went to Turin and the Brydges brothers to The Hague in search of cures for their 'bashfullness'. Indeed, the Ingrams employed Haccius partly because of his ability to introduce Irwin to good company abroad, 'the only way to form a Young Nobleman'.[228] Metcalfe Robinson, though of lower standing, had been delighted by assemblies in the Dutch capital, where women and men talked and gambled for three hours every night. They gathered 'the best bred in the world, & the most open & free in their access & conversation, which are not the last to be desir'd qualifications, a young traveller should hope to meet with'.[229] Unfortunately, such occasions set him up for disappointment in Italy and Vienna. Many used the Habsburg court for the purposes of socialisation by the late eighteenth century, however, with its accommodating diplomats and a nobility apparently more welcoming than in Robinson's day.[230] In 1774, Francis Rawdon arrived in search of 'polish' and 'ease', lacking 'nothing but a little confidence'. He soon discovered that 'Society is upon a better footing than I ever found it' in Italy.[231] Throughout the period, the acquisition of

225 Bodleian, MS Carte 243, fo. 87: James Forbes to Ormond, 7 Aug 1673; BL, Add MS 46955A, fo. 52v: Sir Robert Southwell to Lady Perceval, 3 Feb 1678/9; HL, mssST57, vol. 18, fo. 346: James Brydges to Charles Stuart, 3 Jan 1721/2; vol. 27, fo. 102: same to Andrew Robinson, 15 Nov 1725.

226 SBT, DR18/20/18b: William Leigh to Theophilus Leigh, 23 Jul 1712.

227 SHC, 1248/6/303: Alan Brodrick II to Sir Alan Brodrick, 11 Sep 1725.

228 WYAS, WYL100/C/9A: William Thompson to [John Roades], 18 Jul [1702].

229 WYAS, WYL150/6005/23: Metcalfe Robinson to Sir William Robinson, 24 Apr [1704].

230 Goldsmith, 'Social Challenge', pp. 69–70.

231 PRONI, D2924/1/190: Huntingdon to Earl of Moira, 18 Oct 1774; fo. 203: Francis, Lord Rawdon to Countess of Moira, [April/May 1775]; T3765/M/2/14/100: Huntingdon to Countess of Moira, 22 Oct 1774.

social skills and behavioural codes, through emulation of the best examples, was central to how travel made sons into men and gentlemen. Evidently, it added to the educational significance of Paris, Brussels, The Hague, Vienna and Turin over the rest of Italy.

A less visible group of foreigners, at least in the archives, are travellers from other countries. Although historians have long known that the British were part of an international phenomenon that drew in elites from northern, western and central Europe, they rarely consider it as an international experience.[232] Young men from Britain and Ireland must have rubbed shoulders with foreign travellers, however. Chapter 3 showed that they attended universities and academies alongside French, Dutch and German counterparts. All read the same guidebooks to Italy, meanwhile, shaping similar encounters with the country.[233] The problem is that the British rarely mention foreign travellers in letters and journals, but is this because they never met or because they did not think those meetings important?

A few sources refer to Germans, with whom the British shared French institutions, Italian tours and a reputation for educational travel.[234] Kingston, a bad influence on Sir Philip Perceval at Angers, entertained 'even the hard drincking germans', presumably fellow students at the academy.[235] Irwin borrowed £200 from two unnamed Germans at Augsburg, travelled with them to Venice and would have continued in their company had finances allowed.[236] Cornbury and Leigh similarly entered Italy with a German nobleman, the son of Count Metternich. Leigh thought him 'one who had made good use of his Travells', which had included England, and they 'formed a strict acquaintance' during two weeks of quarantine near Verona. The connection did Leigh 'good offices' when he met the count, a Prussian minister, during the Congress of Utrecht, but there is no evidence of a long-term friendship.[237] These anecdotes in fact suggest that socialising with Germans was not normal: it was evidence of Kingston's depravity and Irwin's financial straits, while Leigh only befriended Metternich's son in unusual circumstances. Robinson suggests more regular contact, but

232 Stoye, *English Travellers Abroad*, p. x; Black, *British Abroad*, p. 6. Brief exceptions include Bignamini, 'Grand Tour: Open Issues', p. 32; Verhoeven, Sweet and Goldsmith, 'Introduction', in Verhoeven, Sweet and Goldsmith, eds, *Beyond the Grand Tour*, pp. 6–8.

233 Ansell, 'Reading and Writing Travels'.

234 De Ridder-Symoens, 'Mobility', p. 433; Leibetseder, 'Across Europe', 430–3.

235 BL Add MS 46954A, fo. 154: Jean Gailhard to Sir Robert Southwell, [*c.* 13 May 1677].

236 WYAS, WYL100/C/9A: Irwin to [John Roades], 23 May [1705]; same to same, 30 May [1705]; same to same, 31 Jul [1705].

237 SBT, DR18/20/18b: William Leigh to Theophilus Leigh, 8 Feb 1712.

similarly short-lived acquaintance, in his complaints of Vienna. The British were so isolated there that 'not so much as the Germans I have known abroad & have had a familiarity with will look upon us', presumably referring to travellers in Italy.[238] Sources occasionally mention Frenchmen, but they are similarly distant. Perceval reported the opinion of 'The french heer in Rome' on the prospect of war and Frederick Frankland left Naples later than planned after hearing that 'a dozen & an half of French men were to do the same'.[239] Isolation from foreign counterparts seems to have been another aspect of British cliquishness. Of these French and German examples, only Metternich's son is an identifiable individual, and only then by his useful parentage.

Conclusion

The lasting connections that emerged from educational travel were among gentlemen and nobles from Britain and Ireland, as continental experience fed into elite formation. They travelled in groups that stressed conformity in daily socialising, excessive consumption and personal honour, continuing the unifying effect of school and university. Alexandra Shepard's work on misrule at Oxford and Cambridge distinguishes between short-lived 'comradeship' and enduring friendship, finding that collective assertions of independence rarely forged personal links between participants, but travel promoted both kinds of connection.[240] On the Continent, shared experiences took young men across the threshold of adulthood and built networks of patronage and mutual support that were crucial in later life. If travel offered an education in national difference, it also confronted them with what they held in common. Travellers encountered foreign people and destinations in each other's company, coming together to speak their own language, eat their own food and host 'partys agreeable to younger Persons'. They developed emotional bonds in the process. When Cornbury died soon after his return, William Leigh was 'sensibly afflicted' by the loss of 'one of the best Friends I could boast of', a patron and fellow traveller for whom he had 'the greatest love and respect'.[241] In these circumstances, some worried about misguided

238 WYAS, WYL150/6005/37: Metcalfe Robinson to Sir William Robinson, 11 Jul [1705].

239 BL Add MS 46955A, fo. 162v: Sir Philip Perceval to Sir Robert Southwell, 21 May 1678; NYCRO, ZON13/2/32: Frederick Frankland to [Thomas Worsley III], 16 Apr 1740.

240 Shepard, *Meanings of Manhood*, pp. 96, 110, 113, 125.

241 SBT, DR18/20/18b: William Leigh to Theophilus Leigh, 15 Mar 1713.

attachments. Not only was Kingston a bad influence on Sir Philip Perceval in France, but their families had also fought bitter land disputes in Cork, leading Sir Robert Southwell to warn that 'a freindship begun abrode' would 'entayle and perpetrate a very inconvenyent one at the returne home'.[242] For good or ill, links forged in travel were hard to break.

While bonds among fellow travellers developed into longer-lasting networks, dealings with local hosts tended to be fleeting. If families such as the Ingrams or the Leighs maintained few international relationships, to judge by surviving correspondence, it was because contemporaries thought of foreign company as an opportunity for learning rather than friendship. Even Southwell and Perceval connections to Florence were of limited use at home, where Robert Southwell 'never thought fit' to deliver the Grand Duke's letter to Charles II.[243] The company of continental elites was instead crucial to a time-bound process of immersion and emulation that, alongside institutional study, developed the minds and bodies of English gentlemen abroad. Educational travel brought young men closer to French and Italian culture, then, but not necessarily to French and Italian individuals. Gentlemen from Britain and Ireland participated in a European travel culture whose main outcomes—the French language, genteel deportment, shared experiences and a network of countrymen—held their greatest relevance at home.

Gerald Newman is right that travel was an accessible way in which families could aim at 'educational and social superiority', an international factor in 'the stagecraft of local power'.[244] Although high spending was important in travelling groups that insisted on the performance of honour and status, social boundaries were more flexible abroad than at home. Those who could afford some form of travel could therefore create links with patrons and peers, whose esteem affirmed membership of the elite. At the same time, they acquired the language and bearing to win and keep that esteem. This process of buying into the elite is clearest for those who began on its fringes, with two recent studies concluding that English Catholics used travel in this way. Liesbeth Corens and Alexander Lock both find them sharing experiences and cultivating friendships with Protestant countrymen abroad, demonstrating their abilities in an environment where standing and reputation counted for more than religion. Travel offered Catholic gentlemen a route into elite society at home.[245]

[242] BL, Add MS 46954A, fo. 158**: Sir Robert Southwell to Jean Gailhard, [mid-May 1677].
[243] BL, Egerton MS 1634: Le Grand, 'Life of Southwell', fo. 16.
[244] Newman, *Rise of English Nationalism*, pp. 42, 44.
[245] Corens, *Confessional Mobility*, p. 103; Lock, *Catholicism, Identity and Politics*, pp. 77–8, 82–4.

Parts of this chapter have found the same process at work among Protestants from Ireland. They travelled in English company, presented themselves as Englishmen and were accepted as such by fellow travellers, expatriates and local hosts. Since identity was a two-sided affair, in which others corroborated or rejected an individual's claims, countrymen and foreigners contributed to the identification of Irish landowners as part of the English elite.[246] Some travellers did reflect the English tendency to speak of the native Irish and the Protestant gentry in similar terms, something that grew more common over the eighteenth century.[247] While Spence thought Arthur Rawdon 'too modest for an Irishman', accounts of Sir John's misunderstandings seem to draw on a tradition of Irish 'bulls'.[248] Throughout the period, however, young men such as the Rawdons remained integral members of travelling English communities, who found far more to unite than divide them abroad. Protestant landowners from Ireland used their residence to access Irish Catholic networks, just as English Catholics used confessional ties, but both groups bought into the English elite through travel. In the process, they shed light on the unifying power of continental experience for all who travelled as English gentlemen.

246 John-Paul A. Ghobrial, 'Moving Stories and What They Tell Us: Early Modern Mobility between Microhistory and Global History', in Ghobrial, ed., *Global History and Microhistory* (Oxford: Oxford University Press, 2019), pp. 243–80, at p. 248.

247 Hayton, *Anglo-Irish Experience*, pp. 1–24.

248 *Spence: Letters*, p. 284: Spence to Mirabella Spence, 1 Jun 1740.

5

Returns from Travel

When Robert Southwell returned from his travels of 1659–62, he immediately set about building a life in London. He made slow progress, however, and retreated to Ireland after a few months, occupying himself with the business of his father's estate. Back home at the 'well advanced' age of 26, Southwell became introspective. He was 'asshamed of my years for the ill account they afford', considering most of them 'blankes wherein I could write nothing', but he recognised that 'the latest number of them', those spent abroad, were 'in this respect of a better edition'.[1] His father wanted to find him work in Dublin with Ormond, the family patron, but Southwell felt that the duke could offer nothing 'agreeing with the method of my education'. His '3 yeares spent abrode' would 'signify as little in order to employment [in Ireland] as if I had beene learning hebrew'. In London, however, 'all the forraine part of improovement, which indeed is all I have, will be serviceable to carry me unto my desired ends'. His travels would do him 'little good in an other place', but the English capital promised 'a station sutable to the method of my Education'. Only in failure would he come home to 'the plow and a privite coarse of life'.[2]

Southwell wanted to make the most of his travels, worrying lest 'all my fathers care and expence, and something of my owne observation abrode should be all buried with me in a corner'.[3] In a note, 'If Proposed', he practised rejecting marriage until he had properly established himself, arguing that 'my education abrode has unshapt me from following the plough or a solitary abode'.[4] Southwell returned to London in late 1663, as his efforts to buy a clerkship of the privy council progressed, but he found no

[1] BL, Add MS 38015, fo. 42: Robert Southwell to [Richard Southwell], 18 Sep 1662; 72852, fo. 14v: same to elder Robert Southwell, [*c.* 1662/3].

[2] BL, Add MS 38015, fo. 64: Robert Southwell to elder Robert Southwell, [*c.* 1662–3]; fo. 82: same to Edward Synge, Bishop of Cork, 10 May 1664; 72852, fos 13–14v: same to elder Robert Southwell, [*c.* 1662–3].

[3] BL, Add MS 38015, fo. 90: Robert Southwell to Helena Southwell, [Mar 1665].

[4] BL, Add MS 38015, fo. 26: Robert Southwell, notes, [*c.* 1662–5].

relief from his anxieties. Amid pages of notes on self-made men, court life and the rewards of hard work, he reminded himself that 'this is the time to Labour, elder time is for play'.[5] His mother hardly helped, reporting that 'wee dayly here of severall of our cuntrey men that cum to good preferment that have neither mony nor kindred nor greate Abillities only theyr dilligenc'. Since Robert had his share 'of the 3 first', she 'should bee sorry you should faile in the last'.[6] Continental travel had prepared him for social advancement and public life, but there was much work still to do.

Was Southwell right to think that his recent travels would help him to move up in the world? The question is not an easy one to answer. Previous chapters have shown that families habitually justified travel with reference to the skills and connections that young men would bring back from the Continent, but historians rarely study the process of returning home in any detail. Few look beyond the clichés of 'cosmopolitanism' and 'polish' in assessing the effects of travel on the individual, while only artistic and architectural influences receive more than cursory examination. In part, historians are hostages to unpromising sources. On the one hand, between the sixteenth and eighteenth centuries, printed criticism of foreign encounters spoke more to anxieties about the corruption of English culture than to any familiarity with actual returned travellers.[7] On the other, manuscript correspondents who comment on the outcomes of travel tend to be those who had funded or encouraged it; they almost always proclaim success. When Edward Dering came home from his uncle Southwell's posting to Iberia, his father found him 'returnd such as I wish him & to my full content'.[8] Sir John Perceval thought that the younger Edward Southwell had 'returned from his travells very much improved, and is indeed a young man of as many good qualities as any I know'.[9] James Brydges similarly found his nephew Leigh 'very much improv'd' and his son Henry 'not a little improved by his Travels'.[10] His grandson and namesake admitted that, 'however agreable my Travels have been to me, the Thoughts of my returning a more worthy member of Society has been no small Joy to me'.[11] Few expand upon these

5 BL, Add MS 38015, fos 66–70: Robert Southwell, notes, [*c.* 1662–5].
6 BL, Add MS 38015, fo. 88: Helena Southwell to Robert Southwell, 1 Mar 1664.
7 Sara Warneke, *Images of the Educational Traveller in Early Modern England* (Leiden: Brill, 1995); Newman, *Rise of English Nationalism*, pp. 98–9.
8 Kent Archives, U1713/C1/3: Edward Dering to Sir Robert Southwell, 23 Aug 1669.
9 BL, Add MS 47031, fo. 210: Sir John Perceval to [Philip] Perceval, 21 Nov 1726.
10 HL, mssST57, vol. 9, fo. 265: James Brydges to Theophilus Leigh, 23 Nov 1713; vol. 30, fo. 302: James Brydges to Edmund Chamberlayne, 9 Oct 1727.
11 SBT, DR18/20/18: James Brydges II to Barbara Blackley, 15 Feb 1752.

formulations, but what did they mean? How had continental experience changed these young men? And with what consequences for the wealth, standing and self-perceptions of their families?

These questions remain challenging ones, but they are central to working out whether—and how—investments in travel paid off. They require deep and varied source material for the years after travel, once again making it a strength, rather than a source of frustration, that so much evidence of continental experience is embedded within family archives. They also involve a group of young men who rarely appear in histories of travel, but who are central to assessing its role and efficacy in family strategy: those who did not go abroad. This chapter approaches these issues in two ways. First, it looks at how families discussed the effects of travel at points that tended to occur soon afterwards: entry into public life for those of sufficiently high standing, more widespread marriage negotiations and, for some Irish landowners, the opportunity to move to England. What difference did it make at these milestones that a young man had—or had not—travelled? Secondly, as Robert Southwell suggests, it was not enough simply to have been abroad; the work of travel continued at home. This chapter also examines the ways in which families and individuals tried to turn continental encounters into cultural capital in the years afterwards. It finds many venues in which travellers displayed their improvements, including but beyond the walls and furnishings of country houses. While historians are drawn to material manifestations, less tangible kinds of performance deserve equal billing. The 'afterlives' of travel, in the sense of how families remembered, reworked and reused their foreign experiences, have received unprecedented attention in the last few years, and they were central to the success or failure of travel as an aspect of family strategy.[12]

Public Life

Landed families were already thinking about entry into public life when young men came home, as the demographic crisis of the late seventeenth and early eighteenth centuries created an 'exceptionally youthful society'.[13] Educational travellers often inherited their estates before or during their

[12] Ansell, 'Reading and Writing Travels'; Mark R. F. Williams, 'The Inner Lives of Early Modern Travel', *Historical Journal*, 62:2 (2019), 349–73; Gallagher, *Learning Languages*, pp. 157–207.

[13] Stone and Stone, *Open Elite?*, p. 103.

time abroad, as in the cases of the Earl of Derby, Sir Philip and Sir John Perceval, Viscount Irwin and the third Edward Southwell. Such men also held office from a young age, an undertaking that met social expectations but also promised power and reasonable financial gain.[14] Many still upheld the Renaissance idea that travel offered the best preparation for public life.[15] Sir Robert Southwell reported that the Duke of Beaufort would provide his son with 'a share in travel, that so his good parts may have full improvement, and give him title henceafter to any thing that is considerable in the Government'.[16] When Sir Philip Perceval died a year after his return, Southwell and his relations similarly argued that 'lively Impressions' of the Continent had left his nephew 'very Acceptable, and Qualified for Publick service'.[17] For Thomas Brodrick, 'seeing the world, and becoming acquainted with the different interests in Europe, quallifies a man for every sort of buisnesse, which alone enables him to stand uppon his own leggs, & will allwayes have more weight then the favour of Courteirs'.[18] Much of the gentry—let alone the nobility—already possessed the wealth, kinship and acquaintance necessary for advancement in government or diplomacy before they crossed the Channel.[19] Equally, as continental experience reached beyond the noblemen of the Tudors and early Stuarts, proportionally fewer travellers could hope for prominence in later decades. For those who aimed at public life, however, travel might still prove decisive.

Skills and connections acquired abroad had favoured Southwell's own rise. Andrew White, an Irish Catholic double-agent encountered in Italy, had given prescient advice on how 'to be taken notice of as for a proper man for businesse', suggesting the friendship of great men and a good knowledge of French and Italian.[20] Although Ormond was already the family patron, the duke recommended Southwell to Charles II on the strength of his 'travels in foreign parts; by which he has so profited'.[21] Like the Earl of

14 Michael J. Braddick, *State Formation in Early Modern England c. 1550–1700* (Cambridge: Cambridge University Press, 2000), pp. 83–4, 347; G. E. Aylmer, *The Crown's Servants: Government and Civil Service under Charles II, 1660–1685* (Oxford: Oxford University Press, 2002), pp. 240–1.

15 Parks, 'Travel as Education', p. 265.

16 TNA, SP31/3, fo. 174: Sir Robert Southwell to Joseph Williamson, 2 Jul 1687.

17 BL, Add MS 46960B, fos 171–4v: draft epitaphs for Sir Philip Perceval, [*c.* 1683–4].

18 SHC, 1248/5/333: Thomas Brodrick to Sir Alan Brodrick, 12 Nov 1723.

19 Aylmer, *Crown's Servants*, pp. 170–1; Horn, *British Diplomatic Service*, pp. 144–81.

20 BL, Egerton MS 1632: Southwell, Italian notebook, fos 10–10v.

21 Hervey Redmond Morres, Viscount Mountmorres, *The History of the Principal Transactions of the Irish Parliament, from the Year 1634 to 1666*, 2 vols (London: 1792), i, p. vi.

Arlington, who was soon convinced of Southwell's 'merit & good parts', Ormond had spent the 1650s with the king in exile.[22] After the Restoration, both noblemen promoted a cohort of educational travellers into different branches of government.[23] The good impression that Southwell had made in Paris on the diplomat Sir Richard Fanshawe, who spread his opinion to influential patrons, equally encouraged his first posting to Lisbon a few years later, alongside linguistic skills that he had demonstrated in work for the privy council.[24] He received a knighthood before his departure for Portugal. Others similarly reaped the rewards of language-learning abroad, securing positions in government, at court and in the diplomatic service.[25] Southwell was right, then, to believe that his travels would help him towards establishment in England.

As Southwell suggests, provincial landowners often connected social standing and public life with residence in London.[26] At the higher reaches of office-holding, whether in crown service or parliament, they were of course correct. Returning travellers with hopes of advancement therefore tended to stay in the capital rather than go home, at least as they attempted to establish themselves. James Brydges started from a more favourable position than Southwell, as the son of a baron and prominent diplomat, but he came back from the Continent in the mid-1690s with similar ambitions to 'strike in with the Government before it is too late'. Straight after his return from Wolfenbüttel, Brydges presumed that his father would want him to live in London rather than 'the Countrey'.[27] His journal tracks rounds of networking and the observation of government business, a circuit reminiscent of Southwell's earlier efforts.[28] Published writers often argued that the tastes and habits of travellers suited them to metropolitan rather than country life,

[22] BL, Stowe MS 744, fo. 99: Earl of Arlington to Sir Edward Dering, 6 Feb 1665/6; Williams, *King's Irishmen*.

[23] Aylmer, *Crown's Servants*, pp. 195–6; Chaney, *Grand Tour and the Great Rebellion*, p. 353. Including William Godolphin, with whom Southwell had lodged in Paris (BL, Add MS 58219, fo. 52), and Joseph Williamson, his Italianate correspondent.

[24] BL, Add MS 58219: Sir Richard Fanshawe to elder Robert Southwell, 6 Aug 1659; 34336, fos 1–2: Arlington to Fanshawe, 1 Nov 1665; fo. 16: Earl of Clarendon to same, 20 Dec 1665.

[25] For example, Barbara C. Murison, 'Blathwayt, William (*bap.* 1650, *d.* 1717)', *ODNB*, online edn, 2008 (https://doi-org/10.1093/ref:odnb/2626).

[26] Rosenheim, *Emergence of a Ruling Order*, pp. 215–52.

[27] HL, mssST57, vol. 1, fo. 20: James Brydges to James Brydges, eighth Baron Chandos, 8 Jul 1694.

[28] BL, Egerton MS 1634: Le Grand, 'Life of Southwell', fos 21–21v; HL, mssST28, vols 1–2: James Brydges, 'Journal of my Daily Actions', 1697–1702.

a source of greater regret for the Earl of Clarendon than for Jean Gailhard.[29] The Huguenot tutor felt that 'Countrey breeding, which is opposed to the Courts, to the Cities, or to Travelling', was a 'clownish' way of raising sons, wondering 'what good employment is such a one fit for?'[30] Men such as Southwell and Brydges clearly agreed. But practically, too, London offered them the best chance to turn continental experience towards personal and family advancement.

Some returning travellers set their sights on parliament. A healthy proportion of Commons seats were held by young men, as were other offices in the period, and many families saw the House as another step towards elite male adulthood. Electoral success required deep pockets and solid connections, such that a seat in the Commons—'the most certain measure of rank apart from a heritable title'—was more the result of wealth and standing than a stepping-stone towards them. It was therefore the sons of an established group of families who timed returns from the Continent in order to stand.[31] Cornbury came home for the 1713 parliament, planning before his death to 'enter into the world with great advantages on his side'.[32] James Brydges likewise brought 19-year-old Henry back for the 1727 election and paid £700 to silence questions about his age, hoping that sitting for Hereford would give him 'an Early turn to Business'.[33] In 1761, the third Edward Southwell took Bridgwater on the interest of his Perceval kinsman, the Earl of Egmont, only a few months after returning from Italy; his trunk arrived while he was away canvassing.[34] Such men formed a larger proportion of the Commons than would appear from the *History of Parliament* series, which estimates that between 5 and 13 per cent of members had travelled for

[29] Edward Hyde, Earl of Clarendon, *A Collection of Several Tracts of the Right Honourable Earl of Clarendon* (London: 1727), pp. 295–6.

[30] Gailhard, *Compleat Gentleman*, ii, pp. 3–5.

[31] Ellis Archer Wasson, 'The House of Commons, 1660–1945: Parliamentary Families and the Political Elite', *English Historical Review*, 106:420 (1991), 635–51, at 637; B. D. Henning, 'The Composition of the House', in Henning, ed., *The History of Parliament: The House of Commons 1660–1690* (London: Secker & Warburg, 1983); D. W. Hayton, 'The Members', in Hayton, Cruickshanks and Handley, eds, *House of Commons 1690–1715*.

[32] SBT, DR18/20/18b: William Leigh to Theophilus Leigh, 8 Feb 1712; same to same, 15 Mar 1713.

[33] HL, mssST57, vol. 30, fo. 151: James Brydges to Andrew Robinson, 23 Jul 1727; fo. 207: same to Mr Brydges of Tybberton, 12 Aug 1727; Romney R. Sedgwick, 'Brydges, Henry, Mq. of Carnarvon (1708–71)', in Sedgwick, ed., *House of Commons 1715–1754*.

[34] Bristol Archives, 45317/2/5/6/17: Katherine Southwell to Edward Southwell III, 21 Mar 1761; Mary M. Drummond, 'Southwell, Edward (1738–77), of Henbury, Glos.', in Lewis Namier and John Brooke, eds, *The History of Parliament: The House of Commons 1754–1790* (London: HMSO, 1964).

their education between 1660 and 1754.[35] If the 19 MPs in this study are a representative sample, the entries for most travellers ignore their time abroad.[36] The only volume on the Lords, for 1660–1715, reports travel by over 35 per cent of peers, usually as part of education, but continental experience among the gentry once again passes under the radar.[37]

Travel was important to the preparations of those destined for parliament. Young men received a rhetorical education at school and university throughout the period, modelled on Roman public speaking, but they moved beyond formal settings during time abroad.[38] They honed speech and deportment in the company of foreign elites, looking to shed the very 'bashfulness' that observers regularly criticised in unsuccessful maiden appearances before the House.[39] Most MPs never spoke in debates, however, and much business was conducted outside the chamber, when observations made abroad surely informed discussions of diplomacy and trade.[40] Sir John Perceval gives some insight into such gatherings, although he entered the Irish parliament before travel amid the disruption caused by Southwell's death. Perceval and his fellow MPs had grown 'so us'd to Parliament language' that it shaped their drinking sessions and 'spoild all intreagues with the Ladies, for it has given us a grave Air of 60 years, & introduc'd so new a manner of speech & addressing, that they think us all Mad'.[41] Shared experience of travel prepared men for these occasions, which enhanced the 'club-like' atmosphere of the Commons, and during

[35] Henning, 'Composition of the House'; Hayton, 'The Members'; Sedgwick, 'Appendix VII', in Sedgwick, ed., *House of Commons 1715–1754*.

[36] Those whose educational travel is mentioned (9): George Brodrick, fourth Viscount Midleton (1754–1836); Hon. James Brydges (1674–1744); George Brydges (1678–1751); John Brydges, Marquis of Carnarvon (1703–27); Henry Brydges, Marquis of Carnarvon (1708–71); Sir Robert Southwell (1635–1702); Edward Southwell (1705–55); John Perceval, Viscount Perceval (1683–1748); Thomas Worsley (1710–78). Those for whom it is omitted (10): George Brodrick, third Viscount Midleton (1730–65); James Brydges, Marquis of Carnarvon (1731–89); Arthur Ingram, third Viscount Irwin (1666–1702); Hon. Henry Ingram (1691–1761); Arthur Ingram of 'Borrowby' [*sic*] (d. 1742); Edward Southwell (1671–1730); Edward Southwell (1738–77); John Perceval, second Earl of Egmont (1711–70); Sir John Walter (*c.* 1674–1722). Robert Molesworth (1656–1725) and Frederick Meinhardt Frankland (*c.* 1694–1768) travelled in later life.

[37] Ruth Paley, ed., *The History of Parliament: The House of Lords, 1660–1715*, 5 vols (Cambridge: Cambridge University Press, 2016), i, pp. 77–8.

[38] Christopher Reid, *Imprison'd Wranglers: The Rhetorical Culture of the House of Commons 1760–1800* (Oxford: Oxford University Press, 2012), pp. 113–55.

[39] Reid, *Imprison'd Wranglers*, p. 148.

[40] Henning, 'The Politics of the Members', in Henning, ed., *House of Commons 1660–1690*; Reid, *Imprison'd Wranglers*, p. 148.

[41] BL, Add MS 47025, fo. 64: Sir John Perceval to Arthur Hill, 30 Oct 1703.

which continental exploits must have filled the conversation of a youthful membership.[42]

The underestimation of travel among the gentry makes it hard to draw firm conclusions about its meaning for major office-holding, but time abroad clearly helped to create as well as reflect the political elite. It can be tempting to downplay the significance of travel, since offices more often rewarded prior standing than they contributed to a family's rise.[43] Many gentlemen and noblemen enjoyed funds and connections that would have brought them into public life even if they had never been abroad, while even the most generous estimates would show that travel was not a requirement for the Commons. From the other side, too, widespread travel among the gentry meant that most did not enter parliament upon their return. But if travel was a product of wealth, it also helped to turn money into skills and networks that maintained the social and political standing of the most prominent families. At the same time, those who went abroad were not limited to an established, national elite. Families who stretched to afford foreign experience, such as the early Southwells and Rawdons, believed that it would furnish sons with accomplishments and patrons necessary to positions in government or diplomacy. If travel helped to prepare the most prominent for inevitable public lives, it also offered hope of advancement to lesser men; it played a role in building as well as manifesting the elite.

Marriage

Marriage was a far more widespread milestone after travel. Several noblemen married before travel in the early seventeenth century, but the practice was fading when the Earl of Derby did so in the 1670s, by which point men of his rank tended to marry late.[44] The mean age of noblemen at first marriage rose from 27 in the mid-seventeenth century to 30 in the mid-eighteenth. Figures for the gentry would probably be lower, as men in the general population married for the first time at 27 or 28 throughout the period. If noblemen tended to go abroad at an earlier age but increasingly married later, finding a wife straight after travel may have been more of a gentry

[42] Paul Langford, *Public Life and the Propertied Englishman, 1689–1798* (Oxford: Clarendon, 1991), p. 189. Barnard calls the Irish parliament a 'national Protestant club': *New Anatomy*, p. 51.

[43] Aylmer, *Crown's Servants*, p. 239; Hayton, 'The Members'; Barnard, *New Anatomy*, p. 51.

[44] Bodleian, MS Carte 50, fo. 98: Ormond to Mr Banks of Winstanley, 24 Jun 1673; Brennan, ed., *Origins of the Grand Tour*, p. 21.

concern. Even so, many noblemen married before the age of 24, almost half in the mid-seventeenth century and a third by the mid-eighteenth.[45] Since a successful match might follow years of financial negotiation, noble and gentry families alike tended to see it as the next priority once a young man had returned from the Continent. If travel and marriage were thus linked chronologically, they also informed each other as successive stages in the construction of elite male adulthood.[46]

Families regularly discussed marriage during travel, not least parental fears that sons might ruin family strategies by marrying abroad. Irwin played upon these concerns by teasing his mother with the prospect of finding a Dutch wife or bringing 'a Pretty Italian Lady Irwin into England, who would charm you by the Sweetness of her manners and Language', though he admitted that 'marriage is never heard of here betwixt strangers and Italians'. The elder Thomas Worsley trusted that the current Lady Irwin had warned her son against marrying abroad, which '(After the care of his Person) is the only Injury he can doe Himselfe & ffamily'.[47] Certainly Katherine Southwell was relieved that Edward was not 'much charm'd with Italy, for excepting the Joys of Classic Ground a la Lytleton, the Modern joys of that Country are not what a Mother Woud wish her Son fond of; a Signora, & all her familly, whoud not be an agreable present at your return'.[48] Here, as with the connections forged with fellow travellers, parents saw the benefits of foreign travel in a national system: they sent sons away to gather skills and networks that would attract an advantageous match at home, rather than to find a bride abroad.

Parents and guardians worried about marriage because many were in the process of arranging one at home. Several cases, from Henry Ingram in the 1650s to John Brydges in the 1720s, have already shown parents recalling their sons in order to secure desirable matches. Both men went back to the Continent when negotiations failed, but were married within

[45] T. H. Hollingsworth, *The Demography of the British Peerage* (London: Population Investigation Committee, London School of Economics, 1964), pp. 15, 25; E. A. Wrigley and R. S. Schofield, *The Population History of England 1541–1871: A Reconstruction* (Cambridge: Cambridge University Press, 1989), p. 255.

[46] French and Rothery, *Man's Estate*, pp. 185–233.

[47] WYL100/C/9A: Thomas Worsley to Dowager Viscountess Irwin, 4 Apr 1705; WYL100/C/20, fo. 190: Irwin to same, 18 Jun [1704]; fo. 160: same to same, 9 Apr [1705]; WYL100/C/9B: same to same, 2 Oct 1705. Marriage in Italy was indeed rare: Black, *Italy and the Grand Tour*, p. 131.

[48] Bristol Archives, 45317/2/5/2/31: Katherine Southwell to Edward Southwell III, 25 Aug 1758. George Lyttelton had praised Ancient Roman virtue: *An Epistle to Mr. Pope, from a Young Gentleman at Rome* (London: 1730).

months of a permanent return.[49] The Southwell, Perceval and Brodrick families all sought lucrative matches while young men were away, while Theophilus Leigh wanted one 'ready cookt' before William came home, 'lest he should match himself, to his own & family's ruin'.[50] Negotiations might take years, but contemporaries thought of marriage as the next step after travel.[51] The minister of the Percevals' parishes in north Cork, Christopher Vowell, marked the returns of Sir Philip in 1679 and Sir John almost 30 years later by recycling questionable verse. Since 'travell was in Order unto rest', both men should 'Secure your vertues in a Married brest'.[52] Once their cousin, the younger Edward Southwell, had returned from Italy, Archbishop King similarly trusted that his father's 'next care will be to buckle him fast to some faire, good Humoured great Fortune'.[53] The prospect of nuptials strengthened the attraction of London immediately after travel, as provincials such as Sir Philip Perceval and William Leigh stayed with relations in order to access a national marriage market.[54] Even where negotiations took time, they shaped how travellers returned from the Continent.

Despite the rise of repeated, shorter journeys in the late seventeenth and early eighteenth centuries, many continued to imply that marriage ended the stage of life at which lengthy voyages were appropriate. Samuel Waring thought of his wedding to Grace Holt as a 'fatal knot' that promised 'noe more of these rambls but in Imagination'.[55] Sir John Perceval also remembered that, before his travels of 1705–7, friends had 'proposed to

[49] John Brydges married Lady Catherine Tollemache in 1724, the year of his return, and Henry Ingram married Lady Essex Montagu in 1661, a few months after his return.

[50] BL, Add MS 46937, fo. 151: William Dobbins to Sir John Perceval, 26 Jul 1659; 46954A, fo. 24: elder Robert Southwell to Sir Robert Southwell, 24 Jan 1676[/7]; SHC, 1248/6/141: Thomas Brodrick to Sir Alan Brodrick, 9 Feb 1724/5; HL, mssST58, vol. 7, fo. 57: Theophilus Leigh to James Brydges, 13 Nov 1710.

[51] After their returns, Sir John Perceval waited three years to marry Catherine Parker, Alan Brodrick II four years for Lady Mary Capel and William Leigh almost a decade for Mary Lord.

[52] BL, Add MS 46956A, fo. 103: Christopher Vowell to Sir Philip Perceval, 16 May 1679; 47025, fo. 92v: same to Sir John Perceval, 24 Jul 1708.

[53] TCD, Lyons MS 750/8, fo. 91: William King to Edward Southwell, 5 May 1726. He found Katherine Watson, daughter of Viscount Sondes, and her £10,000, three years after his son's return: Shirley Matthews, 'Southwell, Edward (1705–55), of King's Weston, nr. Bristol. Glos.', in Sedgwick, ed., *House of Commons 1715–1754*.

[54] BL, Add MS 46954A, fo. 24: elder Robert Southwell to Sir Robert Southwell, 24 Jan 1676[/7]; HL, mssST57, vol. 10, fo. 65: James Brydges to Theophilus Leigh, 15 Feb 1713/14; Rosenheim, *Emergence of a Ruling Order*, pp. 26–9.

[55] PRONI, D695/227: Samuel Waring, notebook, 1687[–1714], p. 55; E. M. Johnston-Liik, *History of the Irish Parliament 1692–1800: Commons, Constituencies and Statutes*, 6 vols (Belfast: Ulster Historical Foundation, 2002), vi, pp. 501–2.

him that this was the only time of life for seeing foreign Countrys', as he would soon 'marry or engage in publick business'.[56] James Brydges policed this boundary in his family. His cousin, George, had travelled at 18 to the Netherlands and Wolfenbüttel, but was later warned of the 'Censure' he would attract when people saw 'a Gentleman easy in his Circumstances and happy in his Family & Domestick Concerns, quit the Pleasures of both for the sake of a Voyage'.[57] Brydges probably wanted to avoid any suggestion of Jacobitism, but his choice of argument is significant. An older rhetoric of one-off travels, clear in seventeenth-century manuals that presumed young men would settle down on their return, persisted alongside the newer reality of repeated journeys to the Continent among gentry and noble families.[58] Brydges was in fact ready to acquiesce to his cousin's travel on health grounds and considered a journey to Spa when his own wife fell ill.[59] But for travellers throughout the period, longer voyages between domestic education and marriage remained an experience set apart, on the threshold of adulthood.

If the inevitability of marriage shaped the ways that families thought about travel, recent continental experience featured prominently in marriage negotiations. Elite families held love and money in balance throughout the period, weighing them according to circumstances. Stereotypes of romantic individuals and pragmatic elders would miss that money and standing are not so easily separated from attraction, but parents and relations generally chose those whom a son or daughter should meet based on rank and wealth. Young people then made their choices from restricted options, considering additional questions of character, reputation, compatibility and mutual affection.[60] The parents or guardians of travellers intended that continental experience should bear plenty of weight in deliberations. The elder Robert Southwell, for instance, hoped that Sir Philip Perceval's English and continental education would encourage the family of a potential bride to overlook the fact that his grandson's finances could not yet 'maintaine

[56] BL, Add MS 47072: Perceval, Autobiography, p. 14.

[57] HL, mssST57, vol. 16, fos 174–6: James Brydges to George Brydges, 31 May 1719.

[58] Alison Games, *The Web of Empire: English Cosmopolitans in an Age of Expansion, 1560–1660* (Oxford: Oxford University Press, 2008), p. 44.

[59] HL, mssST57, vol. 31, fos 212–13: James Brydges to William Leigh, 22 Apr 1728.

[60] Amanda Vickery, *The Gentleman's Daughter: Women's Lives in Georgian England* (New Haven, CT, and London: Yale University Press, 1998), pp. 40–58; Ingrid Tague, *Women of Quality: Accepting and Contesting Ideals of Femininity in England, 1690–1760* (Woodbridge: Boydell, 2002), pp. 73–6; O'Day, 'Matchmaking and Moneymaking'.

a gentleman of good quallity'. Travel, and above all his acquisition of French, stood alongside the purchase of an English estate, which would also 'sound very well'.[61] James Brydges similarly advised his son to spend his remaining time abroad perfecting his French and frequenting assemblies in preparation for an 'interview' with a young woman at home.[62] Parents who had invested heavily in travel unsurprisingly hoped that it would improve the marriageability of their sons, but it is necessary to look at young women and their relations in order to establish whether they were successful.

Where the families of travellers stressed the positive impact of time abroad, those of potential brides saw travel as a test of character that could go wrong as well as right. It is difficult to say whether travelling families tended to intermarry, given the underestimation of gentry travel by available sources, but this wariness about time abroad often emerged from continental experience that the parents of potential brides had gathered themselves or through their own sons. Sir George Rawdon and the Earl of Conway, for instance, took time to arrange a match for Rawdon's daughter, Mary, in the mid-1670s. Conway had part-funded the travels of Arthur Forbes, son of the future Earl of Granard, and suggested him as a 'very fine Gentleman' straight after his return from France. Rawdon's sons were also then in Paris with Conway's support, although reports of John's drinking and 'other dishonest actions' at Foubert's academy had recently scandalised his parents. In this context, Rawdon worried about why Forbes had left debts behind him in France, 'when his Governor was shaken, and he was in his owne disposall'. Conway's enquiries found no evidence of spending on 'either Drinking, Gaming, or upon Woemen', though the young man did seem 'extreamly Wilfull'.[63] The Rawdons remained dubious, such that they could still entertain Lord Coloony's proposal of his son, Richard Coote, almost two years later. Coote, by contrast, seemed to have passed the test of his travels, which had added an Italian tour to similar time with Foubert. He proved 'very winning upon everybody' for his person, humour and education. Coote's prospects had been good when he ruined them by killing a rival at the Rawdon home, leaving Mary to wed Forbes. The unhappiness of their marriage, during which Arthur regularly disappeared to France, suggests that

[61] BL, Add MS 46952, fo. 240: elder Robert Southwell to Sir John Champante, 7 Dec 1675; 46956A, fo. 159v: Sir Robert Southwell to Sir Philip Perceval, 28 Jun 1679.

[62] HL, mssST57, vol. 20, fos 15–16: James Brydges to Charles Stuart, 19 Nov 1721.

[63] HL, mssHA 14541: Conway to Sir George Rawdon, 23 Oct 1675; 14542: same to same, 30 Nov 1675; 14544: same to same, [21 Dec 1675].

Sir George had been right to consider the evidence of travel.[64] Continental experience was not everything, dramatically so alongside a fatal duel, but conduct abroad merited scrutiny among families who had afforded similar educations to their sons. They valued travel, but they also knew what could go wrong.

The Rawdons found themselves on the other side of the equation two generations later, when stories of Sir John's promiscuity came home from Italy. His uncle, Sir Richard Levinge, realised that the Percevals were running 'so considerable a Risque' in letting him marry their daughter, Helena. For the sake of 'your vows, your virtue & all your honor', as well as the 'satisfaction, & content' of his bride and her family, Levinge urged his nephew 'to be faithful always as well as affectionate'.[65] If Rawdon's actions in Italy gave his future in-laws pause, however, there is little trace in the diary of Sir John Perceval, by then elevated to the earldom of Egmont. The earl could think of no 'young gentleman of a more even temper, and better bred' than Sir John, who seemed a 'sincere Christian'. Either tales of Rawdon's behaviour had been exaggerated or Egmont was immensely gullible; alternatively, perhaps, continental exploits did not always harm reputations at home. In any case, Sir John's travels helped to win the approval of the Percevals over a month spent in their company. Amid discussion of Rawdon's standing, fortune and personality, Egmont approved that he 'gives a good account' of his travels. The earl, who had been to Italy in his youth and corresponded with later travellers, would have been a discerning audience.[66] Rawdon escaped judgement of his behaviour abroad, but his father-in-law still tested him on whether he had profited from his travels. Again, family experience of the Continent helped parents to assess their daughters' suitors.

The parents or guardians of potential brides weighed travel alongside other factors. Time abroad could be part of an attractive package, boding well for a young man's personal improvement and public advancement, but it often lost out to other considerations. An earlier Sir John Perceval, who had turned down the chance to travel, placed greater value on other aspects when arranging a match for his sister, Kate, in the 1680s. He sized

[64] *CSPD, Charles II*, xix, pp. 458–9: Sir George Rawdon to Conway, 19 Nov 1677; pp. 479–80: same to same, 30 Nov 1677; xxiv, p. 248: same to same, 12 May 1683; HL, mssHA 14526: Conway to Rawdon, 17 Apr 1674; TNA, SP93/1, fos 255–6: 'Engl[ish] Men passing to Rome', [early 1675].

[65] PRONI, D2924/1/88-9: Sir Richard Levinge to Sir John Rawdon, 2 Nov 1741; 89–90: same to same, 11 Nov 1741.

[66] *Egmont Diary*, iii, pp. 227–9.

up the wealth of neighbouring Irish landowners, but also considered the appearance, education and morals of their sons. Perceval recognised that travel could threaten as well as make a character. One young man was 'well natured' and had 'not seen more of the World' than Trinity College Dublin and 'his own Neighborhood; & addicted to noe vice'. Another, who had studied at Oxford and Middle Temple, perhaps before travel, had 'seen more of the world as well bad as good'. There were 'Circumstances in both to make either acceptable to us'.[67] Perceval eventually settled on a third candidate, William Moore, whose 'good breeding', 'Vivacity' and 'Ingenuity', and above all his father's lands, compensated for his short stature and round shoulders. They also encouraged Perceval to overlook a lack of continental experience: Moore had 'never travel'd, nor appears to have wanted it, having had the good Example of his Father att home allways before him'.[68] Perceval extolled the advantages that his late brother, Sir Philip, had drawn from time on the Continent, but travel—or its absence—was rarely enough to make or break a match. When the Brodricks sought a husband for Alice, a few years before sending Alan to Italy, they similarly weighed travel among other factors. One suitor, John Monckton, 'was newly returnd from travailing' and 'had brought with him … the charecter of sobriety' but 'nott that of being well bred'. Nor would his 'person … please a nice pallett'. Although Monckton had been to Italy, these negatives ruled out a meeting with Alice.[69] If travel was rarely enough by itself, parents who sent their own sons abroad also knew that young men could fail to take advantage of the opportunity.

Suitors equally needed to convince young women themselves, who represented an important audience for returning travellers. While gentlemen's daughters enjoyed a veto rather than free choice, this was the time of life at which their preferences were most respected.[70] In the cases of Mary Rawdon, Kate Perceval, Helena Perceval and Alice Brodrick, all parties agreed that a happy marriage could only begin with their satisfaction. The role of continental experience, and the changes that it could work on mind

67 BL, Add MS 46960A, fos 93–4: Sir John Perceval to Sir Robert Southwell, 3 Apr 1683. The first candidate was probably Henry Ingoldsby: George Dames Burtchaell and Thomas Ulick Sadleir, eds, *Alumni Dublinenses: A Register of the Students, Graduates, Professors and Provosts of Trinity College, in the University of Dublin* (London: Williams and Norgate, 1924), p. 426. The second was Samuel Maynard: Joseph Foster, ed., *Alumni Oxonienses: The Members of the University of Oxford, 1500–1714*, 4 vols (Oxford: Parker, 1891–2), iii, p. 995.

68 BL, Add MS 46960A, fos 169–169v: Sir John Perceval to Sir Robert Southwell, 26 Jun 1683.

69 SHC, 1248/4/259v: Thomas Brodrick to Sir Alan Brodrick, 2 May 1720; Ingamells, *Dictionary*, p. 667.

70 Vickery, *Gentleman's Daughter*, p. 82.

and body, must have influenced young women's opinions in different ways, but it is not always easy to work out why daughters accepted or rejected suitors. Why, for instance, did Mary Rawdon take against Richard Coote even before he had killed a man in her home, when the effects of his travels had won over the rest of her family? These were complex issues, and the residence of daughters in the parental home means that their input does not always appear in letters. James Brydges's correspondence provides one instance, in his attempts to match his niece, Betty, with his cousin's nephew. Sir Charles Lloyd was a Cardiganshire gentleman of £1,100 a year, 'very agreeable in his face, very good Parts, and a vast deal of life and Spirit'. It was his 'misfortune', however, 'to have seen so little of the world that his carriage is awk[w]ard, and his manner of speaking not so polite'. Lloyd had studied at Oxford and Lincoln's Inn, but had evidently not been to the Continent. Betty took 'such a disgust to him', on account of his speech and deportment, 'that she cannot as yet be brought to think of it'. Brydges hoped that his wife would persuade her, but to no avail.[71] Betty, whose male relations had travelled extensively in recent decades, sought the same qualities in a husband that drove so many young men across the Channel.

The examples here have everything but a case of a man having travelled, a woman and her family approving of him on that basis and the conclusion of a successful match. Young women and their relations tended to discuss travel if a suitor had not been abroad or had failed to make the most of the experience. A lack of travel could be something to explain away in a preferred husband or a reason to reject a less attractive option, suggesting that it was rarely a decisive factor. The very need for those explanations nevertheless makes clear that families saw travel as a standard part of elite male upbringing. The rejections, meanwhile, indicate how parents could control some parts of marriage negotiations, in which one family sat in judgement over another. There was little to be done about finances or looks, but parents decided whether sons had travelled and, by extension, whether they had had opportunities to hone their speech and deportment for encounters in London or Bath. Travel alone could not drive rapid rises through marriage, as landed families tended to wed their equals. Since wives took the rank of their husbands, moreover, it is usually daughters rather sons who can be found marrying upwards.[72] When it came to choices among

[71] HL, mssST57, vol. 29, fos 42–3: James Brydges to Dr [Henry] Brydges, 23 Oct 1726; Joseph Foster, ed., *Alumni Oxonienses: The Members of the University of Oxford, 1715–1886*, 4 vols (Oxford: Parker, 1891), iii, p. 860; O'Day, 'Matchmaking and Moneymaking', 281.

[72] Stone and Stone, *Open Elite?*, pp. 247–51; Cannon, *Aristocratic Century*, pp. 71–92.

equals, however, it was best for a young man to have travelled. Preparation for marriage usually features only in accounts of female upbringing, but continental experience helped to make sons acceptable. The attendant risk, of course, was that young men might endanger matches through their conduct abroad. For those who had crossed the Channel, travel loomed over negotiations as a test of character and an opportunity for improvement. In both respects, suitors could succeed or fail. Parents and daughters drew on family experience of the Continent, gathered by fathers, sons and brothers, to frame opinions about travel—or a lack thereof—by potential husbands. Marriage negotiations were thus an early stage in the afterlives of travel, and an initial opportunity for a family to present itself as one that travelled.

Irish Landowners and England

If continental experience played different roles in the searches for employment and a wife, some Irish landowners bound these efforts up, in the years after travel, with the desire for an English 'footing'. Sir Robert Southwell observed in 1675 that 'as soon as any Protestant [in Ireland] finds himself enabled for a journey into England here he resorts either to show his fortune or to strengthen his interest; here he spends his Irish estate and his principal design is how to purchase some thing on this safe side from whence he lately went or is anciently descended'.[73] There is a risk of overgeneralising his comments, not least because the papers of many families survive precisely because they moved to England. Even those who had arrived in Ireland since the early seventeenth century, such as the Southwells, Percevals or Brodricks, were capable of ambiguous, situational pronouncements on their relationship with the country.[74] At the very least, though, Southwell described his own actions over the previous decade. He had not hidden his contempt when forced home from London, writing 'through an Irish mist' to a man he knew from Oxford and complaining to another met in Paris that Ireland was 'most insipid and voyd of curiosity', betraying 'an unexpected narrownesse in the thoughts of most'. 'This Country', he wrote, 'is a meere Scramble, and the belle maniere is not knowne.'[75] Some of this frustration emerged during a

[73] UCC, Boole Library, MS U/20, vol. 2, 1665–75: Robert Southwell to Sir Harry Capel, Aug 1675.

[74] Barnard, 'Crises of Identity', 47; Hayton, *Anglo-Irish Experience*, pp. 37–9.

[75] TNA, SP63/310, fo. 198: Robert Southwell to Joseph Williamson, 24 Sep 1662; *The Correspondence of Henry Oldenburg*, ed. A. Rupert Hall and Marie Boas Hall, 13 vols (Madison:

dispute over poor relief in Kinsale, when Southwell warned an opponent that 'it were better for he and I to be friends for I had been in Italy and knew how to vindicate my concernments'.[76] A successful return to London set him on a course of administrative, diplomatic and parliamentary offices that, alongside his marriage into the Derings of Kent, eventually allowed him to solidify his position in England. He settled at King's Weston, near Bristol, in the late 1670s.

For some landowners from Ireland, English residence accordingly featured among desirable criteria for a potential bride. When Sir John Perceval received a proposal from a wealthy Irish family a year after his return from the Continent, he was 'so far from intending to marry in this Country that I never so much as ask'd the persons name'.[77] The family remembered that his uncle, Sir Philip, would also have married in England, 'where there was greater Choice' of wealthier families 'more suitable to the Dignity of his own'.[78] Thomas Brodrick similarly decided that Alan should marry in England once he was back from Italy, because he was 'nott of a common stamp, wherefore his candle should nott bee putt under a Busshell'.[79] Finding an English wife was never easy because, while elites tended to marry equals, Irish landowners generally had lower incomes than their English counterparts.[80] Marriage in England stood alongside other markers such as travel and public employment, in combinations that were mutually reinforcing. Only once he was 'involved in the Publique Affairs', knighted and settled with Elizabeth Dering in London, could Sir Robert Southwell think of himself as 'a Man who is climbing up in the World'; at that point, too, friends in Ireland began to notice 'the good progresse you make'.[81]

A couple of years before his death, Southwell worked decades of research into a family history of upward and downward social mobility. Ascent 'to a high Degree' and 'a bulke of fortune' between the thirteenth and sixteenth centuries had been marked by crown service, a tradition to which Sir Robert and his son consciously added, but their lowest point had come with Irish residence. Under James I, the Southwells had left Norfolk for 'sanctuary

University of Wisconsin Press, 1965–86), i, p. 434: same to Henry Oldenburg, 19 Sep 1661 [i.e. 1662]; p. 476: same to same, 10 Dec 1662; ii, p. 22: same to same, 11 Feb 1662/3.

76 UCC, Boole Library, MS U/20, vol. 1, 1662–5: Robert Southwell to Thomas Southwell, Jun 1663.

77 BL, Add MS 47025, fo. 106: Sir John Perceval to William Lewis Le Grand, 7 Dec 1708.

78 Anderson, *Genealogical History*, ii, p. 386.

79 SHC, 1248/6/141v: Thomas Brodrick to Sir Alan Brodrick, 9 Feb 1724/5.

80 Barnard, *New Anatomy*, pp. 60, 74.

81 BL, Add MS 38015, fos 107–108v: Sir Robert Southwell to elder Robert Southwell, 4 Apr 1668; 46942, fo. 99: Sir John Perceval to [later Sir] Robert Southwell, 6 May 1665.

in the South of Ireland; not a foot being left on this side, when in 1678 I purchased again in the West of England'.[82] Where Sir Robert's English education and foreign travel had helped him along the way, his father portrayed himself as an unrefined provincial. The elder Robert claimed that his 'debillityes & want of education for discourse' had kept him 'from all companyes & societies espectially from persons of quallity and parts', leaving him reluctant 'to appear on any stage or to act any parte in the face of the people'.[83] He was in fact a prominent administrator of the Munster plantation, whose efforts had 'stemm'd the Tyde, & restor'd his Family', and his self-deprecating letter responded to efforts to get him a seat on the Irish privy council. He too had been thinking of buying an English estate just before his death. But where the elder Robert had lived in Ireland from the age of 5 and never travelled further than London, his son's education had deliberately prepared him for public office and the company of powerful men.[84] The Southwells transformed themselves in a generation, as Toby Barnard puts it, 'from Kinsale Protestants to English landed gentlemen who happened to enjoy a useful Irish rental'.[85] The elder Edward, a 'little Irish man' during a couple of childhood years in Cork, was mourned as 'a friend of Ireland'.[86] While his son entered himself as 'Hibernus' in the matriculation book at Padua, the family retained a sense that their education, marriages and lifestyle, of which successful travel was a prominent part, suited them to English residence.[87] The younger Edward's widow, Katherine, warned that their son's wayward fellow traveller, Francis Annesley, the 18-year-old future first earl, was conversely doomed to return 'as Errant a Teague as he went out, & spend the remainder of his Life, in Drinking & Tyrranising over the poor Irish in the Mountains of Morne'.[88]

The Southwells shared these attitudes with their kinsmen. Sir Robert promised to find an English 'footing' for Sir Philip Perceval but was forced to abandon arrangements, including entry into the London parliament and

[82] Bristol Archives, 45317/16/2/1: Southwell, Narrative, pp. 29–32, 70.

[83] BL, Add MS 38015, fo. 127: elder Robert Southwell to Sir Robert Southwell, 31 Oct 1671.

[84] BL, Add MS 46954A, fo. 135: Sir Robert Southwell to Sir Philip Perceval, 30 Apr 1677; Bristol Archives, 45317/16/2/1: Southwell, Narrative, p. 32; *Boyle Correspondence*, iv, pp. 407–8: elder Robert Southwell to Robert Boyle, 30 Mar 1676; Sean Kelsey, 'Southwell, Robert (*bap.* 1608, *d.* 1677)', *ODNB*, online edn, 2008 (https://doi-org/10.1093/ref:odnb/26065).

[85] Barnard, *Irish Protestant Ascents and Descents*, p. 71.

[86] BL, Add MS 46955A, fo. 136: Sir Philip Perceval to Sir Robert Southwell, 7 May 1678; *Byrd Correspondence*, i, p. 439: Sir John Perceval to William Byrd II, 28 Dec 1730.

[87] Brown, ed., *Inglesi e Scozzesi*, p. 205.

[88] Bristol Archives, 45317/2/5/2/27: Katherine Southwell to Edward Southwell III, 10 Aug 1758.

the purchase of a neighbouring estate outside Bristol, when his nephew died only a year after travel.[89] The plans did not carry over to Perceval's brother, John, who had settled into his Irish inheritance by the time of his own death six years later.[90] A decade earlier, John had agreed with Southwell that his mother's suggestion of a year at Saumur, with its 'brave Academy & many advantages for the education of Youth', amounted to 'noe less than throwing by' his legal studies.[91] In the long term, though, the decision to stay in London left him ill prepared to replace his brother in English public life. The transition to England, and eventually English politics, came in the next generation, when Southwell took John's family into his home and began the domestic and foreign education of another John, the future Earl of Egmont. Soon after his elevation in the 1730s, Egmont commissioned a genealogical history that aimed to prove his ancient nobility, in part by playing down his connections to Ireland. Altogether, it claimed, he and his ancestors 'had never passed … full five and twenty Years' in the country. The earl's lineage and lands, however, as well as 'his Habits, Education, and Acquaintance', tied him strongly to England.[92]

As with questions of marriage and public life, it can be difficult to pin down the role of travel in residential change. After all, the Southwells and Percevals were subject to wider economic forces that were encouraging many Irish landowners into absenteeism.[93] By their own estimation, however, educational mobility was an integral factor in 'the design to live like Englishmen'. They had been to England for schooling, university and preparatory domestic tours. They had seen the Continent as Englishmen in English company, forging connections with English elite. When they returned, they felt entitled to English wives, careers and estates. The Southwells and Percevals stood among other Protestants from Ireland, including the Molesworths, who secured employment and residence in England after time on the Continent.[94] At the very least, travel represented an important way, alongside marriage, office-holding and the purchase of land, in which these families used their wealth to buy into the English elite.

[89] BL, Add MS 46956A, fo. 159: Sir Robert Southwell to Sir Philip Perceval, 28 Jun 1679; Anderson, *Genealogical History*, ii, pp. 385–6.
[90] Anderson, *Genealogical History*, ii, pp. 389–96.
[91] BL, Add MS 46955A, fo. 110: Lady Perceval to Sir Robert Southwell, 13 Apr 1678; 46955B, fo. 127: Southwell to John Perceval, 24 Sep 1678.
[92] Anderson, *Genealogical History*, ii, p. 444.
[93] Toby Barnard, 'Protestantism, Ethnicity and Irish Identities, 1660–1760', in Tony Claydon and Ian McBride, eds, *Protestantism and National Identity: Britain and Ireland, c. 1650–c. 1850* (Cambridge: Cambridge University Press, 1998), pp. 206–35, at pp. 217–19.
[94] Barnard, *New Anatomy*, p. 146.

If young men were not earmarked for England, by contrast, travel began to seem less necessary. Sir Alan Brodrick similarly valued English and foreign education, sending his sons to Eton, Cambridge and in one case Italy.[95] He nevertheless decided that his 22-year-old stepson, Arthur Hill, should not spend time abroad because he was destined to stay in Ireland. Brodrick's untravelled eldest son, St John, ran through some of the negotiating methods open to the younger generation, arguing that Hill was 'of that age and spirit which must be employed otherwise then staying at home'; if not, he might leave for London and 'be soon undone'. Hill's mother and stepfather were resolute, however. Travel would not 'much conduce to his being an useful man at home', where he was already an MP, and Brodrick hoped to buy his stepson a clerkship of the Irish privy council.[96] Robert Southwell had felt that time abroad directly prepared him for the equivalent job in England, leaving traces of agreement from friends and patrons. In this way, attitudes towards travel reflected contemporary understandings of education in general, which was only justifiable as far as it prepared young people for their future roles in life. In some cases, travel itself could represent wasted time at a critical juncture.

'All Men Talke of Their Travells'

Any power that travel held in employment, marriage or other areas of society ultimately depended upon the ways in which travellers cultivated and displayed their experiences in the years afterwards. Yet it can sometimes seem that travellers were better advised to hide their memories and accomplishments at home. Between the sixteenth and eighteenth centuries, 'Italianate' and 'Frenchified' returning travellers were mocked for their outlandish accents, manners, clothes and vices in print and on the stage.[97] At the same time, across Europe, authors told young men to leave behind habits and affectations acquired abroad.[98] These recurring themes must have informed some family warnings against ostentatious returns. When Vowell welcomed home Sir Philip Perceval, for example, he called on him not only to marry but also to

[95] Hayton, *Anglo-Irish Experience*, pp. 95–6.

[96] SHC, 1248/3/396: Sir Alan Brodrick to Thomas Brodrick, 10 Nov 1716; fo. 416: same to same, 30 Mar 1716.

[97] Warneke, *Images of the Educational Traveller*, pp. 105–37, 217–48; Cohen, *Fashioning Masculinity*, pp. 37–41; Gallagher, *Learning Languages*, pp. 199–200.

[98] Stagl, *History of Curiosity*, p. 77; Warneke, *Images of the Educational Traveller*, pp. 239–40.

> Leave that Gallick Art
> Of speaking what they never think: nor write
> As speake. as if they ow'd to truth a spight.
> Tis misbecoming of an English min'd [*sic*]
> Which is to truth and Justice still confin'd[.][99]

Although such exhortations were common, however, they ought not to overshadow the fact that travel was supposed to change young men, whom families deliberately sent abroad at an impressionable age. In the first place, literary and dramatic tropes of the 'Italianate' or 'Frenchified' traveller probably bore little relation to the behaviour of individuals. After all, would families really have spent so much time and money on travel only to guarantee abuse at home? Gallagher has recently used John North, a late sixteenth-century traveller, to offer an alternative 'view from inside' Anglo-Italian encounters. Although Italian fashions, behaviours and language were controversial in Elizabethan England, attitudes and practices learned abroad contributed to North's advancement in London. He bought Italian books, learned French, socialised with fellow Italophiles, including a noble patron, and chose clothing, food and music that advertised his continental experience to an audience of fellow travellers as much as those who had remained at home.[100] Controversy in the late seventeenth and eighteenth centuries centred on France rather than Italy, as print amplified fears of the disruption that 'Frenchified' language, manners, dress, consumption and ideas threatened to society, culture, politics and gender roles. Aristocratic, effeminate 'fops' and 'macaronis', among them stereotypical returning travellers, offered a convenient foil for middling-sort English nationalism, itself suffused with notions of the moral and the masculine.[101] This 'patriotic backlash' should not, however, obscure the truth at its heart: that much of the British elite was indeed attracted by French models.[102] In the seventeenth and eighteenth centuries, as in the sixteenth, it is worth looking beyond the stereotypes to examine how real people performed the changes wrought by travel.

[99] BL, Add MS 46956A, fo. 103: Christopher Vowell to Sir Philip Perceval, 16 May 1679.
[100] John Gallagher, 'The Italian London of John North: Cultural Contact and Linguistic Encounter in Early Modern England', *Renaissance Quarterly*, 70:1 (2017), 88–131.
[101] Stedman, *Cultural Exchange*, pp. 109–254; Philip Carter, *Men and the Emergence of Polite Society, Britain 1660–1800* (Harlow: Longman, 2001), pp. 137–56; Cohen, *Fashioning Masculinity*, pp. 37–41; Newman, *Rise of English Nationalism*, pp. 63–84.
[102] Conway, *Britain, Ireland and Continental Europe*, p. 127; Black, *British Abroad*, p. 334.

Many families acted as if cultural exchange was instead a force for social stability, a bolster to their prestige, and they embraced travel in order to perfect foreign languages and manners. James Brydges confessed an 'unspeakable Concern' that John would return from the Continent 'to turn rake about Town & run into all the extravagancies of that race of People'. He nevertheless urged his son to improve his French and 'naturall parts', lest he throw away his fatherly 'Care & expence' and become 'a Downright simple Country Gentleman spending yor time amongst Horses and Dogs, & Bowling with the meanest of the Company'.[103] Brydges confirms the widespread view that travel prepared young men for life in London rather than the country, through inclination as much as the desire for public office. Crucially, too, he shows that foppishness was not the only way in which returning travellers could prove a poor investment; they might take on too little, as well as too much, of their host cultures. The Rawdons, for instance, had had high hopes for Arthur, but found that his 'mien discovers little of his travels'. Sir George was forced to employ a French dancing master at home to teach the posture and bearing that his son should have acquired abroad.[104] From the sixteenth to the eighteenth centuries, authors and parents alike advised 'critical imitation' of foreign behaviour.[105] If Katherine Southwell noted Edward's intention while abroad 'to learn to Love your own Country' and hoped that he would 'preserve the good sence & probity of a true Englishman', she equally wished for him 'the embellishments of other Countrys; the Scavoir Vivre of the french, the Complaisance of the Italian, the Honesty of the Swiss, but not the Stupidity of the German'.[106] Despite the stereotypes, families saw benefits as well as dangers in direct contact with foreign cultures.

The most obvious way in which young men sought to capitalise on their travels was by talking about them, but here too a balance needed to be struck. Given the importance of conversation to personal reputation, the question of how best to present oneself in company occasioned plenty of reflection and self-scrutiny.[107] Contemporary authors regularly complained about travellers shoving their experiences down people's throats, to the point that having spent time on the Continent almost seems to have been a liability

103 HL, mssST57, vol. 21, fo. 34: James Brydges to John Brydges, 22 Jul 1722.
104 *CSPD, Charles II*, xix, p. 374: Sir George Rawdon to Conway, 22 Sep 1677; p. 511: same to same, 15 Dec 1677.
105 Gallagher, *Learning Languages*, pp. 196–205; Cohen, *Fashioning Masculinity*, pp. 54–63.
106 Bristol Archives, 45317/2/5/2/31: Katherine Southwell to Edward Southwell III, 25 Aug 1758.
107 Carter, *Men and the Emergence of Polite Society*, pp. 163–208.

in conversation by the eighteenth century.[108] Certainly, as educational travel proliferated, families were conscious that young men should not push things too far. Sir Alan Brodrick trusted that his son's 'modestye as well as prudence will render you not inclined to make your travails the subject of your discourse'. He warned Alan against all but the most careful comparisons between England and abroad, and only then when 'urged to give your opinion', for fear of giving off 'an air of liking things abroad better then those at home'. Company would otherwise 'be apt to impute what you say to a vanitye in you of shewing you have seen and observed things which they never did', leading Sir Alan to conclude that 'there is no subject you ought to be lesse fond of discoursing upon then of your travails'. He and his son similarly recognised the need to praise the English constitution rather than the 'glittering shew' of the 'despotick' regimes Alan had seen abroad.[109] Such views were widespread and of long standing, with Black finding that many young men returned as 'better-informed xenophobes'.[110]

At the same time, the degree to which travellers should reminisce was a subject of debate rather than consensus, even within families. While Sir Alan Brodrick warned his son against an ostentatious return, his brother, Thomas, looked forward to the younger Alan coming home a 'master of what some whoe hold their heads high, are wholly ignorant of', from 'the Religious matters of Germany' to the 'taste which you appeare to have of painting'. The latter would make his nephew 'more then ordinarily acceptable to the lovers of itt, and the more because few of our young gentlemen bring back with them much knowledge of what is soe generally admird here, though they more affect being thought knowing then they are'.[111] As the Brodricks indicate, behind complaints about discussion of continental experience lay the implication that everyone was doing it. As in France and Germany, foreign encounters provided 'life-long material to dine out on'.[112] Sir Robert Southwell certainly took it for granted, in a letter to his son's tutor, that 'all men talke of their travells, and illustrate their present discourses by what is past'.[113]

108 Cohen, 'Grand Tour: Language, National Identity and Masculinity', 132.
109 SHC, 1248/6/88v: Alan Brodrick II to Sir Alan Brodrick, 8 Nov 1724; fo. 334: Sir Alan Brodrick to Alan Brodrick II, 31 Oct 1725.
110 Black, *British Abroad*, p. 256.
111 SHC, 1248/6/42: Thomas Brodrick to Sir Alan Brodrick, 7 Jul 1724; fo. 97: same to Alan Brodrick II, 19 Nov 1724.
112 Neil Kenny, *The Uses of Curiosity in Early Modern France and Germany* (Oxford: Oxford University Press, 2004), p. 251.
113 Blackburn CL, Southwell Papers, vol. 1: Sir Robert Southwell to William Webster, 1 Sep 1686.

Southwell had used his own foreign experiences to cement relationships, gaining a reputation for sound travel advice. He had cultivated the brothers of the Duke of Norfolk in Italy, reminding them of their sixteenth-century connections, and continued the project as he headed home. Southwell wrote at length to Edward Howard 'Of Such English as Your Honour will meete in Italy, and of some few others', recommending the sights and contacts of his own journey.[114] Despite the celebrated Italian travels of Howard's grandfather, the 'Collector Earl' of Arundel, up-to-date knowledge was a valuable commodity.[115] Southwell also contributed to experience pooled by his Oxford college, Queen's, telling a contemporary soon after his return that he stood 'obleiged unto you to write something concerning my travaile abrode, and the improovements to be acquired in such an Enterprise'. He began with justifications of travel as a mind-broadening experience, but his fragmentary draft presumably moved on to practical matters.[116] As the years passed, Southwell also shared advice among circles at Whitehall and the Royal Society. Nathaniel Coxe, son of Charles II's physician, was on his way to Italy in 1670 and hoped to see Spain and Portugal, though Southwell feared that his father would 'run with the Streame and tell me those are places dry and melancholy to the view'. He recommended Paris, 'for the last Impression of the businesse of the world', to render Nathaniel 'usefull' for 'the service of your Prince & Country'.[117] Southwell dug into the papers of his own travels to share his 1658 compilation of advice with Isaac Newton, who used it to frame a letter to another traveller in 1669.[118] He may have lent other writings to Sir William Petty, who asked for his friend's 'Routier, Remarques and Advices, with allowance for the present face of things' in 1686. Petty was then planning his son's time in the Low Countries, which Sir Robert knew well from his homeward journey in 1661–2 and his diplomatic posting to Brussels a decade later.[119] Southwell remained an

[114] BL, Add MS 38015, fos 36–9: Robert Southwell to [Edward Howard], [1662].

[115] David Howarth, *Lord Arundel and His Circle* (New Haven, CT, and London: Yale University Press, 1985).

[116] BL, Add MS 38015, fo. 34: Robert Southwell to Gabriel Towerson, [1662].

[117] BL, Add MS 38015, fo. 124: Sir Robert Southwell to [Nathaniel] Coxe, 12 Sep 1670.

[118] King's, Keynes MS 152: Southwell, 'Manuscript concerning travelling'; *The Correspondence of Isaac Newton*, ed. H. W. Turnbull, J. F. Scott, A. Rupert Hall and Laura Tilling, 7 vols (Cambridge: Cambridge University Press, 1959–77), i, pp. 9–12: Newton to Francis Aston, 18 May 1669; Iliffe, 'Foreign Bodies', 366, n. 16. Southwell's manuscript is still among Newton's papers.

[119] *Petty–Southwell Correspondence*, p. 187: Sir William Petty to Sir Robert Southwell, 1 Apr 1686.

authority for those who were about to travel, surely in conversation as much as in surviving letters.

Published critics and wary parents assumed that returned travellers would speak to those who had not—or had not yet—crossed the Channel, but conversation also sustained communities of mutual assistance and esteem among those who had seen the Continent together. When the Earl of Conway, who had attended a Parisian academy in 1640, had a bill of exchange rejected a decade later, he 'went presently to my Lord Mayor [of London], whose eldest sonne, hath marryed my wifes cousin-german, and his yonger sonne, was my great acquaintance in France, where I fownd very much favor and there upon Mr. Gawden accepted the bill'.[120] Robert Southwell kept track of courtiers with whom he had travelled and received help with his father's estate business from John Finch, the brother of the solicitor-general, 'on the Score of their former acquaintance in Italy'.[121] These practical benefits were exactly how families hoped that travel would work, but the papers of other young men are rarely so explicit on the everyday operation of networks forged abroad, not least because a common function of shared travel was to strengthen existing ties of friendship or patronage.[122] Beyond such immediate connections, however, the ability to talk about foreign sights and experiences bound a larger community of travellers. In the 1730s and 1740s, for instance, Egmont included in his diaries several 'pleasant stories' of continental travel that others had shared in his company, sometimes decades after their journeys.[123]

Returning travellers advertised their membership of this in-group through their speech and deportment, as well as their topics of letter-writing and conversation. Plenty is now known about early modern language-learning and the anxieties that surrounded it, but surprisingly little has been written on how people used their newly acquired tongues.[124] If British men and women learned languages for their prestige, how did they make sure that some of that prestige reflected on them? Contemporary authors criticised the elite for peppering their speech with French and Italian phrases, seeing it as 'breaking' the English tongue, and the phenomenon indeed leaves

120 HL, mssHA 14350: Conway to Sir George Rawdon, 1 Sep 1651; Kelsey, 'Conway', *ODNB*.

121 BL, Add MS 46941B, fo. 52: Robert Southwell to Sir John Perceval, 30 Aug 1664; Egerton MS 1634: Le Grand, 'Life of Southwell', fos 10v, 22.

122 For similar effects among the Dutch and Flemish, see Verhoeven, *Europe within Reach*, pp. 39–40.

123 *Egmont Diary*, i, pp. 191–2; ii, p. 207; iii, p. 228.

124 Cohen, *Fashioning Masculinity*; Gallagher, *Learning Languages*.

traces in the letters of returning travellers.[125] Robert Southwell promised his father that he would finish his journey 'alegramente' (cheerfully) and advised Howard of contacts who were 'garbato' (courteous) and 'fidato' (trustworthy). Back home, he quoted Italian proverbs to Joseph Williamson, an Oxford friend and near-contemporary abroad, whom he addressed as 'Mio Caro Signior'.[126] Months after his return, Southwell even labelled a domestic letter from an army officer in Wexford as 'Litera dal Cavaglier'.[127] Such decisions may speak to his flight from a provincial lifestyle, as small self-definitions against a father who occasionally used Irish-language proverbs and terminology, but they were also more widespread.[128] Levinge, who had been in Italy a decade before his nephew Rawdon, exhorted marital fidelity from his 'Caro Nipote', while the well-travelled Molesworths sprinkled French words into everyday financial discussions.[129] Contemporary criticism implies that many travellers did the same in conversation, as an affectation that marked them out to their peers and to those who had stayed at home.

But did a smattering of phrases really justify the time and money spent on learning French, at home or abroad? Returned travellers would continue to read and write the language, as will become clear, but the emphasis that families placed on accent and deportment, learned from the best models, suggests that they had fluent speech in mind. It is less easy, however, to work out the occasions for which they were preparing. Why, for instance, was James Brydges so adamant that French had become 'almost as necessary as one's Mother Tongue' in Britain as much as during travel? Contemporaries had long thought travellers well suited for ambassadorial work, with their languages and courtly experience, but the supply of diplomatic careers was finite. At any rate, noblemen such as Brydges, the Duke of Chandos, mostly

125 Cohen, 'Grand Tour: Language, National Identity and Masculinity', 133–9.

126 BL, Add MS 38015, fo. 22: Robert Southwell to elder Robert Southwell, [*c.* 26 Apr 1661]; fos 36–9: same to [Edward Howard], [1662]; TNA, SP63/310, fo. 198: same to Joseph Williamson, 24 Sep 1662. Williamson had recently been to France as tutor to one of Ormond's sons: Alan Marshall, 'Williamson, Sir Joseph (1633–1701)', *ODNB*, online edn, 2018 (https://doi-org/10.1093/ref:odnb/29571). Southwell's language suggests that Williamson had also been to Italy.

127 NLI, MS 7863, fo. 17: Sir John Boys to Sir John Perceval, 24 Dec [1662].

128 BL, Add MS 38015, fo. 127: elder Robert Southwell to Sir Robert Southwell, 31 Oct 1671: 'our Countrj proverbe will hould good vizt. (bee magh agust ber magh) bee good and have good'; 46054A, fo. 24: same to same, 24 Jan 1676[/7]: an explanation of 'cullup' (i.e. collop, from the Irish *cológ*, the grazing needed for one cow).

129 PRONI, D2924/1/89: Sir Richard Levinge to Sir John Rawdon, 11 Nov 1741; Ingamells, *Dictionary*, p. 599; NLI, microfilm P.3753: Richard Molesworth to Robert Molesworth, 17 Jun 1719; same to John Molesworth, 4 Nov 1724.

saw foreign posts as unattractive prospects.[130] It may be that he had the court in mind, but little work exists on its linguistic history, even under Charles II. It is nevertheless suggestive that French exiles could spend years among the cream of Restoration London without ever needing English.[131] Decades later, George I, who spoke far more French than English, used the former tongue when forced to converse with courtiers.[132] The wider possibility remains that, by the eighteenth century, the elite conducted more day-to-day sociability in French than can usually be recovered from textual sources.[133] Englishmen had certainly spoken Italian together in sixteenth- and seventeenth-century London, where foreign-language in-jokes generated community among returned travellers.[134] Cohen suggests that linguistic knowledge 'might never be tested or verified' at home, with contemporary idealists holding that 'the true gentleman's achievements [should] never be displayed', but this seems an unlikely basis for investment by careful families such as the Rawdons and Southwells.[135] It is equally challenging to find out how returning travellers bore themselves, away from published caricatures of fops and macaronis. So many had learned deportment as well as speech abroad, however, that it must have motivated at least some of the approving comments on a young man's 'person' by the families of potential brides.[136]

If John North chose clothing, food and music that advertised his travels to others in Elizabethan London, young men continued to reflect their experiences in changing patterns of consumption as well as speech and deportment. British elites already ate and drank continental—especially French—food and wine by the eighteenth century and many sought fashionable Parisian clothing, but travellers often came home with renewed enthusiasm.[137] Irwin kept an Italian servant and recruited a French cook when he set up a household in London soon after his return, no doubt

[130] Horn, *British Diplomatic Service*, pp. 87–9.
[131] Lambley, *Teaching and Cultivation*, pp. 366–8.
[132] Hannah Smith, *Georgian Monarchy: Politics and Culture, 1714–1760* (Cambridge: Cambridge University Press, 2006), p. 204.
[133] When one woman met a Frenchman at a London ball in 1771, she took him 'for some Englishman, for French is the common masquerade talk': *A Series of Letters of the First Earl of Malmesbury, His Family and Friends, from 1745 to 1820*, ed. James Howard Harris, Earl of Malmesbury (London: R. Bentley, 1870), p. 228.
[134] Gallagher, *Learning Languages*, pp. 187–8; Gallagher, 'Italian London'.
[135] Cohen, *Fashioning Masculinity*, pp. 59–60.
[136] On physical maturity acquired during travel, see Sarah Goldsmith, 'Imagining Youth: Epistolary Representations of the Eighteenth-Century Adolescent and Youthful Body, *c.*1740–1800', in Sarah Goldsmith, Karen Harvey and Sheryllynne Haggerty, eds, *Letters and the Body, 1700–1830* (London: Routledge, 2022).
[137] Conway, *Britain, Ireland and Continental Europe*, pp. 120–1, 204–5.

remembering his enjoyment of dishes prepared 'after the french fashion' by Huguenot hosts in Delft.[138] Before his untimely death, he subscribed to foreign prints and imported continental wines.[139] Henry Ingram similarly sent wine and books ahead of him as he ended his French travels a few years later.[140] His brother, Rich, wondered whether 'the defferent & pleasing entertainments you've meet with in France, has not alter[ed] your Goust', but nevertheless offered 'an opertunity of fox hunting with two extrodinary good packs of hounds' at Temple Newsam.[141] The sense persisted that travel suited young men to metropolitan rather than country living. Changed tastes probably also favoured repeated trips across the Channel: Irwin saw France in 1712–13, having missed it on his first voyage, and Rich had planned a similar journey a decade after his time in the Netherlands, as long as he could 'get an agreeable company'.[142]

By the 1730s, some young men had institutionalised the performance of recent travel and the sustenance of its connections through the Society of Dilettanti. These impulses were of long standing, but the Society passed them through the lens of mid-eighteenth-century London 'clubbability'. Members relived the connoisseurship and excesses of travel at 'semi-secret' gatherings, while rumours about their conduct only strengthened bonds forged abroad. George Knapton's portraits of the Dilettanti, enjoying the pleasures of Italy in fancy dress, hung in their public meeting room as a projection of their shared identity. Sir John Rawdon, who returned from the Continent in 1740, was among the earliest members. His picture commemorates his promiscuity, as he points from behind at the Callipygian Venus ('la Vénus aux belles fesses') in Rome (Figure 5).[143] It offers a reminder that the behaviour of returning travellers, within and beyond the Society, was never cut off from published stereotypes. While the negative role models of the fop and the macaroni lay behind many parental warnings, the Dilettanti revelled in

138 WYAS, WYL100/C/9A: John Haccius to John Roades, 18 Jun [1704]; WYL100/C/10: Matteo Bianchi to same, 9 Dec 1708; same to same, 29 Mar 1709.

139 WYAS, WYL100/C/10: L. Yersin to [Robert Hopkinson], 28 Jul 1711; WYL100/C/11: William George Pennington to Dowager Viscountess Irwin, 18 Oct 1718.

140 WYAS, WYL100/C/11: Walsh and Clarke to Henry Ingram, 27 Dec 1717.

141 WYAS, WYL100/C/20, fo. 340: [Rich, fifth Viscount Irwin] to Henry Ingram, 19 Oct [1718].

142 WYAS, WYL100/C/20, fo. 326: [Rich, fifth Viscount Irwin] to Henry Ingram, 25 Jun [1717].

143 Jason M. Kelly, *The Society of Dilettanti: Archaeology and Identity in the British Enlightenment* (New Haven, CT, and London: Yale University Press, 2009), pp. 38–9, 42, 56, 59; Bruce Redford, *Dilettanti: The Antic and the Antique in Eighteenth-Century England* (Los Angeles, CA: J. Paul Getty Museum, 2008), pp. 36–7.

Figure 5 Portrait of Sir John Rawdon, fourth Baronet (1720–93), later Earl of Moira, by George Knapton. © The Society of Dilettanti. Not to be reproduced without the permission of the Society of Dilettanti.

them as a rejection of middling sociability and an assertion of their standing, unbound by conventional morality.[144] Histories of the Society might take care to distinguish the activities of a small group, founded only in the 1730s, from the wider experiences of returning travellers.[145] Nor should they assume that educational travellers were already a rarefied elite, since the requirement of having seen Italy ruled out so many with continental experience. For Rawdon, membership indicated how far his family had come since his grandfather's French travels and the financial problems of his father's childhood. The Dilettanti do, though, reflect the wider truth that performance was as much for fellow travellers as those who had stayed at home.

[144] Stedman, *Cultural Exchange*, p. 254; Kelly, *Society of Dilettanti*, pp. 19–21, 34–6, 56.
[145] Kelly, *Society of Dilettanti*, p. 18: 46 members in 1736.

Souvenirs, Commissions and Gifts

The one area in which returning travellers are well studied is the collection of art, particularly from Italy.[146] This tendency is understandable, as pictures and statuary offer more tangible results of travel than speech, deportment or topics of conversation, and more obvious artefacts than the networks forged or strengthened abroad. It must now be clear, though, that art history can only offer a partial view. Redford exaggerates his notion of an elite 'defined' by the antiquities, copies, portraits, landscapes, prints and drawings acquired on the Grand Tour, given the skills and connections also brought back from the Continent, let alone the many other ways, beyond travel, in which families could turn money into social and cultural capital. He is nevertheless right that the 'cultural power' of travel relied on display, perhaps even that 'travel abroad existed for the sake of ostentation at home'.[147] But this recognition demands that souvenirs, like the French language or social connections, be treated as an intended outcome of family strategy: they were subject to the means and ambitions of travellers, and acquired and distributed for social purposes.

Most historians focus on domestic architecture and interiors, presenting country houses shaped by foreign inspiration and filled on 'shopping expeditions' that had 'more to do with home decoration' than education.[148] These methods of advertising travel were important ones, of course, and enhanced the role of the country residence as a 'powerhouse' of landowning hegemony.[149] They find a prime example in Irwin, who spent large sums abroad on 'pictures Books and the like' for 'the ornament of my house'. His Venetian spree would furnish 'my great rambling house with excellent paintings', enlarging a collection still held at Temple Newsam.[150] These purchases did not include a portrait, but he had already commissioned

146 Chaney, *Evolution of the Grand Tour*; Edward Chaney, ed., *The Evolution of English Collecting: Receptions of Italian Art in the Tudor and Stuart Periods* (New Haven, CT, and London: Yale University Press, 2003); Wilton and Bignamini, eds, *Grand Tour*; David R. Marshall, Susan Russell and Karin Wolfe, eds, *Roma Britannica: Art Patronage and Cultural Exchange in Eighteenth-Century Rome* (London: British School at Rome, 2011); Sánchez-Jáuregui and Wilcox, eds, *English Prize*.

147 Redford, *Venice and the Grand Tour*, pp. 16, 65, 81.

148 Brewer, 'Whose Grand Tour?', p. 55; Jon Stobart, 'Introduction: Travel and the British Country House', in Stobart, ed., *Travel and the British Country House: Cultures, Critiques and Consumption in the Long Eighteenth Century* (Manchester: Manchester University Press, 2017), pp. 1–18, at pp. 5–8.

149 Stobart, 'Introduction', in Stobart, ed., *Travel and the British Country House*, p. 2.

150 WYAS, WYL100/C/10: Irwin to Dowager Viscountess Irwin, 22 Jan 1706/7; James Lomax, *Temple Newsam Paintings* (Leeds: Leeds Museums & Galleries, 2000), pp. 5, 28–9.

paintings of himself and his tutor, Haccius, against Italian-style backdrops in the Netherlands (Figures 6 and 7). The younger Edward Southwell started from a lower base, as the first in his family to visit Italy in over 60 years, but he bought freely in Rome. In 1726, he 'spent 150£ in 5 marble tables, 2 Landscipes of Ruins, a little Suite of Brass Medals ... 50£ worth of Prints of Modern and Antique Rome and of the Cheif Paintings, 2 or 3 fans, 2 or 3 Cameo's &ca'.[151] His purchases reflected a significant rise in family means, from an annual rental of around £1,000 during Robert Southwell's travels of 1659–62, and a total income of £2,000 at his death in 1702, to an estate of £6,000 a year when the younger Edward was in Rome.[152] Sir John Rawdon similarly showed augmented spending power around 1740, when observers thought he made most purchases only 'because some body or other wanted to sell'.[153] At his return, his uncle wished to 'tie you up from buying Pictures, which make more Bubbles, & ruin more young men, then all their other expences'.[154]

Italian portraits are usually a staple of work on the Grand Tour, as proof of its 'transformative effect' on the status of returning travellers.[155] Their acquisition was surprisingly patchy, though, with no evidence that Irwin, Rawdon or Edward Southwell sat for an artist in Italy, despite their high spending. Italian portraits were relatively rare in the mid-seventeenth century, and not yet draped in classical imagery, but Robert Southwell employed the Medici court painter, Justus Sustermans, and still displayed the result at home some 30 years later.[156] His nephew, Sir Philip Perceval, had two portraits painted in Rome by the Flemish artist Jacob Ferdinand Voet, sending the smaller one to his mother and presenting the larger to his uncle Southwell.[157] Commissions grew from the 1720s, when Alan Brodrick sat for Rosalba Carriera in Venice, and peaked with Pompeo Batoni, who painted at least 175 British visitors to Rome between the 1750s and the 1780s, of whom 150 were men and around half in their early twenties. The same

[151] BL, Add MS 47031, fo. 140v: Edward Southwell II to Sir John Perceval, 9 Apr 1726.

[152] *Egmont Diary*, i, p. 119; Barnard, *New Anatomy*, p. 61. Said to be £12,000 a year when the third Edward was in Italy: *Montagu Letters*, iii, pp. 223–5: Lady Mary Wortley Montagu to Lady Bute, 26 Sep 1759.

[153] *Montagu Letters*, ii, p. 458: Lady Mary Wortley Montagu to Lady Bute, 22 Jun 1752.

[154] PRONI, D2924/1/89–90: Sir Richard Levinge to Sir John Rawdon, 11 Nov 1741.

[155] Redford, *Venice and the Grand Tour*, p. 81.

[156] BNCF, MS Gal. 254, fo. 192: Robert Southwell to Vincenzio Viviani, 1 Aug 1661; YCBA, MS 8: Inventory of King's Weston, 1695; Stoye, *English Travellers Abroad*, p. 146, 149.

[157] BL, Add MS 46955A, fos 176–83v: Sir Philip Perceval, financial accounts, 1677–8; Add MS 46955B, fo. 36: Kate Perceval to Helena Southwell, 21 Jul 1678; YCBA, MS 8: Inventory of King's Weston.

Figure 6 Portrait of Edward Ingram, fourth Viscount Irwin (1686–1714) by Thomas van der Wilt, Temple Newsam House. © Leeds Museums and Galleries/Bridgeman Images.

Figure 7 Portrait of John Haccius by Thomas van der Wilt, Temple Newsam House. © Leeds Museums and Galleries/Bridgeman Images.

proportion were already peers or destined to become so.[158] While Batoni developed a form that is familiar from any publication on the Grand Tour, its ubiquity is misleading. Previous chapters have shown the inadequacy of seeing the typical educational traveller as a young man who has ventured abroad only in the mid-eighteenth century; who poses on a tour of Italy, often against a classical backdrop; who betrays little of his preparations or additional destinations; who is isolated from the hopes and fears of his family, from the local population and often from his fellow travellers; and who is unquestioningly able to afford such a portrait in the first place.

To begin placing souvenirs in the context of family strategy, it is important to recognise the true range of objects that travellers acquired. Journals and paintings, the standard sources for the Grand Tour, stood alongside 'various accumulations of books, manuscripts, engravings, coins, medals, and other curiosities', but the passage of time and modern collecting practices have shattered these coherent representations of personal experience.[159] Robert Southwell sat for a portrait, but why should it stand apart from the books and elixirs that he sent back from Florence in the same set of boxes?[160] Southwell and his family used a 'Nerve oyl' given by the Grand Duke for years afterwards, while a 'Turky stone' (turquoise) and other items from the Italian chemist Francesco Borri were still on display at King's Weston three decades later.[161] Each of these objects had a role to play in documenting his travels for himself and those at home. Souvenirs might even plug gaps in experience, as in the case of Southwell's son, who bought pictures in Antwerp that 'well represented' those Brussels landmarks too damaged by the French bombardment of 1695 for him to visit the following year.[162] Travel journals are limited sources, as proof of productive 'tours' rather than comprehensive accounts of time abroad, but an additional reason for their limitations is that their authors assumed the parallel display of souvenirs at home.

158 SHC, 1248/6/245: Alan Brodrick II to Thomas Brodrick, 1 Jun 1725; Redford, *Venice and the Grand Tour*, pp. 82, 85; Edgar Peters Bowron and Peter Björn Kerber, *Pompeo Batoni: Prince of Painters in Eighteenth-Century Rome* (New Haven, CT: Yale University Press, 2007), pp. 37–88.

159 Brennan, ed., *Origins of the Grand Tour*, pp. 37–47.

160 BNCF, MS Gal. 254, fos 191–191v: Robert Southwell to Vincenzio Viviani, 1 Aug 1661; fos 195v–196: same to same, 10 Aug 1661.

161 BL, Egerton MS 1634: Le Grand, 'Life of Southwell', fos 11v–12v, 19v; Add MS 46941A, fo. 91v: Sir John Perceval to Robert Southwell, 6 Apr 1664; 46941B, fos 51–52v: Southwell to Perceval, 30 Aug 1664; YCBA, MS 8: Inventory of King's Weston.

162 BL, Add MS 58220: Edward Southwell, Journal of a visit to the Netherlands, i, fos 6v–7.

This range of acquisition meant that travellers could fit purchases to the family budget. Most scholars do not think of souvenirs in this way, but they were part of a wider investment in travel, which also involved decisions over itineraries or the timing of a voyage, and they similarly varied according to means and ambitions.[163] Beyond the most eye-catching purchases, much collecting abroad was, as Stoye puts it, 'the accumulation of small odd things' by 'unexceptional people'.[164] Samuel Waring is a prime example, buying prints and engravings to frame at home rather than paintings or statuary.[165] The Roman souvenir industry had recognised such lighter purses since the mid-seventeenth century, offering books of views and other affordable options.[166] In addition, a range of objects, from stones collected in the catacombs to a licence for eating meat during Lent, attested to Waring's experiences in Italy.[167] He shows how text and object could work together to document travel, ending an account of popular Italian games by confessing that 'I can neither know theire cards nor theire way of playing with them: I have purchased some of them.'[168] Similarly, Waring had already written of his interest in German confessional relations when he noticed a Dominican in Mannheim giving each of his pupils 'a litle knack, a crucifix or sort of lockett with the virgin Marys Picture & our Saviours in it'. He paid a boy two pence for one of the objects but the catechist 'bidd him gett it from me for I would burne it being a protestant, soe that the child crying I was forced to restore it'.[169] Little now remains of Waring's miscellaneous souvenirs, even those he managed to get home, but he offers the perspective of a traveller who only reached Italy through patronage and, of necessity, memorialised his journey on a budget.[170]

Even for the wealthiest travellers, it was difficult to bring purchases home. Besides customs duties, which stood at 20 per cent for prints in 1725, tempting Martha Courthope to write a 'hard word', it was particularly challenging to convey souvenirs in wartime.[171] It took three years for

163 Some recognise that lavish purchases required extraordinary budgets: Chaney, *Evolution of the Grand Tour*, p. 213; Black, *British Abroad*, p. 302.

164 Stoye, *English Travellers Abroad*, p. 163.

165 Barnard, *Making the Grand Figure*, pp. 177, 317.

166 Collins, 'Sites and Sightseers', p. 580.

167 PRONI, D695/214: Ridolfo Guidarelli to Samuel Waring, 7 Mar 1688; D695/226: Waring, travel journal, p. 11.

168 PRONI D695/228: Waring, travel notebook, pp. 22–3.

169 PRONI, D695/225: Samuel Waring, travel journal, 1687, pp. 30–1; 226: Waring, travel journal, p. 11; 228: Waring, notebook, p. 23.

170 Brennan, ed., *Origins of the Grand Tour*, pp. 37–47, for others of 'relatively modest means'.

171 SHC, 1248/6/205: Martha Courthope to Alan Brodrick II, 23 Apr 1725.

Irwin to get his possessions back from Italy during the War of the Spanish Succession, having sent two boxes of books via Livorno and London and another shipment of books and 40 pictures through Rotterdam and Hull. A few paintings were lost and another 'throwune about as of no Vallue' during long delays, while two harpsichords only arrived in Yorkshire five years after Irwin had bought them in Delft.[172] Sir John Perceval lost many of his souvenirs during the same conflict, when French privateers took two ships loaded with 'things that he took much delight in and enterd into the study of in Italy'. Perceval later wrote that he had planned to donate this 'valuable collection of books paintings statues antiquities drafts & musick' to Sir Godfrey Kneller's academy of painting or use them to establish an Irish equivalent, although these claims were never tested.[173] His losses, and Irwin's delays, serve as a reminder that such men also displayed their travels in less tangible ways.

Travellers bought items as gifts and commissions, as well as personal souvenirs, using them to enhance the prestige of their families and strengthen connections among the elite. In this way, foreign purchases were part of family strategy not only in how they were funded, but also in what they were supposed to do. Travellers, like the diplomats discussed in Chapter 4, could give friends and relations access to prestigious foreign markets, not least in art. William Leigh sent James Brydges a sale catalogue from Paris, but his uncle initially refused to buy any pictures, citing uncertain quality, likely cost and the difficulty of sending them home. Brydges was already buckling a week later, though, confessing that 'the names of the masters would tempt me to endeavor to be owner of some of them', 'provided it is certain they are originals'. He sent a list of pictures by Poussin, Raphael, Schiavone, Michelangelo and Domenichino, on which he was willing to spend a total of around £425, before lowering his offers and adding several more by Reni and Carracci in the next letter. The Reni, at least, was in the Cannons library 12 years later, alongside a few unattributed works that may also have come from Paris; Brydges certainly bought more than one. Notably, he depended upon Leigh's judgement of both the pictures themselves and an appropriate

[172] WYAS, WYL100/C/9A: Irwin to John Roades, 8 Aug 1704; WYL100/C/10, Thomas Palmer to same, 29 Jun 1706; [same] to same, 28 Sep 1706; R. Graves to same, 23 Dec [1708]; Richard Hey to same, 1 Nov 1709; WYL100/EA/12/17/5: Joseph Hey to same, 20 May 1709.

[173] BL, Add MS 47025, fo. 70v: Philip Perceval to Sir John Perceval, 7 Feb 1706/7; 47072: Sir John Perceval, Autobiography, p. 23, fo. 103v. Kneller did not open St Luke's Academy until 1711, so Perceval may not have planned to donate his souvenirs.

gift for the auctioneer.[174] Thomas Brodrick similarly instructed Alan to buy art in Italy, looking to expand his impressive collection at Wandsworth House 'because I shall never have such another opertunity'. Brodrick initially promised that 'My ambition must give way to my purse', but soon admitted that 'if the pictures must bee payd for through the nose, I must submitt to itt'. He was willing to spend up to £300, ideally on portraits by masters, but would take 'moderns', 'extraordinary' copies or histories that were 'cheape and very good'. Like Brydges, Brodrick trusted his nephew on what to buy. Alan spent the majority, £260, on five paintings by Poussin, Carracci, Balestra, Luti and Pasinelli, using a standard combination of the dealer Owen Swiney and the merchant Joseph Smith. Alongside the costs of transportation and duties, he may have bought more before leaving Italy. For himself, Alan sent home more modest prints alongside his own portrait and a copy of another by Carriera.[175] Brydges and Brodrick both used travelling nephews, as Martha Courthope put it, to set themselves up in 'picture affairs'.[176]

Objects acquired abroad could enhance a traveller's prospects if distributed wisely at home. Sir John Perceval recognised an opportunity when Robert Southwell sent him 'the fairest Parmizan Cheese I ever saw', giving the 'token' to the Duchess of Ormond.[177] Southwell used very different souvenirs of Italy to strengthen his connection to the well-travelled natural philosopher Robert Boyle, a Cork neighbour whom he must have met at Oxford.[178] When Henry Oldenburg, tutor to Boyle's nephew, encountered Southwell in Paris, he deemed him 'very ingenious and well inclined, apt to make good use of travelling'.[179] Their circle, which would soon promote the intellectual value of travel through the Royal Society, sought to profit from

174 HL, mssST57, vol. 9, fos 36–7: James Brydges to William Leigh, 24 Mar 1712/13; fo. 64: same to same, 3 Apr 1713; fo. 103: same to same, 26 May 1713; fo. 128: same to same, 10 Jul 1713; fo. 199: same to same, 23 Sep 1713; '"An Inventory of His Grace the Duke of Chandos's Seat att Cannons Taken June the 19th 1725" by John Gilbert', ed. Susan Jenkins, *The Volume of the Walpole Society*, 67 (2005), 93–192, at 126. Livres converted according to rates in Black, *British Abroad*, p. xii.

175 SHC, 1248/5/370: Thomas Brodrick to Alan Brodrick II, 28 Jan 1724[/5]; 1248/6/82v: same to same, 6 Nov 1724; fo. 97: same to same, 19 Nov 1724; fo. 110: Alan Brodrick II to Thomas Brodrick, 22 Dec 1724; fo. 219: same to same, 4 May 1725; Ingamells, *Dictionary*, p. 131. For Thomas's collection, see George Vertue, 'Vertue Note Books: Volume III', *The Volume of the Walpole Society*, 22 (1933–4), 9; Hayton, *Anglo-Irish Experience*, p. 95. On Smith and Swiney, see Ingamells, *Dictionary*, pp. 869–71, 919–20.

176 SHC, 1248/6/205: Martha Courthope to Alan Brodrick II, 23 Apr 1725.

177 BL, Add MS 46938, fo. 148: Sir John Perceval to Robert Southwell, 22 Nov 1661.

178 Michael Hunter, *Boyle: Between God and Science* (New Haven, CT, and London: Yale University Press, 2009), pp. 43–56, 92–103.

179 *Oldenburg Correspondence*, i, p. 252: Henry Oldenburg to Robert Boyle, 17 May 1659.

Southwell's time abroad.[180] The young man sent Boyle Italian samples and experimental methods, tried to set up a formal correspondence with the Tuscan court, searched for manuscripts by Galileo and collected a range of curiosities, from a magnifying glass and a handful of Pope Urban's gallstones in Rome to products from Florence's famous apothecaries.[181] Southwell hoped that these efforts would 'testify something of my progresse abroad' and promised Boyle a 'full account' in London. Alongside a thermometer presented in person, they surely encouraged his new patron to sponsor him as a fellow of the society a few months after his return.[182] This endorsement favoured Southwell's general progress, given the overlap between the fellowship and the servants of the restored monarchy, and demonstrates the role that international connections among natural philosophers could play in individual advancement.[183]

More widely, returning travellers distributed a range of consumer goods. Gailhard recommended buying 'some of the Rarities of Towns and Countries, which are less cumbersome and easier to be carried' in order 'to shew and keep them when they are come home', but such purchases also found their way to relations, friends and patrons as tokens of specific destinations.[184] Tea often seems to have represented the Netherlands, amid travellers' interest in the Dutch East India trade. When Edward Southwell was invited to visit the Dowager Countess of Gainsborough after his return from the Low Countries in 1696, he realised while still abroad that he should find 'some good Tea or something of that kind to carry with mee to my Lady to shew I have been travelling'.[185] A few years later, probably during his time in Delft, Rich Ingram also promised a cousin to 'send her some Tee as soon as I can gett some mony'.[186] Neighbouring Flanders meanwhile offered a range of fabrics. While Irwin filled Temple Newsam with Venetian artwork,

180 Carey, 'Compiling Nature's History'; Iliffe, 'Foreign Bodies'.

181 BL, Egerton MS 1632: Southwell, Italian notebook, fo. 44v; BNCF, MS Gal. 161, fo. 307: Robert Southwell to Vincenzio Viviani, 18 Mar 1661; fo. 330v: same to same, Oct 1661; 254, fos 195v–196: same to same, 10 Aug 1661; *Boyle Correspondence*, i, p. 430: same to Robert Boyle, 10 Oct 1660; p. 459: same to same, before Sep 1661; Boschiero, 'Southwell and Viviani'.

182 BL, Add MS 38015, fos 23–23v: Robert Southwell to Robert Boyle, draft letter, [Jan 1661]; W. E. Knowles Middleton, *A History of the Thermometer and Its Use in Meteorology* (Baltimore, MD: Johns Hopkins University Press, 1966), pp. 38–9; Toby Barnard, 'Southwell, Sir Robert (1635–1702)', *ODNB*, online edn, 2004 (https://doi-org/10.1093/ref:odnb/26066).

183 Aylmer, *Crown's Servants*, p. 162; Boschiero, 'Southwell and Viviani', 90, 98, 108.

184 Gailhard, *Compleat Gentleman*, ii, pp. 141–2.

185 Koninklijke Bibliotheek, MS 133c.6: Edward Southwell to Sir Robert Southwell, received 30 Jul 1696.

186 WYAS, WYL100/C/20, fo. 203: Rich Ingram to Dowager Viscountess Irwin, [late 1704/early 1705].

his former companion, Thomas Worsley, used the same merchant to send home muslin and lace from his less ambitious voyage.[187] James Brydges similarly used Henry's time in the Low Countries to buy various cloths, noting that the kinds of cambric 'proper for my wear' were 'those made at Cambray'.[188]

Travellers to Paris received requests for more fashionable items. Brydges asked Leigh to find 'Silk Brocades of the finest & newest fashion of two different sorts, for Manteau & Petticoats', as well as 'a couple of the best & newest fashioned scarfs' neither 'too gay & airy, nor on the other side too grave, they being design'd for a Lady neither very young nor old'. His nephew correctly inferred that he was about to remarry.[189] Arthur Ingram asked his cousin, Henry, to buy a 'very fine' fan for his wife, since 'the women you know think those things are more in perfection att Paris then elsewhere', but he also requested 'two or three pair of beads to wear with neckcloaths' for himself.[190] Work on the Grand Tour gives the impression of Rome, meanwhile, as a marketplace for art and antiquities, which it obviously was, but travellers and their correspondents also knew cheaper products for which the city was renowned. Robert Southwell used Roman gloves to build up his connections, giving them to two Italian academics and an unnamed Englishman, while Sir Philip Perceval sent two dozen to his mother and sisters in France.[191] Orrery probably also bought his 'one Dozen and ½ of womens gloves' for relations, while Charles Butler and Samuel Waring made a trip to buy 'gloves, snuff & washballs Roman'.[192] Southwell's daughter, Helena, even asked Sir John Perceval to buy 'a couple of razors' in Rome for an acquaintance, 'that being a famous place for 'em'.[193] These small purchases—tea, lace, fans, gloves—spread the prestige of travel to those at home, notably including female friends and relations, and they demonstrated

[187] NYCRO, ZON13/1/286: Richard Hey to Thomas Worsley, 2 Feb 1706.

[188] HL, mssST57, vol. 25, fo. 226: James Brydges to Andrew Robinson, 13 Feb 1724/5.

[189] HL, mssST57, vol. 9, fo. 36: James Brydges to William Leigh, 24 Mar 1712/13; fo. 64: same to same, 3 Apr 1713; fos 102–3: same to same, 26 May 1713; SBT, DR18/20/18b: William Leigh to Theophilus Leigh, 20 Sep 1713.

[190] WYAS, WYL100/C/12: Arthur Ingram of Barrowby to Henry Ingram, 31 Mar 1718; same to same, 5 Jun 1718.

[191] BNCF, MS Gal. 254, fo. 178: Robert Southwell to Vincenzio Viviani, 18 Jan 1661; 161, fo. 299: same to same, 19 Feb 1661; BL, Add MS 46956A, fos 122–48v: Alexandre de Rasigade, accounts, 5 Jun 1679.

[192] NLI, MS 13215, fo. 48: Orrery, travel expenses, 1685–9; PRONI, D695/228: Waring, travel notebook, p. 49.

[193] BL, Add 47025, fos 66v–67: Helena Le Grand to Sir John Perceval, 17 Jan 1706/7.

continental experience to potential patrons. In everyday situations, they projected the image of a family that travelled.

The Afterlives of Travel: Reading and Remembering

Books have always featured prominently in treatments of the Grand Tour, but recent work has confirmed their importance.[194] Previous chapters have shown young men reading in preparation for travel and learning from books while abroad, especially when under the eye of tutors. During more mobile periods, print structured both itineraries and representations: Alan Brodrick only went to Frankfurt because 'every Book of travels thro' this part of the world, makes mention of it', while Sir John Rawdon referred correspondents to published descriptions of Versailles.[195] Books also functioned as souvenirs, standing alongside journals and other purchases as ways in which travellers documented their time abroad for themselves and for posterity.[196] Robert Southwell received two small volumes on Rome from a local mathematician and he asked Viviani to get him Italian books on ballet, comedy and glass-making.[197] The financial accounts of his nephew, Sir Philip Perceval, reveal that he commemorated visits by buying 'the book of the Place', whether a guidebook or a history, from Saint-Denis, Richelieu and La Rochelle to Rome, Naples and Loretto.[198] His contemporary, Sir Thomas Isham, brought home 358 books, mostly in Italian and mostly on art and architecture.[199] Recent work has charted the impact of such purchases on country-house

[194] Abigail Brundin and Dunstan Roberts, 'Book-Buying and the Grand Tour: The Italian Books at Belton House in Lincolnshire', *The Library*, 16 (2015), 51–79; María Dolores Sánchez-Jáuregui, 'Books on the Westmorland', in Sánchez-Jáuregui and Wilcox, eds, *English Prize*, pp. 144–53.

[195] SHC, 1248/6/62: Alan Brodrick II to Sir Alan Brodrick, 4 Sep 1724; PRONI, D2924/1, fo. 44: Sir John Rawdon to Thomas Prior, 23 Apr 1738. This point is commonplace: for instance, Sweet, *Cities and the Grand Tour*, pp. 4–5.

[196] Jill Bepler, 'Travelling and Posterity: The Archive, the Library and the Cabinet', in Babel and Paravicini, eds, *Grand Tour*, pp. 191–203, at pp. 195–6.

[197] BNCF, MS Gal. 161, fo. 302v: Robert Southwell to Vincenzio Viviani, 4 Feb 1661; 254, fos 191–191v: same to same, 1 Aug 1661. None were in the family collection by the nineteenth century, though perhaps a 1659 edition of Florio's Italian dictionary had travelled: *Catalogue … Containing the Greater Portion of the United Libraries of Sir Robert Southwell … his Son, the Hon. Edward Southwell, and the Late Lord De Clifford* (London: Thomas Thorpe, 1834).

[198] For Perceval's expenses, see BL, Add MS 46953, fos 190–1; 46954A, fos 206–206v; 46955A, fos 176–183v; 46956A, fos 122–148v. On the role of books in his travels, see Gallagher, 'Language and Education', pp. 122–7.

[199] Keith S. Dent, 'The Informal Education of the Landed Classes in the Eighteenth Century, with Particular Reference to Reading', Ph.D. thesis (Birmingham, 1974), p. 143.

libraries and considered books as one way in which returning travellers displayed their 'intellectual and moral superiority'.[200]

Books were not just for the decoration of library shelves, however; they were working souvenirs. Perceval channelled his tutor, Gailhard, when he recorded that, since every country had a 'genius' for particular kinds of writing, a collection of 'the best and most curious books' from each destination would 'entertaine your selfe' at home 'and keepe your selfe from forgetting what languages you learned in your travells'.[201] Returned travellers throughout the seventeenth and eighteenth centuries similarly used foreign books to maintain their linguistic improvements, a practice that helps to explain how men such as Sir Robert Southwell continued to read, write and speak French and Italian in the years afterwards.[202] Many enlarged their collections of foreign texts after travel, some through London agents and others, including the Southwells and Percevals, by making use of later travellers.[203] Several studies find that such libraries existed in a feedback loop, as foreign books helped to prepare the next generation for travels that would in turn expand a family's collection.[204] Abigail Brundin and Dunstan Roberts argue that libraries spread the benefits of travel more widely, finding that sisters, servants and the local gentry borrowed Italian books at Belton House for language-learning and 'collective enjoyment'.[205] These practices were probably common, as another way in which travellers displayed their cultural capital at home. Print helped to realise the ambitions with which parents sent sons abroad, reinforcing foreign improvements,

200 Mark Purcell, *The Country House Library* (New Haven, CT: Yale University Press, 2017), pp. 119, 128; Sánchez-Jáuregui, 'Books on the Westmorland', p. 151.

201 BL, Add MS 47113: Perceval, Commonplace book, fo. 9v. Gailhard soon made the same point in print: *Compleat Gentleman*, ii, pp. 192–3.

202 Stoye, *English Travellers Abroad*, pp. 136, 162; Gallagher, 'Language and Education', pp. 125–6; Brundin and Roberts, 'Book-Buying', 77. Southwell owned manuscripts in French, Italian and Spanish: BL, Add MS 38536: Catalogue of Sir Robert Southwell's papers, 1692; Royal Society, London, MS 248: Papers collected by Sir Robert Southwell, 17th century; Bodleian, MS Eng. Misc. c 144: Collection of treatises transcribed for and annotated by Sir Robert Southwell, late 17th century. For letters in French: BL, Add MS 38015: Sir Robert Southwell, Correspondence, 1659–97; Egerton MS 2127: Correspondence with Henri Justel, 1684–8. Southwell may still have spoken Italian in 1664: BL, Add MS 46941A, fo. 19v: Lady Perceval to Robert Southwell, 30 Jan 1663[/4].

203 BL, Add MS 46956A, fo. 11: Alexandre de Rasigade to Sir Robert Southwell, 25 Jan 1679; 47025, fo. 78v: Edward Southwell to Sir John Perceval, 30 Sep 1707; Brundin and Roberts, 'Book-Buying', 68–9, 76.

204 Bepler, 'Travelling and Posterity', pp. 193–4; Brundin and Roberts, 'Book-Buying', 70; Dent, 'Informal Education', p. 341.

205 Brundin and Roberts, 'Book-Buying', 53–4, 78–9.

making them available to friends and relations and demonstrating family access to continental knowledge.

Samuel Waring illustrates the role of books in the self-fashioning of a returning traveller.[206] He enjoyed an unusual relationship with print, having gone on the journey that would become Maximilien Misson's famous *New Voyage to Italy*, but he offers a way into the wider uses of reading before, during and after travel. Although little survives of Waring's library, which he kept in a jealously guarded closet, one of his notebooks contains a priced list of 47 books, 12 pamphlets and eight plays. Its military textbooks, architectural works, grammars and well-established travel accounts may represent purchases made in preparation for travel, perhaps even in England under the guidance of Misson, but other entries, such as an assortment of Italian plays and guides to Naples and Pozzuoli, are more likely souvenirs.[207] Under 'Pamphlets', for instance, Waring lists items—'Reliques of Aix le Chapell', 'Habits of Strasburg'—of the kind produced by enterprising locals. He purchased more books after his return, judging by the likely publication dates of some entries. Waring could not have bought the unattributed 'Remark on Italy' before his travels of 1687–8 if it was by either William Bromley (1692) or Joseph Addison (1705), while the 'Delights of Holland' may be William Mountague's 1696 book. The alphabetical list, which leaves gaps for future purchases, includes a long-term record of buying travel-related material before, during and after his journey.[208]

Waring read in order to relive his travels and put them to work at home. Most evidence appears within or alongside draft letters to Misson, as part of a complicated notebook to which Waring returned regularly between 1687 and at least 1714.[209] He got hold of successive editions of *A New Voyage to Italy*, 'runing them over in my chamber where I endulgd my fancie to the utmost', and often read the text to 'refresh my memory in the Ideas I then received'. Indeed, he went through the 'whole booke … yearly'. Waring engaged actively with Misson's texts as the years passed, comparing changes to the 1695, 1699 and 1714 editions.[210] He loosely translated an Italian letter

[206] For a fuller exploration, see Ansell, 'Reading and Writing Travels'.
[207] Toby Barnard, *Brought to Book: Print in Ireland, 1680–1784* (Dublin: Four Courts, 2017), pp. 40–1, 334.
[208] PRONI, D695/228: Waring, travel notebook, pp. 84, 86–90; Bromley, *Remarks in the Grande Tour of France & Italy*; Joseph Addison, *Remarks on Several Parts of Italy, &c. In the Years 1701, 1702, 1703* (London: 1705); William Mountague, *The Delights of Holland: or, A Three Months Travel about that and the other Provinces* (London: 1696).
[209] PRONI, D695/227: Waring, notebook; Ansell, 'Reading and Writing Travels', 1466–7.
[210] PRONI, D695/227: Waring, notebook, pp. 4, 30–1, 51, 55.

added in 1714, surely armed with his copy of Florio's dictionary so long after his travels, and he filled the blank pages of his notebooks with drawings inspired by Misson's plates (Figures 8 and 9).[211] The draft letters represent Waring's engagement with travel books up to around 1700–5, some 15 years after his journey. Although they concentrate on praising *A New Voyage to Italy*, they rely on reading that was transnational, multilingual and chronologically broad. His list of 'Authors that have writt of Italy' contains Gilbert Burnet, Richard Lassels and John Ray in English, but also the Frenchmen Charles Patin, Pierre du Val, Jacques de Villamont and Jean Dumont alongside Franz Schott of Antwerp. While Waring knew early seventeenth-century authors in Schott and Villamont, he continued to engage with writers such as Dumont, first published in 1694, after his own voyage. He offers Misson verdicts that demonstrate his familiarity with each author, even though only Burnet, Lassels and Schott appear in his booklist, and various indications suggest that he read them in a mixture of English, French and Italian.[212] Waring's intensive reading of *A New Voyage to Italy* arose from his relationship with the author, but it was part of a wider engagement with travel writing in the years after his journey.

Although Waring read in his chamber and kept his books in a closet, reading and discussion turned foreign experience into cultural capital. His association with *A New Voyage to Italy*, particularly after his appearance in the 1699 appendix, guaranteed authority in local circles.[213] Waring met with 'inquistors in relation to' Misson's book, presumably in County Down, Belfast and Dublin, and fielded queries as 'an eye wittness of the whole relation'. He found 'soe many varieties of fancies to be pleased as there are mouths to ask questions withall', telling Misson that he had been quizzed on 'matters that neither you nor I thought of while abroad'.[214] Waring failed to answer only one question among the 'reasonable', finding himself unable to offer 'some parallel betwen the old Rome & the new, as to its former & pleasent [*sic*] splendour & glory and to relate what of the old Rome is now standing'. He told Misson that he had 'taken a litle pains by collection, out of yours & other bookes', to provide an answer that 'seemd as gave them

211 PRONI, D695/227: Waring, notebook, p. 39; D695/228: Waring, travel notebook, pp. 6–7, 84, 86–90; Misson, *New Voyage to Italy*, fourth English edn (London: 1714), ii, pp. 598–602; Ansell, 'Reading and Writing Travels', 1471, n. 145.

212 PRONI, D695/227: Waring, notebook, pp. 4, 9, 52–3; 228: Waring, notebook, pp. 84, 86–90; Ansell, 'Reading and Writing Travels', 1468–70.

213 Maximilien Misson, *A New Voyage to Italy*, second English edn, 2 vols (London: 1699), ii, sigs Hh2–Ii4: letter 'To Samuel Waring, Esq'.

214 PRONI, D695/227: Waring, notebook, pp. 6, 8, 23, 51–2.

Figure 8 'Divers habillemens de Femmes de Nuremberg', Maximilien Misson, *Nouveau voyage d'Italie*, fourth French edn, 3 vols (The Hague, 1702), i, p. 89. Bodleian Libraries, University of Oxford, Douce M 245. The plate first appeared in Misson, *Nouveau voyage d'Italie*, third French edn, 2 vols (The Hague, 1698).

Figure 9 Samuel Waring's sketches of women. Misson had included plates of women before, including one mislabelled as Nuremberg, but Waring's sketches most closely resemble those in the 1698 French edition onwards. PRONI, D695/228, pp. 6–7. By permission of the Deputy Keeper of the Records, Public Record Office of Northern Ireland.

& my selfe some satisfaction, tho I could not at the same time justify the truth or exactness of it either'.[215] The resort to books in order to sustain and embellish memories of travel was common across the seventeenth and eighteenth centuries, from a law student refreshing his account of France to a Cambridge don pursuing his interest in Germany, and the English translation of Germain Brice's 1684 description of Paris offered itself to returned travellers in precisely these terms.[216] Recollections of travel stood alongside conversation and reading as the years passed, uniting personal experience and the printed word in the maintenance of prestigious expertise.

Travel in the Family Archive

If returned travellers read books in order to relive their foreign encounters and transform them into cultural capital, they also came back to their own manuscripts. Travel represents a richly documented time of life, increasingly exploited for all kinds of social and cultural history, because families such as the Southwells and Percevals guarded its traces with care. Others, including the Ingram, Leigh and Brydges families, lack documentation of the first generation of travel but afterwards afford it an equally privileged position. The meticulous archiving of correspondence, financial records and other material relating to travel is another indication of the value that these families placed on successive journeys to the Continent.[217] Few historians recognise, however, that archives relating to travel were built and used before their employment for modern research. These papers offer insights into the afterlives of travel as resources for their creators, as well as for friends and relations in search of vicarious encounters with the Continent. They also provided descendants with materials for planning their own journeys, contributing to the virtuous circle in which successful travel favoured its repetition in future generations.[218] Previous chapters have found the Ingrams comparing Irwin's spending abroad with the records of his predecessors and erasing sexually explicit phrases from the letters of his brother, Henry, presumably because they too would be read by future generations. The

215 PRONI, D695/227: Waring, notebook, pp. 5–6, 8.
216 Germain Brice, *A New Description of Paris*, trans. James Wright (London: 1687), sig. A2r; Claydon, *Europe and the Making of England*, pp. 22–3; D. R. Woolf, *Reading History in Early Modern England* (Cambridge: Cambridge University Press, 2000), pp. 124–5; Conway, *Britain, Ireland and Continental Europe*, p. 201.
217 Williams, 'Inner Lives', 354.
218 Bepler, 'Travelling and Posterity', p. 193.

Southwells and Percevals equally delved into earlier travel correspondence for biographical and genealogical projects. These practices seem to have been common, but the most prominent documents in the years after travel were the diaries, journals and notebooks that had been kept abroad.

As Waring's drawings suggest, travel journals could be open-ended documents on which travellers continued to work in the years afterwards.[219] Three generations of Southwells produced at least 11 texts between 1658 and 1726, with varying evidence of their later usage. There is little indication that Robert returned to his rough commonplace books for France and Italy, compiled in 1659–61, but he evidently shared his preparatory notes with Isaac Newton.[220] If fulfilled, Petty's request for a Dutch 'Routier' of 1661–2 would explain why such a document has left no other trace. Robert's son, the elder Edward, produced at least four diaries over two trips to the Low Countries in the 1690s. The journal for 1691 and two from 1696 are 7 × 2- or 7 × 3-inch notebooks that are pocket-sized but difficult to read, especially the later two with their vertical text, suggesting that they were produced abroad but not designed for sharing.[221] A fourth text, however, covering the beginning of the second journey, is a neat copy in the hand of a family clerk, containing a few corrections by Edward himself, and of a user-friendly 9.5 × 7 inches.[222] Unlike his father, Edward clearly prepared some of his notebooks for posterity or a circle of readers. The family reached a pinnacle of presentation with his own son, the younger Edward, who produced at least four larger, neater diaries of foreign travel, from the 'Journal of Our Journey' with his father in 1723 to 'A Journal of My Journey' in 1725–6.[223] A lack of errors, as well as the consistency of Edward's hand, indicate that these are copies made at home, perhaps from discarded notes. Histories of papal bulls or the Jesuits seem inserted into a reshaped narrative, while gaps left

219 For travel further afield, see Eva Johanna Holmberg, 'Writing the Travelling Self: Travel and Life-Writing in Peter Mundy's (1597–1667) *Itinerarium Mundii*', *Renaissance Studies*, 31:4 (2017), 608–25.

220 King's, Keynes MS 152: Southwell, 'Manuscript concerning travelling'; BL, Add MS 58219: Southwell, French diary; Egerton MS 1632: Southwell, Italian notebook.

221 Beinecke, Osborn MS b66: [Edward Southwell], Holland voyage; BL, Add MS 58220: [Southwell], Journal of a visit to the Netherlands.

222 BL, Add MS 21495: Edward Southwell, Journal of a tour in Holland, 1696; Fremantle, 'Visit to the United Provinces and Cleves', 40–1, n. 4.

223 Clark Library, MS.1975.002: Edward Southwell II, Journal of Our Journey; BL, Add MS 34753: Southwell, Journal of travels in Lorraine and France; BNF, MS Anglais 194: Southwell, Travels; BL, Egerton MS 3805: Southwell, Journal of travels in France and Italy. The last ends in Naples, so a journal of Southwell's return through Germany is probably missing.

for forgotten details speak of ongoing work.[224] These changes may suggest a trend towards formalisation, moving from Robert's miscellaneous notes on conversation to his grandson's bland daily entries, but others had written up their journals at the time of the older man's travels.[225] More likely, the Southwells had self-consciously begun a travel archive in the decades since Robert had first crossed the Channel.

Privileged young men may not have published their journals, a practice that they left to tutors or other middling-sort writers, but work on the vibrancy of manuscript culture offers a reminder that print was not the only way in which travellers could share their experiences.[226] Original or copied journals circulated among select groups of family and friends throughout the seventeenth and eighteenth centuries, a phenomenon that must have shaped how travellers wrote their narratives.[227] If the Southwells only offer hints of circulation, their Perceval relations certainly shared manuscripts within and beyond the family. Although Sir Philip's 'diary' does not survive, John looked forward to borrowing it at home, venturing that 'an elder Brother shall travell for the service & advantage of the younger'.[228] John's son, Sir John, shared his manuscript beyond the family, commending the younger Edward Southwell for keeping 'a Journal which will hereafter [be] usefull to you as it will also be entertaining to those you shew it to'. Sir John recalled that he had done 'the same 19 years ago', but had 'frequently regretted my giving it to a Person who took the same Rout, and after composing out of it any Account of his own travells made away with mine'.[229] The common practice of lending journals may help to explain why historians have underestimated travel among the gentry, if more diaries once existed than now survive in family archives. Regardless, manuscripts by friends or relations stood alongside books in helping readers to plan itineraries or enjoy vicarious encounters with the Continent.

224 BNF, MS Anglais 194: Southwell, Travels, fos 16, 16v, 38, 45, 47v, 57.

225 Williams, 'Inner Lives', 350, 358.

226 Turner, *British Travel Writers*, pp. 16–17; Harold Love and Arthur F. Marotti, 'Manuscript Transmission and Circulation', in David Loewenstein and Janel Mueller, eds, *The Cambridge History of Early Modern English Literature* (Cambridge: Cambridge University Press, 2002), pp. 55–80.

227 Bepler, 'Travelling and Posterity', pp. 191–2; Black, *British Abroad*, pp. v–vi; Brennan, ed., *Origins of the Grand Tour*, pp. 77–9, 158–60, 235–6.

228 BL, Add MS 46955B, fo. 125v: John Perceval to Sir Robert Southwell, 23 Sep 1678. The book in which he recorded Gailhard's teachings contains nothing of his itinerary or impressions: Add MS 47113: Sir Philip Perceval, Commonplace book.

229 BL, Add MS 47031, fo. 72: Sir John Perceval to Edward Southwell II, 12 Jan 1725/6.

The Southwells and Percevals shared expertise most freely with immediate family, who benefited from correspondence and discussion as well as journals. To begin with, Sir Robert Southwell grouped and preserved his 'Copies of letters writt while I was travelling'.[230] He also drew on his experiences to formulate advice for the next generation, writing to Sir Philip Perceval of 'my owne practice when I was in Travell'. He warned of the dangers of conversation as his nephew entered Italy, but also imparted 'the methods that I used to Ingratiate my selfe with the best sort of People'.[231] Letters also hint at everyday discussion, such as Perceval's comment that his uncle had 'very often tould mee' to avoid countrymen abroad.[232] Years later, when his only son crossed the Channel, Southwell wrote letters on how to interact with great men and what to 'fix' in his mind; after all, people would 'from this very Crisis pronounce of you for the Rest of your life'. Edward dutifully endorsed his 'fathers advice on my going to Holland'.[233] Sir Robert had such an influence over the travels of his great-nephew, Sir John Perceval, that his death forced the young man put them off for several years, and Perceval in turn sustained the connection with Southwell's grandson, the younger Edward. While Perceval shared knowledge of Italy gleaned in 1706–7, especially the 'unexpressible pleasure' he had taken in the remains of Ancient Rome, Southwell wrote to 'refresh' his cousin's 'memory' and update him on the latest archaeological 'Discoverys'. Reading his letters, Perceval 'dreamed more of Travelling than ever I did in my Life'. In time, he planned to use Southwell's writing as a 'patern' for his son's correspondence and hoped that his cousin would 'live to instruct my Grand Children how to travell advantageously'.[234] Edward stayed with Sir John immediately after his return, no doubt indulging his cousin's hope that he would 'recollect the Million of things you have seen and heard', and he brought his brother, William, for a similar debrief after travels that have otherwise left little trace.[235] Perceval had surely had many such discussions during his upbringing in the house of William's grandfather, Sir Robert

[230] BL, Add MS 38015, fo. 20.

[231] BL, Add MS 46955A, fos 59v–60: Sir Robert Southwell to Sir Philip Perceval, 11 Feb 1678.

[232] BL, Add MS 46954B, fo. 5v: Sir Philip Perceval to Sir Robert Southwell, 20 Jul 1677.

[233] BL, Add MS 38015, fos 388–389v: Sir Robert Southwell to Edward Southwell, 5 Jan 1690/1; fos 390–390v: same to same, 19 Jan 1690/1.

[234] BL, Add MS 47031, fo. 39: Sir John Perceval to Edward Southwell II, 1 Dec 1725; fo. 72: same to same, 12 Jan 1725/6; fos 115–115v: same to same, 12 Mar 1725/6; fo. 139: Southwell to Perceval, 9 Apr 1726; fo. 203: Perceval to Southwell, 14 Sep 1726.

[235] BL, Add MS 47031, fo. 203: Sir John Perceval to Edward Southwell II, 14 Sep 1726; *Egmont Diary*, ii, p. 502.

Southwell. Conversation, in homes adorned with souvenirs, transmitted foreign experience across the generations just as much as tangible letters and journals, contributing to the sense of a family as one that travelled.

Samuel Waring shared both his experiences and his reading with family members. One notebook bears an improvised title page, 'Mr Waren traveals throug Germany and Italy 1687', in an adolescent hand that may belong to his nephew, judging by its similarity to 'Wm Close's Book 1697' in the front of the collection of drafts to Misson.[236] William was then in his teens, as planning for his career appears in correspondence the following year, so his uncle's manuscripts featured in the final stages of his upbringing.[237] Waring has written 'ffor Mr Wm Close 1697' on the same page, hinting at a passing down of experience, but the fact that Waring used the notebook to record changes to Misson's 1714 edition suggests that Close returned it afterwards. Uncle and nephew demonstrate that print and personal experience came together in family engagement with travel culture, as Waring provided Close with both reading notes and journals kept abroad. Another annotation, identifying 'a draft of a letter to Max. Misson, who wrote an Account of Italy during a tour made in Company with Mr Waring', appears in a third and possibly later eighteenth-century hand, implying that Waring's notes remained a resource for generations.[238] He may also have shared his journals beyond the family, as he certainly borrowed a manuscript account of the Rhineland from a traveller who had been in Italy at the same time as him, 'tho unknown to us'.[239]

Most studies of European travel writing confine themselves to print and, where they incorporate manuscripts, almost always to letters and journals.[240] These materials played an important role in the afterlives of travel, but a wider range of texts deployed continental experience, often over several generations, to project a prestigious image within and beyond the family. From biographical and autobiographical texts to genealogies and histories, these sources provide an insight into what educational travel meant in the long run, and how families incorporated it into the stories that they told about themselves. They often wrote for themselves, too, as part of an 'internal narrative' of status and identity.[241] Sir Robert Southwell addressed

236 PRONI, D695/225: Waring, travel journal, p. 4; D695/227: Waring, notebook, p. 1.

237 PRONI, D695/73, William Waring to Samuel Waring, 9 Feb 1698/9.

238 PRONI, D695/227: Waring, notebook, pp. 22–3.

239 PRONI, D695/227: Waring, notebook, p. 54.

240 Work on voyages beyond Europe has been quicker to value manuscripts: Holmberg, 'Writing the Travelling Self'.

241 French, 'The "Remembered Family"', 545.

his 'Domestick essay' on family history to his son, though he assumed that his daughters would also read the manuscript. He concluded of their fluctuating fortunes that 'Such Planetary Motions attend Familys, as well as Private Men', but he also left room for individual agency. Southwell found the priorities of a lifetime reflected in the traces of his ancestors, recognising one medieval figure who had given 'his sons such Education, as putt them forward in the world' and another who had travelled to Gascony in order to be 'Educated to Publick Businesse and the Knowledge of that Countrey'. When Sir Robert came to his own life, he implied a similar relationship between foreign experience and crown service: 'having past France, Italy, Germany, Holland and Flanders, I was admitted Clerke of the Councill to K. Charles the 2d in September 1664'.[242] These events were over two years apart, but clearly linked in Southwell's mind.

Educational travel is more prominent in other biographical manuscripts by family members, including a 'Life of Sir Robert Southwell' written by his daughter, Helena Le Grand. Sir John Perceval recorded that she had 'collected' the account from her father, who died in 1702, and given it to him in 1731. Perceval probably meant that Helena had worked with Southwell's archive, as the text is followed by notes on his letters and movements. The account has been a valuable source for earlier chapters, though, because it includes things that do not appear in surviving correspondence, suggesting that Le Grand also drew on family discussions. It begins with Southwell's childhood but focuses on his time abroad, ending with his missions to Portugal. Helena dedicated 12 out of 30 folios to her father's educational travels of 1659–62, apparently agreeing with him that these years were 'of a better edition'.[243] Perhaps she produced the account, which she evidently shared with relations, for use within a well-travelled family. Perceval's own autobiographical texts—two narratives and a chronological summary to 1729—similarly afford detailed treatment to his education, English tour and foreign travels, as well as to later journeys made with his wife and children. An introduction to his great-uncle Southwell emphasises his travels, in retrospect an important moment for a family that had not been abroad within living memory.[244]

Sir John Perceval was collecting materials in the late 1720s and early 1730s with an eye to the genealogical history that he commissioned a few years later as Earl of Egmont, allowing James Anderson and William Whiston

[242] Bristol Archives, 45317/16/2/1: Southwell, Narrative, pp. 14, 23, 28–30.
[243] BL, Egerton MS 1634: Le Grand, 'Life of Southwell'.
[244] BL, Add MS 47072: Perceval, Autobiography, pp. 6–14, 21–3, fos 103–103v, 105–6.

to produce a two-volume work from family archives.[245] Foreign travel fades in their account of successive generations, from a third of Sir Philip Perceval's biography to half a page on Egmont—mainly a providential escape from privateers—to a couple of lines on his son.[246] While Anderson and Whiston may have chosen not to dwell on the voyages of living readers, given their claim to be writing for the family itself, Sir Philip's death a year after his travels provided little other material with which to flesh out an account. Their work is more useful for the light it sheds on the workings of memory in the 60 years since his return. Although the authors consulted manuscripts, they occasionally use living witnesses, both family members and servants, to corroborate aspects of their history.[247] As such, they hint at an oral tradition that survived alongside the family archive. While this evidence has already proved useful for duels, which never appear in contemporary letters, the flexibility of memory is clearest for Sir Philip's audience with the pope in 1678. Perceval's correspondence is muted, reporting that Innocent XI had asked him 'many questions conserning England and att last dismisd mee Laden with his blessings which I did not aske for, butt I gott noe medall as I might have expected'. Rasigade agreed that the pope's questions had been 'fort triviales'.[248] A different story, perhaps the one shared after Perceval's return, emerges from the 1742 history. Innocent XI had treated him 'with very great Distinction, the Pope being well informed of his great Interest in Ireland, which he would have willingly added to that of the Catholic Party, by Arguments of the great Power he would acquire by joining that Party, of which he flattered him he would be the Leader; but he used his Persuasions to that Purpose in vain'.[249] Such myths must often have gathered around educational travel, which was a test of personal fortitude and family standing. They contributed to the two or three generations of 'cumulative oral memory' that was, as French puts it, 'the most powerful source of … collective identity-formation' among the early modern gentry.[250]

245 Francis Espinasse and S. J. Skedd, 'Anderson, James (*bap.* 1679, *d.* 1739)', *ODNB*, online edn, 2006 (https://doi-org/10.1093/ref:odnb/474).
246 Anderson, *Genealogical History*, ii, pp. 379–82, 405, 456.
247 Anderson, *Genealogical History*, ii, pp. 371–2.
248 BL, Add MS 46955B, fo. 58v: Alexandre de Rasigade to Sir Robert Southwell, 13 Aug 1678; fos 64–64v: Sir Philip Perceval to same, 20 Aug 1678.
249 Anderson, *Genealogical History*, ii, p. 381.
250 French, 'The "Remembered Family"', 543.

Conclusion

Parents, travellers, relations and friends saw time abroad within educational strategies that aimed at personal and family advancement. The first half of this chapter gauged their success at the subsequent milestones of elite, male, English adulthood. Travel was extremely useful, though not essential, to young men embarking on public life, furnishing accomplishments and connections that helped to secure positions in government or diplomacy. Here, Renaissance-style preparation for service to one's prince blended with personal and familial benefits. While public life might prove lucrative, or at least confirm existing standing, linked notions of travel and crown service remained powerful components of family identity into the eighteenth century. Marriage negotiations, meanwhile, provided a negative inducement to cross the Channel. Travel was rarely decisive in standard matches between social equals, but suitors had to explain away its absence as the practice spread. A lack of continental experience gave young women and their families a reason to reject an undesirable candidate, but parents could anticipate this objection by sending sons abroad. Finally, for some Irish landowners, travel offered one way in which to buy into the English elite, overlapping with the search for wives, offices and lands in England. Their experiences fit with the wider sense that time abroad prepared young men for metropolitan lives and employment.

The second half of the chapter explored the changeable meanings that continental experience held for travellers and their families. Most work tends to follow seventeenth- and eighteenth-century print by dwelling on what travel implied for the nation, in terms of expense, enlightenment or corruption. Recent contributions are more promising, though, with John Gallagher offering a 'view from inside' Anglo-Italian cultural exchange and Mark Williams exploring 'the inner lives of travel'.[251] Gallagher is right that bodily, behavioural and intellectual performances at home established the social value of travel. Its power depended upon display, including but beyond the paintings and statues that fill much of its historiography. Returned travellers shaped their conversation, speech, deportment and consumption in order to advertise membership of an in-group, to each other as much as to those who had not been abroad. Williams is similarly correct that individuals reframed their foreign experiences according to the anxieties with which they wrestled—or indeed the opportunities with which they met—at home. Returned travellers read books and manuscripts in order to relive their

[251] Gallagher, 'Italian London'; Williams, 'Inner Lives'.

experiences, maintain their accomplishments and turn them towards social ends. The self-fashioning of the traveller extended well beyond the moment of encounter with the foreign. In terms of outcomes, then, much depended upon the abilities of a young man, not only to have gathered improvements abroad, but also to perform them at home. As Robert Southwell realised in Italy, 'some are wiser that stay at home then who travel'.[252]

At the same time, this chapter has followed the afterlives of travel beyond interiority to shared meanings, finding that families were deeply implicated in returns from the Continent. While young men were still abroad or in the days after their return, souvenirs, commissions and gifts contributed to family prestige and spread foreign encounters to parents, siblings and kin. Crucially, they did so at a cost that suited family budgets. In the months and years that followed, the influence of travel on office-holding, marriage and residence meant as much for family prospects as for individual biographies. Over the even longer term, travellers contributed to growing archives of continental experience. Letters, journals and financial records became resources for siblings and descendants, who planned routes, travelled vicariously or pieced together family histories, whether for internal or more public consumption. Travel became a shared experience across generations, as relations inwardly pooled useful knowledge and outwardly presented themselves as a family that travelled. If time abroad might contribute to social stability or advancement in the years afterwards, it assuredly occupied a central position in the self-perceptions of families among the gentry and nobility.

252 BL, Egerton MS 1632: Robert Southwell, Italian notebook, fo. 16.

Conclusion

Before crossing the Channel in 1659, Robert Southwell noted that 'Tis here in Plymouth a great Practice to exchange Children with the French people of Nantes Burdeaux Rochell &c: and to breed them up to the Language for a yeare.'[1] Frustratingly little can be recovered of such traditions, but they offer a reminder that educational travel has taken many forms, depending on time, place and social standing. Students attended institutions on the Continent in the medieval period, during the so-called Age of the Grand Tour and, until very recently, under the Erasmus Programme. Such options have always coexisted and overlapped with more flexible approaches, from Grand Tours to gap years. While the children—presumably only the boys—of early modern Plymouth went on what would now be thought of as an extended French exchange, Southwell recorded their tradition as he began four generations of educational travel in his own family. He and his male descendants went to the Continent in search of the knowledge and connections that would attain and secure their position in English landed society. They first travelled in this way in 1659 and returned from the last recorded instance in 1760, framing a distinctive period of educational travel for them and for the wider elite.

By paying attention to what families did and discussed, rather than to well-worn stereotypes, this book has found that existing work on the seventeenth and eighteenth centuries at once overestimates the Italian Grand Tour and underplays educational travel. Eye-catching journeys to Italy were only one part of a broader culture of educational mobility, within which continental experience was widespread and habitual among the gentry. Average landowning families knew far more of France and the Low Countries, and to a lesser extent Germany and Italy, than it would seem from their histories or those of the Grand Tour. They secured time abroad for their sons in several ways, valuing it highly enough to invest significant time, money and emotional energy. They did so in the belief that travel

[1] BL, Add MS 58219: Southwell, French diary, fo. 4.

would not only help to make their sons men, gentlemen and Englishmen, but also to secure or advance family standing. At the same time, shared skills and experiences encouraged a common identity among returned travellers. Formative encounters with continental Europe helped to identify and define the elite.

A period on the Continent became an integral part of gentlemanly upbringing between the mid-seventeenth and mid-eighteenth centuries not because of any advantages that it might bring to the nation, but because of what it could do for families. As the elite moved further away from lineage as a guarantor of status, the demands of landed identity continued to expand. This was a period of French cultural dominance in Britain and Ireland, as elsewhere in Europe, and young men travelled to draw gentlemanly 'parts' from the most prestigious sources. At the same time, they could display their means and strengthen their connections. But why did they do so in such numbers in the century after 1650? The phenomenon cannot be dismissed as a function of rising means: landed incomes were stagnant over the period, helping to explain why the benchmark costs of travel—around £300 a year and another £100 for a top-class tutor—remained so consistent.[2] Rather, just as the cachet of French culture increased, developments in the second half of the seventeenth century made educational travel a more affordable way to buy into the elite. Improvements in transport drew together a north-western European cluster, within which new academies and changing universities adapted themselves to family practices. Meanwhile, educational travel developed alongside other commitments to the Continent: the expansion of the military and the diplomatic service, especially after 1689, brought opportunities to exploit networks of kin and acquaintance in search of a foreign education. Cumulatively, family strategies lay behind the popularisation of educational travel between the mid-seventeenth and mid-eighteenth centuries.

Family strategy also shaped individual experiences. Although most studies of the Grand Tour pick up their subjects on the Continent and drop them before they return, historians cannot understand these encounters without following family-level thought and action before, during, after and instead of travel. Questions about why young men went abroad, why their experiences mattered and whether they paid off only find answers in a family context. To begin with, travel took its place in longer courses of upbringing. Parents sent sons across the Channel to continue an education that had already featured some combination of tuition, schooling, university,

[2] Rosenheim, *Emergence of a Ruling Order*, pp. 49–58.

a French-run institution in London and a domestic tour. They wanted them to develop the skills and connections upon which they had worked at home, perfecting their French and deepening relationships with schoolmates, university contemporaries, neighbours and patrons. They hoped that this latest step away from home would produce independent and authoritative gentlemen who established and valued their Englishness. Just as importantly, they used travel to demonstrate family spending power.

A young man's travels emerged from his family's financial and social resources, as well as from the need to share them with brothers and sisters. These factors decided who went abroad, where they went, what they did, for how long and with whom. Time abroad was expensive, of course—that was part of the point—but a French academy or Dutch university, perhaps followed by a local tour, was more accessible than a lengthy journey to Italy. So too was the exploitation of military or diplomatic networks. Educational travel was just as complex as domestic upbringing, then, as families assembled courses from a range of options. Even at the highest levels, it was never a case of unthinkingly sending a son through Eton, Oxford and Italy. The varying use of unequal resources produced many instances of what would now be called the Grand Tour, but many more adaptations and alternatives. After all, the base requirements of educational travel—immersion in French culture, opportunities to develop independence—never insisted upon a voyage to Italy. Time abroad was generally the most expensive stage of education, but it was not beyond the means of gentry families. They used it to buy into the elite as far as they were able.

The work of travel continued at home, where family efforts suggest some relationship between geographical and social mobility. The Southwells brought home languages, gifts and strengthened connections to domestic patrons, using them to rise socially and politically in England. Many more, such as the Leighs, crossed the Channel in search of the same benefits. The ability to talk about and manifest travel offered an appreciable edge at the milestones of elite male adulthood, in marriage negotiations, entry into public life and, in an Irish context, efforts to transfer residence to England. In a society that demanded the constant performance of elite status, however, the fear of decline probably motivated more journeys than the hope of success. Travel shored up family positions, a transgenerational obligation for the noble Ingrams and a way for the Brydges family to secure recent advances. It became a worthwhile investment for the ambitious and a requirement for the established. As a result, although social and political success depended upon individual ability and application, time abroad consistently informed how elite families saw themselves and presented to others. Layered memories, souvenirs and archives helped later travellers to plan new journeys and reinforced the identities of families who travelled. Beyond moments of encounter with the foreign, travel found meaning in individual lives and across generations.

Continental travel was part of a culture of educational mobility that awaits further exploration. Provincial British and Irish landowners also sent sons to institutions in southern England, including French-run academies, and undertook domestic tours, with these options preparing some for the Continent and providing others with alternatives. Most of the evidence in this book comes from families who sent at least one son abroad, but many parents thought foreign travel beyond them, whether because of lesser means or greater anxieties. Though the most prestigious methods were to be found abroad, did domestic mobility provide similar outcomes? Families who used Foubert's London academy as an alternative to foreign travel certainly pursued French culture and connections to fellow landowners.[3] A comprehensive account of educational mobility would also incorporate girls. The Rawdons of the north of Ireland and the Leighs of Gloucestershire sent daughters to relations in London, for instance, while the Percevals made use of boarding schools near the capital.[4] How popular were these options, and were they another factor in the coalescence of a national elite? Recent work on Irish and Scottish landowners suggests that girls' educational mobility differed from that of boys only in degree, with London the preferred destination for acquiring polite manners, accent and connections by the eighteenth century.[5] As with boys, these decisions also affected the maintenance or advancement of family standing, since women were bearers of status and had a better prospect of marrying upwards.[6] At the least, stark contrasts between immobile daughters and mobile sons are probably misplaced. Within the elite, the ways in which financial means and gender differences shaped educational mobility need more work.

Like an expensive education today, the use of travel in family strategies could backfire. Success depended on sons, including lazy young men

[3] Ansell, 'Foubert's Academy', pp. 55–7.

[4] WYAS, Bradford, SpSt/5/4/3/34, fo. 22: Sir George Rawdon to John Stanhope, 19 Feb 1680; SpSt/6/1/7: same to same, Jul 1682; SBT, DR18/20/18b: Theophilus Leigh to James Leigh, 11 Dec 1710; BL, Add MS 46954A, fo. 135: Sir Robert Southwell to Sir Philip Perceval, 30 Apr 1677.

[5] Katharine Glover, *Elite Women and Polite Society in Eighteenth-Century Scotland* (Woodbridge: Boydell, 2011), pp. 38–44, 147, 164–5; Rachel Wilson, *Elite Women in Ascendancy Ireland, 1690–1745: Imitation and Innovation* (Martlesham: Boydell, 2015), pp. 54–6.

[6] Vickery, *Gentleman's Daughter.*

such as Henry Brydges, and many could not satisfy the lofty intentions with which they were sent abroad.[7] Others risked family stability through overspending, as Irwin was warned of his predecessors. Travellers were also hostages to fortune: Edward Rawdon, John Rawdon and John Chamberlayne all died abroad; Sir Philip Perceval lasted a year at home. The failure of travel rarely brought about family decline by itself, however, because—as will be discussed below—it always stood alongside other ways in which landowners established their position. Families nevertheless believed in time abroad as a route to social stability or upward mobility, even if the results can be difficult to disentangle from other proceeds of wealth and standing. Debates over travel involved fathers, mothers, uncles, aunts, eldest sons and younger brothers because they concerned the prospects of the whole family. Sometimes, as with the Southwells, it is possible to identify feedback loops in which successful travel encouraged its repetition: skills and connections gained abroad contributed towards rising social and political standing, which in turn made further travel necessary and feasible. It is usually easier to find travel contributing to educational practices and senses of self in landed families who had often, by the mid-eighteenth century, built up three or four generations of continental experience. Grand Tour portraits and biographical dictionaries are misleadingly individualising: over generations, the risks, benefits and meanings of educational travel were made in families.

Family adoption of educational travel was equally an important aspect of elite formation. In a crucial period for the coalescence of a British ruling class, time abroad informed its membership and cultural priorities.[8] Upward mobility and elite formation are not the same, of course, but families rose when they bought into elite practices and risked falling when they could not; in turn, elite culture was the sum of family decisions. Henry French has recently called for 'a more detailed social history of gentility—that is, the active processes by which the landed elite conceived of their identity, and attempted to perpetuate it'.[9] Educational travel must be central: from the mid-seventeenth century, it was far more widespread than existing work would suggest and became a habitual practice over generations of landowning families. They sent sons abroad to acquire 'parts' that slotted into a domestic hierarchy and to create or renew connections among a domestic elite. Encounters with foreigners were comparatively fleeting

[7] Cohen, 'Grand Tour: Constructing the English Gentleman', 247–8.
[8] Rosenheim, *Emergence of a Ruling Order*.
[9] French, 'Gentlemen', p. 283.

and mostly seen as opportunities for learning, because time abroad found meaning as far as it contributed to standing at home.

Above all, young men went abroad in pursuit of the language and manners of France, the priorities of most who sent sons abroad and always the minimum expectations. It is well known that French culture became fashionable across Europe in this period. The investments that gentry and noble families were willing to make, in search of pure accents and refined deportment, show that French models were even more central to British and Irish elite culture than is usually recognised, however. So much educational practice revolved around that fact, from French-speaking tutors—themselves participants in professional mobility or confessional exile—to French-run institutions in London to settlement at continental institutions. Regular war with France after 1689 made options there less accessible, but it did little to harm the cachet of French culture. Indeed, establishments in the Netherlands, Lorraine, Geneva and Turin were popular because they offered Francophone immersion. Parents prepared sons to use the language in the rest of their lives, more than in momentary encounters with foreign hosts, and had in mind courtly occasions, political careers and—perhaps—daily encounters among themselves. Plenty is now known about how and why the British and Irish elite learned French, but the next step will be a fuller exploration of when they used it.

Educational travel informed the membership as well as the culture of the elite, furnishing not only the 'parts' of a 'compleat gentleman', but also his essential connections. Families such as the Leighs travelled in pursuit of patronage, and competition was fierce to go abroad with the influential Ingrams, because time on the Continent offered the chance to establish or strengthen relationships with such benefactors, without whom little could be achieved in seventeenth- and eighteenth-century society. Beyond interpersonal links within parties, common experiences simultaneously bound together a wider cohort of travellers. They thought of themselves as a transient community abroad, with shared understandings of honour, shared domains of English speech, shared celebrations of English food and—for the most part—shared Anglican worship. They encountered the same foreign sights and people in each other's company, receiving parental exhortations to judge them by English standards. And if identity is something imposed as well as projected, they were thought of as 'gentilhommes anglois' and 'milordi inglesi' by their hosts.[10] When they returned home, they reminisced about their experiences, performed the changes that travel had worked on them

10 Sweet, *Cities and the Grand Tour*, p. 13.

and displayed their souvenirs—all to retain the esteem of fellow travellers as much as that of an untravelled audience. With a stress on both words, sons became 'English gentlemen' abroad.

This process is clearest for the Anglo-Irish. Even as continental seminaries provided opportunities to Catholic Irishmen under the Penal Laws, time abroad allowed many Protestant landowners to assert their standing and draw closer to an English elite.[11] Much eighteenth-century historiography deals with the important question of how Protestants came to identify with an Irish 'nation', but opposing forces were also at work in the most prominent families: educational travel helped them to retain their English ethnicity, even to move to England.[12] Recent work has found a similar dynamic at work among English Catholics. While education in France and Flanders allowed Catholic girls to maintain cultural distinctiveness, shared encounters with the Continent offered their brothers a way into the Protestant elite.[13]

What about Scottish families? Whether travel bound them into a British elite or helped to maintain a distinct national identity deserves greater exploration, but it surely depends on chronology. Calvinist connections shaped periods along the Loire, where Scottish travellers and teachers clustered at the reformed academies, but did a distinctive relationship with France survive the Revocation of the Edict of Nantes?[14] Here and elsewhere, did the identifiably Scottish networks of the seventeenth century outlast the union of 1707?[15] They seem to have, at least among travellers, artists, diplomats and exiles in Italy. But it remains to be seen whether these networks shaped appreciably different experiences or, as in the case of Anglo-Irish landowners, facilitated journeys that drew travellers closer to a common British elite. Work on eighteenth-century Scots and the Grand Tour does little to suggest distinctive encounters, focusing instead on the impact of Italy in Scotland, and English and Scottish examples appear interchangeably

[11] Liam Chambers, 'The Irish in Europe in the Eighteenth Century, 1691–1815', in James Kelly, ed., *The Cambridge History of Ireland, Volume III, 1730–1880* (Cambridge: Cambridge University Press, 2018), pp. 569–92, at pp. 575–80. Conway suggests that the prestige of travel was particularly necessary for Irish Protestants, who were in less secure possession of their estates: *Britain, Ireland and Continental Europe*, pp. 129–30.

[12] Hayton, *Anglo-Irish Experience*, pp. 25–48.

[13] Corens, *Confessional Mobility*; Lock, *Catholicism, Identity and Politics*.

[14] *Journals of Sir John Lauder, Lord Fountainhall, with his Observations on Public Affairs and other Memoranda, 1665–1676*, ed. Donald Crawford (Edinburgh: Edinburgh University Press, 1900), pp. 1–166.

[15] Games, *Web of Empire*, p. 36.

in books on British travel.[16] The suspicion is that adding Scottish families to this study would barely have changed its conclusions: extensive work on Scots at the Dutch universities further entrenches a northerly, institutional vision of educational travel, while a recent exploration of its afterlives has focused on the Clerks of Penicuik.[17] Evidently, though, there is more to uncover about the place of continental travel in the social and cultural history of seventeenth- and eighteenth-century Britain and Ireland.

Part of normalising travel—placing it within lives, identities and social history—is recognising that it was not the only way in which families could invest financial and social capital in the hope of stability or advancement. The period offered complementary routes to social reproduction and, perhaps, upward mobility: among other requirements, it was felt that a gentleman should be a generous host, he should collect art, he should wear expensive clothes, he should own a home befitting his standing and he should hold local or national office. How families allocated resources to these priorities depended on their outlooks and circumstances. In the 1670s, Sir George Rawdon considered downsizing his lands and household to keep three sons in France; in the same decade, Sir Robert Southwell prioritised the travels of the Perceval heir and looked to buy him an English estate, dismissing travel by his younger nephews. Property seems longer lasting, but travel could outlive individuals as shared family practice, sustained by oral memory, archives and souvenirs. The younger James Brydges still went to Göttingen and Orleans, though his grandfather's bare house would soon be demolished to satisfy creditors. Travel, with flexible itineraries to suit most landed budgets, was in fact among the most accessible routes to gentility. A performative elite had fuzzy requirements; as Irwin and his trustees found, it left room for disputes over what was best for personal and family honour, and indeed how far they overlapped. Educational travel stood alongside other options for displaying standing but, increasingly, a 'compleat gentleman' had been abroad.

So how much of a problem was it, not to have travelled? Again, most of the families in this book are here because they sent at least one son abroad. When they downplayed travel, it was usually because they wanted to excuse its absence in young men who were otherwise attractive marriage prospects. When they insisted on its necessity, it was either a reason to reject

16 Basil Skinner, *Scots in Italy in the Eighteenth Century* (Edinburgh: Scottish National Portrait Gallery, 1966); Iain Gordon Brown, 'Water, Windows, and Women: The Significance of Venice for Scots in the Age of the Grand Tour', *Eighteenth-Century Life*, 30:3 (2006), 1–50.

17 Mijers, *News from the Republick of Letters*; Williams, 'Inner Lives'.

an unsuitable candidate or a celebration of family tradition. Some caution is therefore needed before pronouncing that time abroad was essential. It was clearly not an insurmountable issue for those who invested in other markers of elite identity, as untravelled younger brothers inherited in stable families such as the Percevals and the Ingrams. Trawling the archives for families who discussed but rejected foreign travel is unlikely to be cost-effective. The papers of those with continental experience nevertheless show that it offered an advantage in marriage negotiations and entry into public life, as well as an enduring source of family status. Whether or not one had been abroad entered what Shepard has, for those lower down the social order, called 'the calculus of esteem'.[18] Educational travel aided the Southwell rise from provincial to metropolitan life and helped the Brydges family to secure rapid gains, but it was unlikely to fuel many meteoric ascents, not least because these were rare in themselves. Travel allowed some to maintain their standing and others to contemplate gradual upward mobility. It also contributed to a powerful familial self-image, reflecting wealth and enhancing status. In these respects, it was worth the investment.

The Grand Tour is usually treated in isolation, but family archives show that educational travel was part of wider European engagement. Not only did the military and the diplomatic corps inform many early experiences, but they offered subsequent careers abroad. Meanwhile, as shorter 'pleasure trips' grew popular over the same period, elite families repeatedly crossed the Channel.[19] By the turn of the eighteenth century, the same families—even the same people—knew the Continent as educational travellers, soldiers, diplomats and travellers for health and leisure. Yet more possibilities existed: William Ingram, Irwin's youngest brother, was apprenticed to an Amsterdam merchant in the 1720s; in the same decade, William Leigh evaded his creditors over six years in Utrecht, which he knew well from his time with Cornbury. Historians of the Grand Tour focus on Italy, which attracted many of those who could afford it, but broader educational travel shared a north-western European centre of gravity with other journeys in the same families. Some still saw time abroad as a one-off, with Samuel Waring reminiscing for decades about 'one of the cheiffest & most satisfactory adventur[es] of my whole life'.[20] Men of greater means gained a taste for travel, which they pursued on visits to France and the Low Countries with wives and children,

18 Alexandra Shepard, *Accounting for Oneself: Worth, Status, and the Social Order in Early Modern England* (Oxford: Oxford University Press, 2015).
19 The term is Verhoeven's: 'Foreshadowing Tourism'.
20 PRONI, D695/227: Waring, notebook, p. 55.

but even they set apart journeys made on the threshold of adulthood. When all these experiences are taken together, however, it becomes clear that landed families were strikingly familiar with Britain's neighbours, with profound consequences for the membership and culture of the seventeenth- and eighteenth-century elite.

The educational component of continental travel began to fall away in the mid-eighteenth century. By 1750, the Ingrams, Worsleys, Franklands, Leighs and Percevals had all undertaken the final educational journey to leave a trace in their archives.[21] The Brydges, Southwell and Robinson families had done so within another decade.[22] The Rawdons and Brodricks persisted into the 1770s, though in the latter case without much of a paper trail.[23] Again, this trend was not a product of simple economics, as landed incomes rose significantly between the 1760s and the 1790s. Families even ploughed much of their growing income into foreign trips, which gained in popularity after the Seven Years War.[24] As is clear from the travels of Huntingdon and his Rawdon nephews, however, young men abroad for their education were increasingly outnumbered by older travellers and family groups. These decades are often called a 'golden age' of British travel to Italy, and they are the period around which stereotypes of the Grand Tour have formed, but they actually saw the decline of educational travel to the Continent.[25] By the late eighteenth century, notions of home-grown masculinity were bringing a century of French cultural hegemony to an end and with it much of the mobility that it had inspired. With the rise of public schools, the gentry slowly abandoned continental models in the few decades before 1800, though they still crossed the Channel for health and leisure.[26] The story of the later eighteenth century is well told by Cohen, Newman and Linda Colley: broadly, French culture went from a central plank of elite identity, a buttress of social division, to an external 'other' against which

[21] Aside from journeys already discussed, see SBT, DR18/31/881: James Leigh, Accounts and description of European tour, 1747–8.

[22] Ingamells, *Dictionary*, pp. 816–17.

[23] George Brodrick, third Viscount Midleton, joined the Dilettanti in 1751, so had been to Italy by the age of 21: Lionel Cust, *History of the Society of Dilettanti*, ed. Sidney Colvin (London: Macmillan, 1898), p. 258. The fourth viscount probably accompanied his younger brother, Thomas, to Italy in 1774: R. G. Thorne, 'Brodrick, George, 4th Visct. Midleton (1754–1836)', in Thorne, ed., *The History of Parliament: The House of Commons 1790–1820* (London: Secker & Warburg, 1986); Ingamells, *Dictionary*, pp. 131–2.

[24] Conway, *Britain, Ireland and Continental Europe*, p. 191.

[25] Bignamini, 'Grand Tour: Open Issues', p. 33; Conway, *Britain, Ireland and Continental Europe*, p. 191.

[26] French and Rothery, *Man's Estate*, pp. 140, 159–60.

a British nation could define itself.[27] Once again, behind these sweeping changes, lay family decisions about how to bring up their children.

Educational travel has changed rather than disappeared, of course. Young gentlemen still travelled through Europe in the nineteenth century, though not as to fonts of civilisation, and they increasingly toured the eastern Mediterranean and the expanding empire. Both kinds of travel still developed elite male independence.[28] Non-elite journeys have become more visible when they have been centrally organised, by middle-class reformers in the nineteenth century and through state initiatives in the twentieth.[29] Educational travel is often part of wider foreign engagement, with deepening European commitments in the late seventeenth century, the imperial turn after 1815 and now the end of UK participation in the Erasmus Programme—amid plans for a global outlook—as part of Brexit. British Erasmus students were disproportionately middle-class and privately educated, though by no means overwhelmingly so. At least lower-income students no longer need to flatter patrons or travel in service to spend time on the Continent, perhaps, while gap years show that the comfortable will travel anyway, as they have for centuries, with or without institutional support. Indeed, a pre-university gap year is a strong predictor of later study abroad.[30] A factor in problems of access to Erasmus seems to have been the enduring connection between European travel and social reproduction, the idea that Paris or Rome were only for wealthy repeat travellers interested in high culture. James Brydges's explanation of his eldest son's travels—that 'the more of the world he sees … the sooner he will become a Citizen of it'—would surprise few today.[31] The fear is that, in the twenty-first century as in the eighteenth, such practices have too often entrenched privilege. The real question is whether what comes next will be any different.

[27] Cohen, *Fashioning Masculinity*; Newman, *Rise of English Nationalism*; Linda Colley, *Britons: Forging the Nation, 1707–1837* (New Haven, CT, and London: Yale University Press, 1992).
[28] French and Rothery, *Man's Estate*, pp. 160–81.
[29] Michele M. Strong, *Education, Travel and the 'Civilisation' of the Victorian Working Classes* (Basingstoke: Palgrave Macmillan, 2014).
[30] Russell King, Allan Findlay and Jill Ahrens, *International Student Mobility Literature Review* (Bristol: Higher Education Funding Council for England, 2010), pp. 27–33.
[31] HL, mssST57, vol. 20, fo. 16: James Brydges to Charles Stuart, 19 Nov 1721.

Bibliography

Manuscript Sources

Archives Nationales, Paris

O/1/915-917: Départements des grands officiers de la Maison du Roi: Académies, manèges, [17th–18th centuries].

Beinecke Library, Yale

Osborn MS b66: [Edward Southwell], Holland voyage, 1691.

Osborn MS b112: Robert Southwell, Commonplace book, [1650s].

Osborn MS fb90: Edward Southwell, Accounts of travels, 1684–1715.

Biblioteca Nazionale Centrale di Firenze, Florence

MS Gal. 158: Familiar letters from Vincenzio Viviani, 1664–71.

MS Gal. 161: Familiar letters to Vincenzio Viviani, 1640–62.

MS Gal. 254: Scientific letters to Vincenzio Viviani, 1642–66.

MS Gal. 258: Letters from, to and relating to Vincenzio Viviani, 1655–1715.

Bibliothèque Mazarine, Paris

MS 4569: [Edward Southwell], Journey from London to Paris, 1714.

MS 4570: Giovanni Battista Gornia, 'Viaggio fatto dal serenissimo Principe Cosimo Terzo di Toscana per la Spagna, Inghilterra, Francia, ed altri luoghi negli anni 1668 e 1669'.

Bibliothèque Nationale de France, Paris

MS Anglais 194: Travels of the Hon. Edw[ard] Southwell [II], 1725.

Blackburn Central Library

Southwell Papers, vol. 1: Correspondence and papers, 1685–7.

Bodleian Library, Oxford

MS Carte 50: Copies of letters from the Duke of Ormond, 1669–87.

MS Carte 70: Papers relating to the public affairs of Ireland, and records of the Butler family of Ormond, 17th–18th centuries.

MS Carte 219: Correspondence and papers of the Duke of Ormond, chiefly on Irish and English public affairs, 1660–83.

MS Carte 243: Correspondence and papers of the Duke of Ormond, 1667–81.

MS Eng. Lett. c 328/3: Correspondence of Francis Parry, 1665–80.
MS Eng. Misc. c 144: Collection of treatises transcribed for and annotated by Sir Robert Southwell, late 17th century.
MS Eng. Poet. f 6: Robert Southwell, Poetic commonplace book, [1650s].

Boole Library, University College Cork
MS U/20, vol. 1: Kinsale Manorial Papers, 1662–5.
MS U/20, vol. 2: Kinsale Manorial Papers, 1665–75.

Bristol Archives
45317/2/5/1: Letters from Katherine Southwell to her son, Edward, 1749–55.
45317/2/5/2: Letters from Katherine Southwell to her son, Edward, 1758.
45317/2/5/6: Letters from Katherine Southwell to her son, Edward, 1760–1.
45317/16/2/1: Sir Robert Southwell, Narrative of Southwell family history, 1700.

British Library, London
Add MS 4293: Warrants, letters etc collected by Thomas Birch, 1536–1765.
Add MS 9714: Collection of papers and letters relative to the town of Kinsale, and to the connection of the Southwell family with that place, 1637–1778.
Add MS 21495: Edward Southwell, Journal of a tour in Holland, 1696.
Add MS 22511-22514: Dispatches and papers relating to Lord Carteret's embassy to Sweden, 1719–20.
Add MS 28724: [Edward Southwell], 'Discourse concerning the securing the Government of the Kingdome of Ireland to the Interest of the English Nation', [*c.* 1695].
Add MS 34336: Sir Robert Southwell, Letter-book during negotiations in Portugal, 1665–9.
Add MS 34753: Edward Southwell II, Journal of travels in Lorraine and France, 1723.
Add MS 35100: Francis Parry, Letter-book as envoy to Portugal, 1668–74.
Add MS 38015: Sir Robert Southwell, Correspondence, 1659–97.
Add MS 38536: Catalogue of Sir Robert Southwell's papers, 1692.
Add MS 46932-46942: Correspondence of Sir John Perceval, first Baronet, 1647–65.
Add MS 46943-46956: Correspondence of Sir Philip Perceval, second Baronet, 1665–80.
Add MS 46957-46962: Correspondence of Sir John Perceval, third Baronet, 1680–6.
Add MS 46963: Correspondence of Sir Robert Southwell and Katherine, Lady Perceval, as guardians of Sir Edward Perceval, fourth Baronet, 1686–90.
Add MS 47025-47033: Letter-books of Sir John Perceval, fifth Baronet, later Baron Perceval (1715), Viscount Perceval (1722) and Earl of Egmont (1733), 1697–1736.
Add MS 47059: Sir John Perceval, later Earl of Egmont, Journals of travels to France and the Low Countries, 1718–23.
Add MS 47072: Sir John Perceval, later Earl of Egmont, Autobiography and notes, [*c.* 1729].

Add MS 47113: Sir Philip Perceval, Commonplace book, *c.* 1669–77.
Add MS 47127: Sir John Perceval, later Earl of Egmont, Commonplace book, 1708–22.
Add MS 58219: Robert Southwell, Diary and commonplace book of travels in France, 1659.
Add MS 58220: Edward Southwell, Journal of a visit to the Netherlands, 2 vols, 1696.
Add MS 61639: Correspondence and papers of Earl of Sunderland relating to Ireland, [1709]–1715.
Add MS 70013: Correspondence and papers of Robert Harley, Earl of Oxford, 1679–86.
Add MS 72852: Letters of Sir William and Lady Petty to elder Robert Southwell and Sir Robert Southwell, 1663–83.
Egerton MS 1627: Sir Robert Southwell, Memoranda and diary, 1674–85.
Egerton MS 1632: Robert Southwell, Notebook during his stay in Italy, 1660–1.
Egerton MS 1634: Helena Le Grand, 'Life of Sir Robert Southwell', [before 1731].
Egerton MS 2127: Sir Robert Southwell, Correspondence with Henri Justel, 1684–8.
Egerton MS 3805: Edward Southwell II, Journal of travels in France and Italy, 1725–6.
Harley MS 7614, fo. 101: Humble Petition of Solomon Foubert, [after 1685].
Stowe MS 744: Miscellaneous letters, many to Sir Edward Dering, first Baronet, and Sir Edward Dering, second Baronet, 1640–67.

Brotherton Library, University of Leeds
YAS/MS328: Edward Southwell II, Account of my Journey Begun 6 Aug[us]t 1724.

Christ Church, Oxford
MS 426: Benson of Oxford's letters relating to Lord Kinsale [i.e. Kingsale], 1676–81.
MS 427: Letters relating to the education of Sir Thomas Southwell at Oxford, 1681–4.
MS 475: Francis Atterbury, 'Regiones, Urbes, Montes, Flumina, &c.', [after 1685].

Christ's College, Cambridge
Fellows' Papers, Box 24: Letters from John Covel to John Roades, 1702–4.

Clark Library, UCLA
MS.1968.006: [John Perceval, future second Earl of Egmont], Journal of a tour in France and the Netherlands, 2 vols, 1726.
MS.1975.002: Edward Southwell II, Journal of Our Journey, 1723.

Huntington Library, San Marino, California
mssEL 9015-9091: Ellesmere Collection: Wrey Papers, 1664–1700.

mssHA: Correspondence of Hastings family, Earls of Huntingdon, 1477–1892, including Hastings Irish Papers, 1583–1751 [including Rawdon family: boxes are in chronological order in collection, but individual items are not].
mssHA, box 8: Hastings Irish Papers, 1640/1–8 (including mssHA 14336).
mssHA, box 9: Hastings Irish Papers, 1648/9–55 (including mssHA 14350).
mssHA, box 22: Hastings Irish Papers, 1667/8–70 (including mssHA 14482, 14486, 14489).
mssHA, box 23: Hastings Irish Papers, 1670–3 (including mssHA 14498, 14499, 14507, 14508, 14512, 14523, 15603).
mssHA, box 24: Hastings Irish Papers, 1673/4–6 (including mssHA 14525, 14526, 14527, 14531, 14535, 14536, 14538, 14539, 14541, 14542, 14543, 14544, 14545, 14546, 14549, 14553, 14554, 15264, 15373, 15604, 15642, 15667, 15700).
mssHA, box 25: Hastings Irish Papers, 1676/7–80 (including mssHA 15242).
mssHAF, box 20: Hastings accounts and financial papers, 1670–2 [Rawdons].
mssHAF, box 22: Hastings accounts and financial papers, 1674/5–6 [Rawdons].
mssHAF, box 23: Hastings accounts and financial papers, 1676/7–82/3 [Rawdons].
mssHAF, box 27: Hastings accounts and financial papers, 1694/5–1699/1700 [Rawdons].
mssHM 30301: Henry Brydges, Journal, 1720.
mssHM 81607-81660: Leigh family papers, 1686/7–1866.
mssST12: James Brydges, first Duke of Chandos, Accounts with brokers etc, 4 vols, 1711-44.
mssST57: Letters from James Brydges, first Duke of Chandos, 57 vols, 1694–1744.
mssST58: Letters to James Brydges, first Duke of Chandos, 14 vols, 1700–12.
mssST109, vol. 1: James Brydges, third Duke of Chandos, Travel diary for Germany and France, 1751.
mssSTB, box 12, folder 53: James Brydges, third Duke of Chandos, Letter-book, 1750–1.

Kent Archives, Maidstone
U1713/C1: Dering-Southwell correspondence, 1664–79.
U1713/C2: Dering-Southwell correspondence, 1680–1.

King's College, Cambridge
Keynes MS 152: 'An Abridgement of a Manuscript of Sr Robert Southwell's concerning travelling. Writt. 1658. Feb. 20', [in Southwell's hand, 1658–61].

Koninklijke Bibliotheek, The Hague
MS 133c.6: Correspondence of E[dward] Southwell, 1696.

Leicester, Leicestershire and Rutland Record Office, Wigston
DG7, Bundle 28: Correspondence of second Earl of Nottingham, 1723–9.

National Art Library, Victoria and Albert Museum, London
Forster MS 47.A.40: Letters and papers of the Duke of Ormond, vol. 2, 1678–84.

National Library of Ireland, Dublin
MS 36: Orrery Papers, Correspondence, 1684–99.
MS 664: Correspondence of Sir William King with Sir Robert Southwell, 1681–4.
MS 2493: Duke of Ormond's correspondence during Oxford chancellorship, from 1669.
MS 7863: Perceval family history, 1646–[99].
MS 13215: Travelling expenses of Lionel, third Earl of Orrery, 1685–9.
Microfilm P.3752: Letters from Robert Molesworth, later first Viscount Molesworth, to his wife, 1689–1724.
Microfilm P.3753: Letters to or from Richard, third Viscount Molesworth, 1727–44.

National Records of Scotland, Edinburgh
GD406/1: Correspondence of the Dukes of Hamilton, 1563–1712.

Northamptonshire Record Office, Northampton
IC: Correspondence of Isham family of Lamport.

North Yorkshire County Record Office, Northallerton
ZON13/1: Personal papers of Thomas Worsley (1649–1715) of Hovingham.
ZON13/2: Personal papers of Thomas Worsley II (1686–1751) of Hovingham.

Public Record Office of Northern Ireland, Belfast
D695: Waring Manuscripts, 1667–1797.
D695/225: Samuel Waring, travel journal, 1687.
D695/226: Samuel Waring, travel journal, 1687–8.
D695/227: Samuel Waring, notebook, 1687[–1714].
D695/228: Samuel Waring, travel notebook, 1687–8.
D2924/1: Copies of letters etc, mainly written by or received by the Rawdon family, 1722–*c.* 1800.
T3765/M/2/14: Granard Papers, 1750–87.

Royal Society, London
MS 248: Papers collected by Sir Robert Southwell, 17th century.

Shakespeare Birthplace Trust, Stratford-upon-Avon
DR18/20/18: Leigh family of Adlestrop, letters and papers, 1630–1800.
DR18/20/18b: Leigh family of Adlestrop, letters and papers, 1630–1800.
DR18/31/881: James Leigh, Accounts and description of European tour, 1747–8.
DR18/31/884a: Typescript copy of Mary Leigh, 'History of the Leigh Family of Adlestrop', 1788.
DR671/59a: William Leigh, Travel journal, 1710–11.
DR671/60: Draft inscriptions for Theophilus Leigh's monument, 1727.
DR671/61-67: William Leigh, Diaries while resident in the Netherlands, 1726–32.

Surrey History Centre, Woking
1248/2: Brodrick manuscripts, *c.* 1680–[1716].
1248/3: Brodrick manuscripts, 1700–17.
1248/4: Brodrick manuscripts, 1703–21.
1248/5: Brodrick manuscripts, 1715–24.
1248/6: Brodrick manuscripts, 1723–7.
1248/7: Brodrick manuscripts, *c.* 1710–1800.

The National Archives, Kew
PROB 11/488/419: Will of third Viscount Irwin, 24 Jun 1706 [died 1702].
SP29/411: Secretaries of State: State Papers Domestic, Charles II: Letters and papers, 1679.
SP31/3: Secretaries of State: State Papers Domestic, James II: Letters and papers, 1687.
SP44/25: Secretaries of State: State Papers: Entry Books: Warrants: Sir William Morice and Sir John Trevor, 1667–71.
SP44/51: Secretaries of State: State Papers: Entry Books: Warrants, with some proclamations: Coventry and Jenkins, 1678–81.
SP44/336: Secretaries of State: State Papers: Entry Books: Warrants and passes: Warrants, Sunderland, 1685–6.
SP44/346: Secretaries of State: State Papers: Entry Books: Warrants and passes, Trenchard and Trumbull, 1694–7.
SP44/390: Secretaries of State: State Papers: Entry Books: Warrants and passes, passes, including post warrants, 1704–6.
SP44/391: Secretaries of State: State Papers: Entry Books: Warrants and passes, passes, including post warrants, 1704–6.
SP63/310: State Paper Office: State Papers Ireland, Elizabeth I to George III, 1558–1782: Letters and papers, 1662.
SP78/127: Secretaries of State: State Papers Foreign, France, 1577–1780: W. Perwich, F. Vernon, St Albans, B. Whiting, R. Petit, R. Montague..., 1669.
SP85/15: Secretaries of State: State Papers Foreign, Italian States and Rome, 1544–1773: John Walton, Rome, 1724–5.
SP85/16: Secretaries of State: State Papers Foreign, Italian States and Rome, 1544–1773: John Walton, Rome, 1726–9.
SP93/1: Secretaries of State: State Papers Foreign, Sicily and Naples: Consuls Browne, Davies, etc, 1584–1674.
SP98/21: Secretaries of State: State Papers Foreign, Tuscany, 1582–1780: Sir Lambert Blackwell and consuls, 1703.
SP98/41: Secretaries of State: State Papers Foreign, Tuscany, 1582–1780: John Walton, 1737–9.
SP98/43: Secretaries of State: State Papers Foreign, Tuscany, 1582–1780: John Walton, 1740–1.

Trinity College Dublin
Lyons MS 750/4: Letters from William King, Archbishop of Dublin, 1712–15.
Lyons MS 750/8: Letters from King, 1725–7.
Lyons MSS 1995-2008/1821-1950: Letters to King, 1717–20.

Lyons MSS 1995-2008/2081-2210: Letters to King, 1724–37.
Lyons MS 2535: Transcripts of letters from King, 1716–18.
Lyons MS 2537: Transcripts of letters from King, 1723–5.

West Yorkshire Archive Service, Bradford
SpSt/5/4/1/1: Richard Rawdon, 'Irish Accompts', 1652–86.
SpSt/5/4/3/34: Bills, receipts and accounts of Rawdon family in England, 1654–86.
SpSt/6/1/7: Letters between Stanhope and Rawdon families, 1640–91.
SpSt/6/1/13: Letters from Elizabeth Rawdon to John Stanhope of Horsforth, 1682–90.

West Yorkshire Archive Service, Leeds
WYL100/C/7: Correspondence of Ingram family and John Roades, Temple Newsam estate steward, 1695–7.
WYL100/C/8: Correspondence of Ingram family and John Roades, 1698–1701.
WYL100/C/9A: Correspondence of Ingram family and John Roades, 1702–6.
WYL100/C/9B: Correspondence of Ingram family and John Roades, 1701–5.
WYL100/C/10: Correspondence of Ingram family, John Roades and his successor, Robert Hopkinson, 1706–13.
WYL100/C/11: Correspondence of Ingram family, 1714–17.
WYL100/C/12: Correspondence of Ingram family, 1718–22.
WYL100/C/20: Correspondence of Ingram family, undated, 17th–18th centuries.
WYL100/EA/10: Bills of exchange for Ingram family, 1661–1745.
WYL100/EA/12/17: Stewards' vouchers for education of Ingram family, 1623–1744.
WYL100/EA/15: Ingram family estate rentals, 1627–1764.
WYL150/6002: Letters from Sir William Robinson to his wife and sons, 1687–1728.
WYL150/6005: Letters from Metcalfe Robinson to his parents, [1704–14].

Yale Center for British Art, New Haven, CT
MS 8: Inventory of King's Weston, 1695.

Printed Primary Sources

'A Little News from Horsham': Some 18th Century Letters from Hills, ed. Susan C. Djabri (Horsham: Horsham Museum Society, 2002).
Addison, Joseph, *Remarks on Several Parts of Italy, &c. In the Years 1701, 1702, 1703* (London: 1705).
Anderson, James, *A Genealogical History of the House of Yvery, in its Different Branches of Yvery, Luvel, Perceval and Gournay*, 2 vols (London: 1742).
Austen-Leigh, R. A., *The Eton College Register, 1698–1752* (Eton: Spottiswoode, Ballantyne & Co., 1927).
Bacon, Francis, *The Essays*, ed. John Pitcher (London: Penguin, 1985).

Barker, G. F. Russell, and Alan H. Stenning, *The Record of Old Westminsters*, 2 vols (London: Chiswick, 1928).

Boyle, Robert, *The Correspondence of Robert Boyle*, ed. Michael Hunter, Antonio Clericuzio, and Lawrence Principe, 6 vols (London: Pickering & Chatto, 2001).

Brice, Germain, *A New Description of Paris*, trans. James Wright (London: 1687).

Brome, James, *Travels over England, Scotland and Wales* (London: 1700).

Bromley, William, *Remarks in the Grande Tour of France & Italy* (London: 1692).

Brown, Edward, *A Brief Account of Some Travels in Divers Parts of Europe* (London: 1685).

Brown, Horatio F., ed., *Inglesi e Scozzesi all' Università di Padova dall' anno 1618 sino al 1765* (Venice: 1921).

Brydges, Cassandra, *Cassandra Brydges, Duchess of Chandos, 1670–1735: Life and Letters*, ed. Rosemary O'Day (Woodbridge: Boydell, 2007).

Burtchaell, George Dames, and Thomas Ulick Sadleir, eds, *Alumni Dublinenses: A Register of the Students, Graduates, Professors and Provosts of Trinity College, in the University of Dublin* (London: Williams and Norgate, 1924).

Byrd, William, *Another Secret Diary of William Byrd of Westover, 1739–1741. With Letters and Literary Exercises, 1696–1726*, ed. Maude H. Woodfin, trans. Marion Tinling (Richmond, VA: Dietz Press, 1942).

Byrd, William, *The Correspondence of Three William Byrds of Westover, Virginia, 1684–1776*, ed. Marion Tinling, 2 vols (Charlottesville: University Press of Virginia, 1977).

Calendar of State Papers, Domestic Series, [CSPD; of the Commonwealth, 1649–60], 13 vols (London: Longman & Co. and Trübner & Co., 1875–86).

Calendar of State Papers, Domestic Series [CSPD], of the Reign of Charles II, 28 vols (London: Longman, Green, Longman & Roberts, 1860–1938).

Camden, William, *Britannia* (London: 1586).

Castiglione, Baldassare, *The Courtier: or, The Complete Gentleman and Gentlewoman* (London: 1729).

Catalogue ... Containing the Greater Portion of the United Libraries of Sir Robert Southwell ... his Son, the Hon. Edward Southwell, and the Late Lord De Clifford (London: Thomas Thorpe, 1834).

Chandos, Elizabeth Brydges, Baroness, 'Lady Chandos' Register', ed. C. H. Collins Baker, *The Genealogists' Magazine*, 10:8 (1948), 255–64.

Clarendon, Edward Hyde, Earl of, *A Collection of Several Tracts of the Right Honourable Earl of Clarendon* (London: 1727).

Conway, Anne, *The Conway Letters: The Correspondence of Anne, Viscountess Conway, Henry More, and Their Friends 1642–1684*, ed. Marjorie Hope Nicolson, rev. Sarah Hutton (Oxford: Clarendon, 1992).

Dalton, Charles, ed., *English Army Lists and Commission Registers, 1661–1714*, 6 vols (London: Eyre & Spottiswoode, 1892–1904).

Dalton, Charles, ed., *George I's Army, 1714–1727*, 2 vols (London: Eyre & Spottiswoode, 1910–12).

[De Blainville, P.], *A Diary of the Journey Through the North of England made by William and John Blathwayt of Dyrham Park in 1703*, ed. Nora Hardwick (Bristol: N. Hardwick, 1977).

[De Blainville, P.], *The Grand Tour: Letters and Accounts Relating to the Travels Through Europe of the Brothers William and John Blathwayt of Dyrham Park, 1705–1708*, ed. Nora Hardwick (Bristol: N. Hardwick, 1985).

Dering, Sir Edward, *The Diaries and Papers of Sir Edward Dering, Second Baronet, 1644 to 1684*, ed. Maurice F. Bond (London: HMSO, 1976).

Dumont de Bostaquet, Isaac, *Memoirs of Isaac Dumont de Bostaquet, a Gentleman of Normandy Before and After the Revocation of the Edict of Nantes*, ed. and trans. D. W. Ressinger (London: Huguenot Society of Great Britain and Ireland, 2005).

Foster, Joseph, ed., *Alumni Oxonienses: The Members of the University of Oxford, 1715–1886*, 4 vols (Oxford: Parker, 1891).

Foster, Joseph, ed., *Alumni Oxonienses: The Members of the University of Oxford, 1500–1714*, 4 vols (Oxford: Parker, 1891–2).

Gailhard, Jean, *The Present State of the Princes and Republicks of Italy* (London: 1668).

Gailhard, Jean, *The Present State of the Republick of Venice* (London: 1669).

Gailhard, Jean, *The Compleat Gentleman: or, Directions for the Education of Youth as to their Breeding at Home and Travelling Abroad*, 2 vols (London: 1678).

Gilbert, John, 'An Inventory of His Grace the Duke of Chandos's Seat att Cannons Taken June the 19th 1725', ed. Susan Jenkins, *The Volume of the Walpole Society*, 67 (2005), 93–192.

Gracián, Baltasar, *The Compleat Gentleman: or a Description of the Several Qualifications, both Natural and Acquired, that are Necessary to Form a Great Man* (London: 1730).

HMC (Historical Manuscripts Commission), *Report on Manuscripts in Various Collections, Vol. VIII* (London: HMSO, 1913).

HMC (Historical Manuscripts Commission), *Manuscripts of the Earl of Egmont: Diary of Viscount Percival, Afterwards First Earl of Egmont*, 3 vols (London: HMSO, 1920–3).

Howell, James, *Instructions for Forraine Travell* (London: 1642).

Hudson, John, *Fabularum Aesopicarum Collectio, Quotquot Graece Reperiuntur. Accedit Interpretatio Latina* (Oxford: 1718).

Lassels, Richard, *The Voyage of Italy, or A Compleat Journey through Italy* (Paris: 1670).

Lauder, Sir John, *Journals of Sir John Lauder, Lord Fountainhall, with his Observations on Public Affairs and other Memoranda, 1665–1676*, ed. Donald Crawford (Edinburgh: Edinburgh University Press, 1900).

Leigh, Edward, *England Described* (London: 1659).

Letters of Eminent Men, Addressed to Ralph Thoresby, F.R.S., 2 vols (London: Henry Colburn and Richard Bentley, 1832).

Lingard, Richard, *A Letter of Advice to a Young Gentleman Leaving the University* (London: 1671).

Locke, John, *Some Thoughts Concerning Education* (London: 1693).

Lyttelton, George, *An Epistle to Mr. Pope, from a Young Gentleman at Rome* (London: 1730).

Macky, John, *A Journey through England. In Familiar Letters from a Gentleman Here, to his Friend Abroad* (London: 1714).

Magalotti, Lorenzo, 'An Unexpected Visitor: Cosimo de'Medici's Visit to Cork in 1669', trans. Anne O'Connor, *Journal of the Cork Historical and Archaeological Society*, 111 (2006), 1–6.

Malmesbury, James Howard Harris, Earl of, ed., *A Series of Letters of the First Earl of Malmesbury, His Family and Friends, from 1745 to 1820* (London: R. Bentley, 1870).

Misson, [Henri], *M. Misson's Memoirs and Observations in his Travels over England. With Some Account of Scotland and Ireland*, trans. John Ozell (London: 1719).

Misson, Maximilien, *A New Voyage to Italy*, first English edn, 2 vols (London: 1695).

Misson, Maximilien, *A New Voyage to Italy*, second English edn, 2 vols (London: 1699).

Misson, Maximilien, *A New Voyage to Italy*, fourth English edn, 4 vols (London: 1714).

Misson, Maximilien, *Nouveau voyage d'Italie*, third French edn, 2 vols (The Hague: 1698).

Misson, Maximilien, *Nouveau voyage d'Italie*, fourth French edn, 3 vols (The Hague: 1702).

Molesworth, Robert, *An Account of Denmark, as it was in the Year 1692* (London: 1694).

Molesworth, Robert, *The True Way to Render Ireland Happy and Secure* (Dublin: 1697).

Montagu, Lady Mary Wortley, *The Complete Letters of Lady Mary Wortley Montagu*, ed. Robert Halsband, 3 vols (Oxford: Clarendon, 1965–7).

Mountague, William, *The Delights of Holland: or, A Three Months Travel about that and the other Provinces* (London: 1696).

Mountmorres, Hervey Redmond Morres, Viscount, *The History of the Principal Transactions of the Irish Parliament, from the Year 1634 to 1666*, 2 vols (London: 1792).

Newcastle, Margaret Cavendish, Marchioness of, *Orations of Divers Sorts Accommodated to Divers Places Written by the Lady Marchioness of Newcastle* (London: 1662).

Newton, Isaac, *The Correspondence of Isaac Newton*, ed. H. W. Turnbull, J. F. Scott, A. Rupert Hall and Laura Tilling, 7 vols (Cambridge: Cambridge University Press, 1959–77).

Nugent, Thomas, *The Grand Tour* (London: 1749).

Ohlmeyer, Jane, and Éamonn Ó Ciardha, eds, *The Irish Statute Staple Books, 1596–1687* (Dublin: Dublin Corporation, 1998).

Oldenburg, Henry, *The Correspondence of Henry Oldenburg*, ed. A. Rupert Hall and Marie Boas Hall, 13 vols (Madison: University of Wisconsin Press, 1965–86).

Osborne, Francis, *Advice to a Son* (Oxford: 1656).

Peacham, Henry, *The Compleat Gentleman Fashioning Him Absolute in the Most Necessary & Commendable Qualities Concerning Minde or Bodie that May Be Required in a Noble Gentlema[n]* (London: 1622).

Peacock, Edward, *Index to English-Speaking Students who have Graduated at Leyden University* (London: Longmans, Green & Co., 1883).

Perceval, Sir John, *The English Travels of Sir John Percival and William Byrd II: The Percival Diary of 1701*, ed. Mark R. Wenger (Columbia: University of Missouri Press, 1989).

Petty, Sir William, and Sir Robert Southwell, *The Petty-Southwell Correspondence 1676–1687*, ed. Henry William Edmund Petty-FitzMaurice, Marquis of Lansdowne (New York: Augustus M. Kelley, 1967).

[Pullen, Tobias], *A Vindication of Sr. Robert King's Designs and Actions, in Relation to the Late, and Present Lord Kingston* (Dublin: 1699).

Raguenet de Saint-Albin, Olivier, 'Livre des pensionnaires et des externes à l'Académie d'équitation d'Angers au XVIIIe siècle (1755–1790), d'après un document inédit', *Revue de l'Anjou*, 68 (1914), 161–78, 339–60.

Sheridan, Thomas, *A Plan of Education for the Young Nobility and Gentry of Great Britain* (Dublin: 1769).

Smith, Thomas, *De Republica Anglorum* (London: 1583).

Spence, Joseph, *Joseph Spence: Letters from the Grand Tour*, ed. Slava Klima (Montreal: McGill-Queen's University Press, 1975).

Sterry, Wasey, *The Eton College Register, 1441–1698* (Eton: Spottiswoode, Ballantyne & Co., 1943).

Stukeley, William, *Itinerarium Curiosum* (London: 1724).

Sturgess, H. A. C., *Register of Admissions to the Honourable Society of the Middle Temple: From the Fifteenth Century to the Year 1944*, 3 vols (London: Butterworth, 1949).

The Adventures of Mr. Loveill, Interspers'd with Many Real Amours of the Modern Polite World (London: 1750).

Theatri Oxoniensis Encaenia (Oxford: 1677).

The Visor Pluckt Off from Richard Thompson of Bristol, Clerk in a Plain and True Character of Him (London: 1681).

Vertue, George, 'Vertue Note Books: Volume III', *The Volume of the Walpole Society*, 22 (1933–4).

Walpole, Horace, *The Yale Edition of Horace Walpole's Correspondence*, ed. W. S. Lewis, 48 vols (New Haven, CT: Yale University Press, 1937–83).

Secondary Sources

Adler, Judith, 'Travel as Performed Art', *American Journal of Sociology*, 94:6 (1989), 1366–91.

Anderson, Jocelyn, *Touring and Publicizing England's Country Houses in the Long Eighteenth Century* (New York: Bloomsbury Academic, 2018).

Andrews, Malcolm, *The Search for the Picturesque: Landscape Aesthetics and Tourism in Britain, 1760–1800* (Aldershot: Scolar, 1989).

Ansell, Richard, 'Educational Travel in Protestant Families from Post-Restoration Ireland', *Historical Journal*, 58:4 (2015), 931–58.

Ansell, Richard, 'Foubert's Academy: British and Irish Elite Formation in Seventeenth- and Eighteenth-Century Paris and London', in Rosemary Sweet, Gerrit Verhoeven and Sarah Goldsmith, eds, *Beyond the Grand Tour: Northern Metropolises and Early Modern Travel Behaviour* (London: Routledge, 2017), pp. 46–64.

Ansell, Richard, 'Reading and Writing Travels: Maximilien Misson, Samuel Waring and the Afterlives of European Voyages, *c.*1687–1714', *English Historical Review*, 133:565 (2018), 1446–77.

Appadurai, Arjun, 'Introduction: Commodities and the Politics of Value', in Appadurai, ed., *The Social Life of Things: Commodities in Cultural Perspective* (Cambridge: Cambridge University Press, 1986), pp. 3–63.

Asch, Ronald G., *Nobilities in Transition 1550–1700: Courtiers and Rebels in Britain and Europe* (London: Arnold, 2003).

Aylmer, G. E., *The Crown's Servants: Government and Civil Service under Charles II, 1660-1685* (Oxford: Oxford University Press, 2002).

Barker, Hannah, *Family and Business during the Industrial Revolution* (Oxford: Oxford University Press, 2017).

Barnard, T. C. [Toby], 'Crises of Identity among Irish Protestants 1641–1685', *Past & Present*, 127 (1990), 39–83.

Barnard, Toby, 'Protestantism, Ethnicity and Irish Identities, 1660–1760', in Tony Claydon and Ian McBride, eds, *Protestantism and National Identity: Britain and Ireland, c. 1650–c. 1850* (Cambridge: Cambridge University Press, 1998), pp. 206–35.

Barnard, Toby, *A New Anatomy of Ireland: The Irish Protestants, 1649–1770* (New Haven, CT, and London: Yale University Press, 2003).

Barnard, Toby, *Irish Protestant Ascents and Descents* (Dublin: Four Courts, 2003).

Barnard, Toby, *Making the Grand Figure: Lives and Possessions in Ireland, 1641–1770* (New Haven, CT, and London: Yale University Press, 2004).

Barnard, Toby, *Brought to Book: Print in Ireland, 1680–1784* (Dublin: Four Courts, 2017).

Bates, Richard, 'The Petit Tour to Spa, 1763–87', in Rosemary Sweet, Gerrit Verhoeven and Sarah Goldsmith, eds, *Beyond the Grand Tour: Northern Metropolises and Early Modern Travel Behaviour* (London: Routledge, 2017), pp. 127–46.

Ben-Amos, Ilana Krausman, 'Reciprocal Bonding: Parents and Their Offspring in Early Modern England', *Journal of Family History*, 25:3 (2000), 291–312.

Bepler, Jill, 'Travelling and Posterity: The Archive, the Library and the Cabinet', in Rainer Babel and Werner Paravicini, eds, *Grand Tour: Adeliges Reisen und Europäische Kultur vom 14. bis zum 18. Jahrhundert* (Ostfildern: Thorbecke, 2005), pp. 191–203.

Bianchi, Paola, 'In cerca del moderno: studenti e viaggiatori inglesi a Torino nel Settecento', *Rivista storica italiana*, 115:3 (2003), 1021–51.

Bianchi, Paola, and Karin Wolfe, eds, *Turin and the British in the Age of the Grand Tour* (Cambridge: Cambridge University Press, 2017).

Bignamini, Ilaria, 'The Grand Tour: Open Issues', in Andrew Wilton and Ilaria Bignamini, eds, *Grand Tour: The Lure of Italy in the Eighteenth Century* (London: Tate Gallery, 1996), pp. 31–5.

Bignamini, Ilaria, 'The Italians as Spectators and Actors: The Grand Tour Reflected', in Clare Hornsby, ed., *The Impact of Italy: The Grand Tour and Beyond* (London: British School at Rome, 2000), pp. 29–47.

Black, Jeremy, *The British Abroad: The Grand Tour in the Eighteenth Century* (Stroud: Sutton Publishing, 2003).

Black, Jeremy, *France and the Grand Tour* (Basingstoke: Palgrave Macmillan, 2003).
Black, Jeremy, *Italy and the Grand Tour* (New Haven, CT, and London: Yale University Press, 2003).
Booker, John, *Travellers' Money* (Stroud: Sutton Publishing, 1994).
Borsay, Peter, 'Children, Adolescents and Fashionable Urban Society in Eighteenth-Century England', in Anja Müller, ed., *Fashioning Childhood in the Eighteenth Century: Age and Identity* (Aldershot: Ashgate, 2006), pp. 53–62.
Boschiero, Luciano, 'Robert Southwell and Vincenzio Viviani: Their Friendship and an Attempt at Italian-English Scientific Collaboration', *Parergon*, 26 (2009), 87–108.
Boutier, Jean, 'Le grand tour: une pratique d'éducation des noblesses européennes (XVIe–XVIIIe siècles', in *Le voyage à l'époque moderne* (Paris: Presses de l'université de Paris-Sorbonne, 2004), pp. 7–21.
Boutier, Jean, 'Le Grand Tour des gentilhommes et les academies d'éducation pour la noblesse: France et Italie, XVIe–XVIIIe siècle', in Rainer Babel and Werner Paravicini, eds, *Grand Tour: Adeliges Reisen und Europäische Kultur vom 14. bis zum 18. Jahrhundert* (Ostfildern: Thorbecke, 2005), pp. 237–53.
Boutier, Jean, 'Compétence internationale, émergence d'une "profession" et circulation des savoirs: le tuteur aristocratique dans l'Angleterre du XVIIe siècle', in Maria Pia Paoli, ed., *Saperi a confronto nell'Europa dei secoli XIII–XIX* (Pisa: Edizioni della Normale, 2009), pp. 151–79.
Boutier, Jean, 'Inventing the "Grand Tour": An Historiographical Success Story, between Anticipation and *Nachleben*', paper delivered at the *Art of Travel* conference, NUI Galway, 7 November 2016.
Bowden, Caroline, 'Convent Schooling for English Girls in the "Exile" Period, 1600–1800', *Studies in Church History*, 55 (2019), 177–204.
Bowron, Edgar Peters, and Peter Björn Kerber, *Pompeo Batoni: Prince of Painters in Eighteenth-Century Rome* (New Haven, CT: Yale University Press, 2007).
Bracewell, Wendy, 'Arguing from Experience: Travelees versus Travelers in Early Modern Exchanges', *Renaissance Studies*, 3:4 (2019), 548–67.
Braddick, Michael J., *State Formation in Early Modern England c. 1550–1700* (Cambridge: Cambridge University Press, 2000).
Brauer, G. C., *The Education of a Gentleman: Theories of Gentlemanly Education in England, 1660–1775* (New York: Bookman Associates, 1959).
Brennan, Michael G., ed., *The Origins of the Grand Tour: The Travels of Robert Montagu, Lord Mandeville (1649–1654), William Hammond (1655–1658) and Banaster Maynard (1660–1663)* (London: Hakluyt Society, 2004).
Brewer, John, 'Whose Grand Tour?', in María Dolores Sánchez-Jáuregui and Scott Wilcox, eds, *The English Prize: The Capture of the Westmorland, An Episode of the Grand Tour* (New Haven, CT, and London: Yale University Press, 2012), pp. 45–61.
Brown, Iain Gordon, 'Water, Windows, and Women: The Significance of Venice for Scots in the Age of the Grand Tour', *Eighteenth-Century Life*, 30:3 (2006), 1–50.
Brundin, Abigail, and Dunstan Roberts, 'Book-Buying and the Grand Tour: The Italian Books at Belton House in Lincolnshire', *The Library*, 16 (2015), 51–79.
Bryant, Margaret E., *The London Experience of Secondary Education* (London: Athlone Press, 1986).

Burke, Peter, *Languages and Communities in Early Modern Europe* (Cambridge: Cambridge University Press, 2004).
Calaresu, Melissa, 'Looking for Virgil's Tomb: The End of the Grand Tour and the Cosmopolitan Ideal in Europe', in Jaś Elsner and Joan-Pau Rubiés, eds, *Voyages and Visions: Towards a Cultural History of Travel* (London: Reaktion, 1999), pp. 138–61.
Calaresu, Melissa, 'Thomas Jones' Neapolitan Kitchen: The Material Cultures of Food on the Grand Tour', *Journal of Early Modern History*, 24:1 (2020), 84–102.
Cannon, John, *Aristocratic Century: The Peerage of Eighteenth-Century England* (Cambridge: Cambridge University Press, 1984).
Capp, Bernard, *The Ties that Bind: Siblings, Family, and Society in Early Modern England* (Oxford: Oxford University Press, 2018).
Carey, Daniel, 'Compiling Nature's History: Travellers and Travel Narratives in the Early Royal Society', *Annals of Science*, 54:3 (1997), 269–92.
Carter, Philip, *Men and the Emergence of Polite Society, Britain 1660–1800* (Harlow: Longman, 2001).
Ceserani, Giovanna, Giorgio Caviglia, Nicole Coleman, Thea de Armond, Sarah Murray and Molly Taylor-Poleskey, 'British Travelers in Eighteenth-Century Italy: The Grand Tour and the Profession of Architecture', *American Historical Review*, 122:2 (2017), 425–50.
Chambers, Liam, 'The Irish in Europe in the Eighteenth Century, 1691–1815', in James Kelly, ed., *The Cambridge History of Ireland, Volume III, 1730–1880* (Cambridge: Cambridge University Press, 2018), pp. 569–92.
Chaney, Edward, *The Grand Tour and the Great Rebellion: Richard Lassels and 'The Voyage of Italy' in the Seventeenth Century* (Geneva: Slatkine, 1985).
Chaney, Edward, *The Evolution of the Grand Tour: Anglo-Italian Cultural Relations since the Renaissance* (London: Routledge, 2000).
Chaney, Edward, *The Evolution of English Collecting: Receptions of Italian Art in the Tudor and Stuart Periods* (New Haven, CT, and London: Yale University Press, 2003).
Chaney, Edward, and Timothy Wilks, *The Jacobean Grand Tour: Early Stuart Travellers in Europe* (London: I.B. Tauris, 2014).
Chard, Chloe, *Pleasure and Guilt on the Grand Tour: Travel Writing and Imaginative Geography, 1600–1830* (Manchester: Manchester University Press, 1999).
Chard, Chloe, *A Critical Reader of the Romantic Grand Tour: Tristes Plaisirs* (Manchester: Manchester University Press, 2014).
Chareyre, Philippe, 'Les protestants de Saumur au XVIIe siècle, religion et société', in *Saumur, capitale européenne du protestantisme au XVIIe siècle* (Fontevrault-l'Abbaye: Abbaye royale de Fontevraud, 1991), pp. 27–70.
Childs, John, *The Army of Charles II* (London: Routledge & Kegan Paul, 1976).
Chodějovská, Eva, and Zdeněk Hojda, 'Abroad, or Still "at Home"? Young Noblemen from the Czech Lands and the Empire in the Seventeenth and Eighteenth Centuries', in Rosemary Sweet, Gerrit Verhoeven and Sarah Goldsmith, eds, *Beyond the Grand Tour: Northern Metropolises and Early Modern Travel Behaviour* (London: Routledge, 2017), pp. 83–107.
Claydon, Tony, *Europe and the Making of England, 1660–1760* (Cambridge: Cambridge University Press, 2007).

Cliffe, J. T., *The Yorkshire Gentry from the Reformation to the Civil War* (London: Athlone Press, 1969).

Cohen, Michèle, 'The Grand Tour: Constructing the English Gentleman in Eighteenth-Century France', *History of Education*, 21:3 (1992), 241–57.

Cohen, Michèle, *Fashioning Masculinity: National Identity and Language in the Eighteenth Century* (London: Routledge, 1996).

Cohen, Michèle, 'Manliness, Effeminacy and the French: Gender and the Construction of National Character in Eighteenth-Century England', in Tim Hitchcock and Michèle Cohen, eds, *English Masculinities, 1660–1800* (London: Longman, 1999), pp. 44–61.

Cohen, Michèle, 'The Grand Tour: Language, National Identity and Masculinity', *Changing English*, 8:2 (2001), 129–41.

Colley, Linda, *Britons: Forging the Nation, 1707–1837* (New Haven, CT, and London: Yale University Press, 1992).

Collins, Jeffrey, 'Sites and Sightseers: Rome through Foreign Eyes', in Pamela M. Jones, Barbara Wisch and Simon Ditchfield, eds, *A Companion to Early Modern Rome, 1492–1692* (Leiden and Boston, MA: Brill, 2019), pp. 564–81.

Connolly, S. J., *Religion, Law, and Power: The Making of Protestant Ireland, 1660–1760* (Oxford: Clarendon, 1992).

Conway, Stephen, *Britain, Ireland and Continental Europe in the Eighteenth Century: Similarities, Connections, Identities* (Oxford: Oxford University Press, 2011).

Corens, Liesbeth, 'Catholic Nuns and English Identities. English Protestant Travellers on the English Convents in the Low Countries, 1660–1730', *Recusant History*, 30:3 (2011), 441–59.

Corens, Liesbeth, *Confessional Mobility and English Catholics in Counter-Reformation Europe* (Oxford: Oxford University Press, 2019).

Corp, Edward T., *The Stuarts in Italy, 1719–1766: A Royal Court in Permanent Exile* (Cambridge: Cambridge University Press, 2011).

Coutu, Joan, *Then and Now: Collecting and Classicism in Eighteenth-Century England* (Montreal: McGill-Queen's University Press, 2015).

Coward, Barry, 'The Social and Political Position of the Earls of Derby in Later Seventeenth-Century Lancashire', *Transactions of the Historic Society of Lancashire and Cheshire*, 132 (1982/3), 127–54.

Cramsie, John, *British Travellers and the Encounter with Britain, 1450–1700* (Woodbridge: Boydell, 2015).

Cressy, David, 'Educational Opportunity in Tudor and Stuart England', *History of Education Quarterly*, 16:3 (1976), 301–20.

Cressy, David, 'Kinship and Kin Interaction in Early Modern England', *Past & Present*, 113:1 (1986), 38–69.

Cust, Lionel, *History of the Society of Dilettanti*, ed. Sidney Colvin (London: Macmillan, 1898).

Delorme, Edmond, *Lunéville et son arrondissement*, 2 vols (Marseille: Laffitte Reprints, 1977).

Dent, Keith, 'The Informal Education of the Landed Classes in the Eighteenth Century, with Particular Reference to Reading', Ph.D. thesis (Birmingham, 1974).

Dent, Keith, 'Travel as Education: The English Landed Classes in the Eighteenth Century', *Educational Studies*, 1:3 (1975), 171–80.

De Ridder-Symoens, Hilda, 'Mobility', in De Ridder-Symoens, ed., *A History of the University in Europe, Volume II: Universities in Early Modern Europe* (Cambridge: Cambridge University Press, 1996), pp. 416–48.

De Schickler, Fernand, *Les églises du refuge en Angleterre*, 3 vols (Paris: Fischbacher, 1892).

De Seta, Cesare, 'Grand Tour: The Lure of Italy in the Eighteenth Century', in Andrew Wilton and Ilaria Bignamini, eds, *Grand Tour: The Lure of Italy in the Eighteenth Century* (London: Tate Gallery, 1996), pp. 13–19.

Devlin, Eoin, 'English Encounters with Papal Rome in the Late Counter-Reformation, *c.*1685–*c.*1697', Ph.D. thesis (Cambridge, 2010).

DIB (*Dictionary of Irish Biography*), Royal Irish Academy, 2009–18; online edn, 2021 (www.dib.ie).

Dickson, David, 'Jacobitism in Eighteenth-Century Ireland: A Munster Perspective', *Éire-Ireland*, 39 (2004), 38–99.

Dickson, P. G. M., and J. V. Beckett, 'The Finances of the Dukes of Chandos: Aristocratic Inheritance, Marriage, and Debt in Eighteenth-Century England', *Huntington Library Quarterly*, 64:3/4 (2001), 309–55.

Dolan, Brian, *Ladies of the Grand Tour* (London: Flamingo, 2002).

Doucet, Corinne, 'Les académies équestres et l'éducation de la noblesse (XVIe–XVIIIe siècle)', *Revue historique*, 628 (2003/4), 817–36.

Eagles, Robin, *Francophilia in English Society, 1748–1815* (Basingstoke: Macmillan, 2000).

Ezell, Margaret J. M., 'John Locke's Images of Childhood: Early Eighteenth-Century Response to *Some Thoughts Concerning Education*', *Eighteenth-Century Studies*, 17:2 (1983–4), 139–55.

Fantoni, Marcello, 'The Grand Duchy of Tuscany: The Courts of the Medici 1532–1737', in John Adamson, ed., *The Princely Courts of Europe: Ritual, Politics and Culture under the Ancien Régime 1500–1750* (London: Seven Dials, 2000), pp. 255–73.

Feingold, Mordechai, 'The Humanities', in Nicholas Tyacke, ed., *The History of the University of Oxford: Volume IV: Seventeenth-Century Oxford* (Oxford: Clarendon, 1997), pp. 211–358.

Firth, C. H., *Modern Languages at Oxford, 1724–1929* (London: Oxford University Press, 1929).

Fletcher, Anthony, *Gender, Sex and Subordination in England, 1500–1800* (New Haven, CT, and London: Yale University Press, 1995).

Fletcher, Anthony, *Growing up in England: The Experience of Childhood, 1600–1914* (New Haven, CT, and London: Yale University Press, 2008).

Fletcher, Anthony, 'The Ambition of a Young Baronet: Sir Thomas Isham of Lamport, 1657–1681', in Angela McShane and Garthine Walker, eds, *The Extraordinary and the Everyday in Early Modern England* (Basingstoke: Palgrave Macmillan, 2010), pp. 31–47.

Fox, Adam, 'Words, Words, Words: Education, Literacy and Print', in Keith Wrightson, ed., *A Social History of England, 1500–1750* (Cambridge: Cambridge University Press, 2017), pp. 129–51.

Fremantle, Katharine, 'A Visit to the United Provinces and Cleves in the Time of William III: Described in Edward Southwell's Journal', *Nederlands Kunsthistorisch Jaarboek*, 21 (1970), 39–68.

French, Henry, 'Gentlemen', in Keith Wrightson, ed., *A Social History of England, 1500–1750* (Cambridge: Cambridge University Press, 2017), pp. 269–89.

French, Henry, 'The "Remembered Family" and Dynastic Senses of Identity among the English Gentry *c.*1600–1800', *Historical Research*, 92:257 (2019), 529–46.

French, Henry, and Mark Rothery, *Man's Estate: Landed Gentry Masculinities, c.1660–1900* (Oxford: Oxford University Press, 2012).

French, Henry, and Mark Rothery, 'Male Anxiety among Younger Sons of the English Landed Gentry, 1700–1900', *Historical Journal*, 62:4 (2019), 967–95.

Frijhoff, Willem, 'Étudiants étrangers à l'académie d'équitation d'Angers au XVIIe siècle', *Lias*, 4 (1977), 13–84.

Frijhoff, Willem, 'Éducation, savoir, compétence: les transformations du Grand Tour dans les Provinces-Unies à l'époque moderne', in Rainer Babel and Werner Paravicini, eds, *Grand Tour: Adeliges Reisen und Europäische Kultur vom 14. bis zum 18. Jahrhundert* (Ostfildern: Thorbecke, 2005), pp. 609–35.

Gallagher, John, 'The Italian London of John North: Cultural Contact and Linguistic Encounter in Early Modern England', *Renaissance Quarterly*, 70:1 (2017), 88–131.

Gallagher, John, *Learning Languages in Early Modern England* (Oxford: Oxford University Press, 2019).

Gallagher, John, 'Language and Education on the Grand Tour of Sir Philip Perceval', in Helmut Glück, Mark Häberlein and Andreas Flurschütz da Cruz, eds, *Adel und Mehrsprachigkeit in der Frühen Neuzeit. Ziele, Formen und Praktiken des Erwerbs und Gebrauchs von Fremdsprachen* (Wiesbaden: Harrassowitz Verlag in Kommission, 2019), pp. 113–33.

Games, Alison, *The Web of Empire: English Cosmopolitans in an Age of Expansion, 1560–1660* (Oxford: Oxford University Press, 2008).

Ghazvinian, John H., '"A Certain Tickling Humour": English Travellers, 1560–1660', D.Phil. thesis (Oxford, 2003).

Ghobrial, John-Paul A., 'Moving Stories and What They Tell Us: Early Modern Mobility between Microhistory and Global History', in Ghobrial, ed., *Global History and Microhistory* (Oxford: Oxford University Press, 2019), pp. 243–80.

Gilbert, Christopher, 'A Nobleman and the Grand Tour: Lord Irwin and Marco Ricci', *Apollo: The International Magazine of Arts*, 51 (1966), 358–63.

Glickman, Gabriel, *The English Catholic Community, 1688–1745: Politics, Culture and Ideology* (Woodbridge: Boydell, 2009).

Glover, Katharine, *Elite Women and Polite Society in Eighteenth-Century Scotland* (Woodbridge: Boydell, 2011).

Goldsmith, Sarah, 'The Social Challenge: Northern and Central European Societies on the Eighteenth-Century Aristocratic Grand Tour', in Rosemary Sweet, Gerrit Verhoeven and Sarah Goldsmith, eds, *Beyond the Grand Tour: Northern Metropolises and Early Modern Travel Behaviour* (London: Routledge, 2017), pp. 65–82.

Goldsmith, Sarah, 'Nostalgia, Homesickness and Emotional Formation on the Eighteenth-Century Grand Tour', *Cultural and Social History*, 15:3 (2018), 333–60.

Goldsmith, Sarah, *Masculinity and Danger on the Eighteenth-Century Grand Tour* (London: University of London Press, 2020).

Goldsmith, Sarah, 'Imagining Youth: Epistolary Representations of the Eighteenth-Century Adolescent and Youthful Body, *c.* 1740–1800', in Sarah Goldsmith, Karen Harvey and Sheryllynne Haggerty, eds, *Letters and the Body, 1700–1830* (London: Routledge, 2022).

Graham, Aaron, 'Auditing Leviathan: Corruption and State Formation in Early Eighteenth-Century Britain', *English Historical Review*, 128:533 (2013), 806–38.

Graham, Aaron, *Corruption, Party, and Government in Britain, 1702–1713* (Oxford: Oxford University Press, 2015).

Graham, Aaron, 'Connoisseurship, Consumption, Company, and James Brydges, First Duke of Chandos, 1705–13', *Huntington Library Quarterly*, 80:4 (2017), 539–57.

Green, Michael, 'A Huguenot Education for the Early Modern Nobility', *Proceedings of the Huguenot Society*, 30:1 (2013), 73–91.

Green, Michael, 'Bridging the English Channel: Huguenots in the Educational Milieu of the English Upper Class', *Paedagogica Historica*, 54:4 (2018), 389–409.

Hague, Stephen, *The Gentleman's House in the British Atlantic World 1680–1780* (Basingstoke: Palgrave Macmillan, 2015).

Hall, R., 'Whig Party Fortunes in the Yorkshire County Election of 1708', *Northern History*, 32:1 (1996), 111–32.

Hareven, Tamara K., 'The History of the Family and the Complexity of Social Change', *American Historical Review*, 96:1 (1991), 95–124.

Harris, Amy, *Siblinghood and Social Relations in Georgian England: Share and Share Alike* (Manchester: Manchester University Press, 2012).

Hayton, D. W., *The Anglo-Irish Experience, 1680–1730: Religion, Identity, and Patriotism* (Woodbridge: Boydell, 2012).

Hayton, D. W., Eveline Cruickshanks and Stuart Handley, eds, *The History of Parliament: The House of Commons 1690–1715* (Cambridge: Cambridge University Press, 2002).

Heal, Ambrose, *The English Writing-Masters and their Copy-Books, 1570–1800* (Hildesheim: Georg Olms Verlagsbuchhandlung, 1962).

Heal, Felicity, *Hospitality in Early Modern England* (Oxford: Clarendon, 1990).

Heal, Felicity, and Clive Holmes, *The Gentry in England and Wales, 1500–1700* (Basingstoke: Macmillan, 1994).

Henning, B. D., ed., *The History of Parliament: The House of Commons 1660–1690* (London: Secker & Warburg, 1983).

Hill, Rosemary, *Stonehenge* (London: Profile, 2008).

Hiscock, W. G., *Henry Aldrich of Christ Church, 1648–1710* (Oxford: Holywell Press, 1960).

Holland, Robert, *The Warm South: How the Mediterranean Shaped the British Imagination* (New Haven, CT: Yale University Press, 2018).

Hollingsworth, T. H., *The Demography of the British Peerage* (London: Population Investigation Committee, London School of Economics, 1964).

Hollingsworth, T. H., 'Mortality in the British Peerage Families since 1600', *Population*, 32:1 (1977), 323–52.

Holmberg, Eva Johanna, 'Writing the Travelling Self: Travel and Life-Writing in Peter Mundy's (1597–1667) *Itinerarium Mundii*', *Renaissance Studies*, 31:4 (2017), 608–25.

Holmes, Geoffrey, *Augustan England: Professions, State and Society, 1680–1730* (London: Allen & Unwin, 1982).

Holmes, Geoffrey, *British Politics in the Age of Anne* (London: Hambledon, 1987).

Horn, D. B., *The British Diplomatic Service, 1689–1789* (Oxford: Clarendon, 1961).

Hornsby, Clare, ed., *The Impact of Italy: The Grand Tour and Beyond* (London: British School at Rome, 2000).

Houlbrooke, Ralph A., *The English Family 1450–1700* (London: Longman, 1984).

Howard, Clare, *English Travellers of the Renaissance* (London and New York: J. Lane, 1914).

Howarth, David, *Lord Arundel and His Circle* (New Haven, CT, and London: Yale University Press, 1985).

Howarth, David, *Images of Rule: Art and Politics in the English Renaissance, 1485–1649* (Basingstoke: Macmillan, 1997).

Hunter, Michael, *Boyle: Between God and Science* (New Haven, CT, and London: Yale University Press, 2009).

Iliffe, Robert, 'Foreign Bodies: Travel, Empire and the Early Royal Society of London. Part 1. Englishmen on Tour', *Canadian Journal of History*, 33:3 (1998), 358–85.

Ingamells, John, *A Dictionary of British and Irish Travellers in Italy, 1701–1800* (New Haven, CT, and London: Yale University Press, 1997).

Jacobsen, Helen, *Luxury and Power: The Material World of the Stuart Diplomat, 1660–1714* (Oxford: Oxford University Press, 2012).

Jenkins, Susan, *Portrait of a Patron: The Patronage and Collecting of James Brydges, 1st Duke of Chandos (1674–1744)* (Aldershot: Ashgate, 2007).

Johnston-Liik, E. M., *History of the Irish Parliament 1692–1800: Commons, Constituencies and Statutes*, 6 vols (Belfast: Ulster Historical Foundation, 2002).

Kelly, Jason M., 'Riots, Revelries, and Rumor: Libertinism and Masculine Association in Enlightenment London', *Journal of British Studies*, 45:4 (2006), 759–95.

Kelly, Jason M., *The Society of Dilettanti: Archaeology and Identity in the British Enlightenment* (New Haven, CT, and London: Yale University Press, 2009).

Kelly, Jason M., 'Reading the Grand Tour at a Distance: Archives and Datasets in Digital History', *American Historical Review*, 122:2 (2017), 451–63.

Kennedy, Máire, *French Books in Eighteenth-Century Ireland* (Oxford: Voltaire Foundation, 2001).

Kenny, Neil, *The Uses of Curiosity in Early Modern France and Germany* (Oxford: Oxford University Press, 2004).

King, Russell, Allan Findlay and Jill Ahrens, *International Student Mobility Literature Review* (Bristol: Higher Education Funding Council for England, 2010).

Kinsley, Zoë, *Women Writing the Home Tour, 1682–1812* (Aldershot: Ashgate, 2008).

Lambley, Kathleen, *The Teaching and Cultivation of the French Language in England During Tudor and Stuart Times* (Manchester: Manchester University Press, 1920).

Langford, Paul, *Public Life and the Propertied Englishman, 1689–1798* (Oxford: Clarendon, 1991).

Langford, Paul, *Englishness Identified: Manners and Character, 1650–1850* (Oxford: Oxford University Press, 2000).

Leibetseder, Mathis, 'Across Europe: Educational Travelling of German Noblemen in a Comparative Perspective', *Journal of Early Modern History*, 14 (2010), 417–49.

Lemmings, David, *Gentlemen and Barristers: The Inns of Court and the English Bar, 1680–1730* (Oxford: Clarendon, 1990).

Levack, Brian P., 'Law', in Nicholas Tyacke, ed., *The History of the University of Oxford: Volume IV: Seventeenth-Century Oxford* (Oxford: Clarendon, 1997), pp. 559–68.

Littlewood, Ian, *Sultry Climates: Travel and Sex since the Grand Tour* (London: John Murray, 2001).

Lock, Alexander, *Catholicism, Identity and Politics in the Age of Enlightenment: The Life and Career of Sir Thomas Gascoigne, 1745–1810* (Woodbridge: Boydell, 2016).

Lomax, James, *Temple Newsam Paintings* (Leeds: Leeds Museums & Galleries, 2000).

Lough, John, *France Observed in the Seventeenth Century by British Travellers* (Stocksfield: Oriel Press, 1985).

Love, Harold, and Arthur F. Marotti, 'Manuscript Transmission and Circulation', in David Loewenstein and Janel Mueller, eds, *The Cambridge History of Early Modern English Literature* (Cambridge: Cambridge University Press, 2002), pp. 55–80.

Lucas, Paul, 'A Collective Biography of Students and Barristers of Lincoln's Inn, 1680–1804: A Study in the "Aristocratic Resurgence" of the Eighteenth Century', *Journal of Modern History*, 46:2 (1974), 227–61.

Mączak, Antoni, *Travel in Early Modern Europe*, trans. Ursula Phillips (Cambridge: Polity, 1995).

Manning, Roger B., *Swordsmen: The Martial Ethos in the Three Kingdoms* (Oxford: Oxford University Press, 2003).

Manning, Roger B., *An Apprenticeship in Arms: The Origins of the British Army 1585–1702* (Oxford: Oxford University Press, 2006).

Marshall, David R., Susan Russell and Karin Wolfe, eds, *Roma Britannica: Art Patronage and Cultural Exchange in Eighteenth-Century Rome* (London: British School at Rome, 2011).

Mascuch, Michael, 'Continuity and Change in a Patronage Society: The Social Mobility of British Autobiographers, 1600–1750', *Journal of Historical Sociology*, 7:2 (1994), 177–97.

Mascuch, Michael, 'Social Mobility and Middling Self-Identity: The Ethos of British Autobiographers, 1600–1750', *Social History*, 20:1 (1995), 45–61.

Matteson, Robert S., *A Large Private Park: The Collection of Archbishop William King, 1650–1729*, 2 vols (Cambridge: LP, 2003).

Mayo, Hugh, 'Robert Molesworth's *Account of Denmark*: Its Roots and Its Impact', Ph.D. thesis (University of Southern Denmark, 2000).

McClelland, Charles E., *State, Society, and University in Germany 1700–1914* (Cambridge: Cambridge University Press, 1980).

McCormack, Matthew, 'Dance and Drill', *Cultural and Social History*, 8:3 (2011), 315–30.

McIntyre, Sylvia, 'Bath: The Rise of a Resort Town, 1660–1800', in Peter Clark, ed., *Country Towns in Pre-Industrial England* (Leicester: Leicester University Press, 1981), pp. 198–249.

McRae, Andrew, *Literature and Domestic Travel in Early Modern England* (Cambridge: Cambridge University Press, 2009).

Mead, William Edward, *The Grand Tour in the Eighteenth Century* (Boston, MA, and New York: Houghton Mifflin, 1914).

Middleton, W. E. Knowles, *A History of the Thermometer and Its Use in Meteorology* (Baltimore, MD: Johns Hopkins University Press, 1966).

Middleton, W. E. Knowles, *The Experimenters: A Study of the Accademia del Cimento* (Baltimore, MD: Johns Hopkins University Press, 1971).

Mijers, Esther, 'Irish Students in the Netherlands, 1650–1750', *Archivium Hibernicum*, 59 (2005), 66–78.

Mijers, Esther, *'News from the Republick of Letters': Scottish Students, Charles Mackie, and the United Provinces, 1650–1750* (Leiden: Brill, 2012).

Mingay, G. E., *English Landed Society in the Eighteenth Century* (London: Routledge & Kegan Paul, 1963).

Moir, Esther, *The Discovery of Britain: The English Tourists, 1540–1840* (London: Routledge & Kegan Paul, 1964).

Molesworth, William, 'John Molesworth (1679–1726) as a Patron of Art: Complacence, Connoisseurship and Commissions', M.Litt. thesis (Trinity College Dublin, 2010).

Morgan, Victor, *A History of the University of Cambridge: Volume 2, 1546–1750* (Cambridge: Cambridge University Press, 2004).

Mori, Jennifer, *The Culture of Diplomacy: Britain in Europe, c. 1750–1830* (Manchester: Manchester University Press, 2010).

Motley, Mark, *Becoming a French Aristocrat: The Education of the Court Nobility, 1580–1715* (Princeton, NJ: Princeton University Press, 1990).

Muir, Rory, *Gentlemen of Uncertain Fortune: How Younger Sons Made Their Way in Jane Austen's England* (New Haven, CT: Yale University Press, 2019).

Namier, Lewis, and John Brooke, eds, *The History of Parliament: The House of Commons 1754–1790* (London: HMSO, 1964).

Newman, Gerald, *The Rise of English Nationalism: A Cultural History, 1740–1830* (Basingstoke: Macmillan, 1997).

O'Day, Rosemary, *Education and Society, 1500–1800: The Social Foundations of Education in Early Modern Britain* (London: Longman, 1982).

O'Day, Rosemary, *The Family and Family Relationships, 1500–1900: England, France and the United States of America* (Basingstoke: Macmillan, 1994).

O'Day, Rosemary, 'Matchmaking and Moneymaking in a Patronage Society: The First Duke and Duchess of Chandos, *c.*1712–35', *Economic History Review*, 66:1 (2013), 273–96.

ODNB (*Oxford Dictionary of National Biography*), Oxford University Press, 2004; online edn, 2004–18 (www.oxforddnb.com).

Ohlmeyer, Jane, *Making Ireland English: The Irish Aristocracy in the Seventeenth Century* (New Haven, CT, and London: Yale University Press, 2012).

Ousby, Ian, *The Englishman's England: Taste, Travel and the Rise of Tourism* (Cambridge: Cambridge University Press, 1990.

Paley, Ruth, ed., *The History of Parliament: The House of Lords, 1660–1715*, 5 vols (Cambridge: Cambridge University Press, 2016).

Parks, George B., 'Travel as Education', in Richard Foster Jones, ed., *The Seventeenth Century: Studies in the History of English Thought and Literature from Bacon to Pope* (Stanford, CA: Stanford University Press, 1965), pp. 264–90.

Peltonen, Markku, *The Duel in Early Modern England: Civility, Politeness and Honour* (Cambridge: Cambridge University Press, 2003).
Perry, Ruth, *Novel Relations: The Transformation of Kinship in English Literature and Culture, 1748–1818* (Cambridge: Cambridge University Press, 2004).
Pincus, Steve, *1688: The First Modern Revolution* (New Haven, CT: Yale University Press, 2009).
Pittion, Jean-Paul, 'Saumur et les voyageurs britanniques au dix-septième siècle', *Annales de Bretagne et des pays de l'Ouest*, 121:3 (2014), 97–117.
Plumb, J. H., 'The New World of Children', in Neil McKendrick, John Brewer and J. H. Plumb, eds, *The Birth of a Consumer Society: The Commercialisation of Eighteenth-Century England* (London: Europa, 1982), pp. 286–312.
Pollock, Linda, '"Teach Her to Live under Obedience": The Making of Women in the Upper Ranks of Early Modern England', *Continuity and Change*, 4, special issue 2 (1989), 231–58.
Porter, Stephen, 'University and Society', in Nicholas Tyacke, ed., *The History of the University of Oxford: Volume IV: Seventeenth-Century Oxford* (Oxford: Clarendon, 1997), pp. 25–104.
Prest, Wilfrid R., *The Inns of Court under Elizabeth I and the Early Stuarts, 1590–1640* (London: Longman, 1972).
Prögler, Daniela, *English Students at Leiden University, 1575–1650* (Farnham: Ashgate, 2013).
Purcell, Mark, *The Country House Library* (New Haven, CT: Yale University Press, 2017).
Redford, Bruce, *Venice and the Grand Tour* (New Haven, CT, and London: Yale University Press, 1996).
Redford, Bruce, *Dilettanti: The Antic and the Antique in Eighteenth-Century England* (Los Angeles, CA: J. Paul Getty Museum, 2008).
Reid, Christopher, *Imprison'd Wranglers: The Rhetorical Culture of the House of Commons 1760–1800* (Oxford: Oxford University Press, 2012).
Richardson, R. C., 'The Generation Gap: Parental Advice in Early Modern England', *Clio*, 32:1 (2002), 1–25.
Roberts, David, *Thomas Betterton: The Greatest Actor of the Restoration Stage* (Cambridge: Cambridge University Press, 2010).
Rosenheim, James M., *The Emergence of a Ruling Order: English Landed Society, 1650–1750* (London: Longman, 1998).
Ross, Alan S., 'Pupils' Choices and Social Mobility after the Thirty Years War: A Quantitative Study', *Historical Journal*, 57:2 (2014), 311–41.
Rothery, Mark, and Jon Stobart, 'Inheritance Events and Spending Patterns in the English Country House: The Leigh Family of Stoneleigh Abbey, 1736–1806', *Continuity and Change*, 27:3 (2012), 379–407.
Sánchez-Jáuregui, María Dolores, 'Educating the Traveler: The Tutors', in María Dolores Sánchez-Jáuregui and Scott Wilcox, eds, *The English Prize: The Capture of the Westmorland, An Episode of the Grand Tour* (New Haven, CT, and London: Yale University Press, 2012), pp. 88–97.
Sánchez-Jáuregui, María Dolores, 'Books on the Westmorland', in María Dolores Sánchez-Jáuregui and Scott Wilcox, eds, *The English Prize: The Capture of the Westmorland, An Episode of the Grand Tour* (New Haven, CT, and London: Yale University Press, 2012), pp. 144–53.

Sánchez-Jáuregui, María Dolores, and Scott Wilcox, eds, *The English Prize: The Capture of the Westmorland, An Episode of the Grand Tour* (New Haven, CT, and London: Yale University Press, 2012).

Schalk, Ellery, *From Valor to Pedigree: Ideas of Nobility in France in the Sixteenth and Seventeenth Centuries* (Princeton, NJ: Princeton University Press, 1986).

Schneider, Gary, *The Culture of Epistolarity: Vernacular Letters and Letter Writing in Early Modern England* (Newark: University of Delaware Press, 2005).

Sedgwick, Romney, ed., *The History of Parliament: The House of Commons 1715–1754* (London: HMSO, 1970).

Shepard, Alexandra, *Meanings of Manhood in Early Modern England* (Oxford: Oxford University Press, 2003).

Shepard, Alexandra, *Accounting for Oneself: Worth, Status, and the Social Order in Early Modern England* (Oxford: Oxford University Press, 2015).

Skinner, Basil, *Scots in Italy in the Eighteenth Century* (Edinburgh: Scottish National Portrait Gallery, 1966).

Smith, Geoffrey, *The Cavaliers in Exile, 1640–1660* (Basingstoke: Palgrave Macmillan, 2003).

Smith, Hannah, *Georgian Monarchy: Politics and Culture, 1714–1760* (Cambridge: Cambridge University Press, 2006).

Stagl, Justin, *A History of Curiosity: The Theory of Travel 1550–1800* (Chur: Harwood Academic Publishers, 1995).

Stedman, Gesa, *Cultural Exchange in Seventeenth-Century France and England* (Farnham: Ashgate, 2013).

Stewart, Gordon M., 'British Students at the University of Göttingen in the Eighteenth Century', *German Life and Letters*, 33:1 (1979), 24–41.

Stobart, Jon, 'Introduction: Travel and the British Country House', in Stobart, ed., *Travel and the British Country House: Cultures, Critiques and Consumption in the Long Eighteenth Century* (Manchester: Manchester University Press, 2017).

Stone, Lawrence, and Jeanne C. Fawtier Stone, *An Open Elite? England 1540–1880* (Oxford: Clarendon Press, 1984).

Stoye, John, *English Travellers Abroad, 1604–1667: Their Influence in English Society and Politics* (New Haven, CT, and London: Yale University Press, 1989).

Stoye, John, 'The Grand Tour in the Seventeenth Century', *Journal of Anglo-Italian Studies*, 1 (1991), 62–73.

Strong, Michele M., *Education, Travel and the 'Civilisation' of the Victorian Working Classes* (Basingstoke: Palgrave Macmillan, 2014).

Sweet, Rosemary, *Cities and the Grand Tour: The British in Italy, c.1690–1820* (Cambridge: Cambridge University Press, 2012).

Sweet, Rosemary, Gerrit Verhoeven and Sarah Goldsmith, 'Introduction', in Sweet, Verhoeven and Goldsmith, eds, *Beyond the Grand Tour: Northern Metropolises and Early Modern Travel Behaviour* (London: Routledge, 2017).

Tadmor, Naomi, *Family and Friends in Eighteenth-Century England: Household, Kinship and Patronage* (Cambridge: Cambridge University Press, 2001).

Tadmor, Naomi, 'Early Modern English Kinship in the Long Run: Reflections on Continuity and Change', *Continuity and Change*, 25:1 (2010), 15–48.

Tague, Ingrid, *Women of Quality: Accepting and Contesting Ideals of Femininity in England, 1690–1760* (Woodbridge: Boydell, 2002).

Thirsk, Joan, 'Younger Sons in the Seventeenth Century', *History*, 54:182 (1969), 358–77.

Thomas, Keith, *The Ends of Life: Roads to Fulfilment in Early Modern England* (Oxford: Oxford University Press, 2009).

Thomas, Keith, *In Pursuit of Civility: Manners and Civilization in Early Modern England* (New Haven, CT: Yale University Press, 2018).

Thorne, R. G., ed., *The History of Parliament: The House of Commons 1790–1820* (London: Secker & Warburg, 1986).

Thrush, Andrew, and John P. Ferris, eds, *The History of Parliament: The House of Commons, 1604–1629* (Cambridge: Cambridge University Press, 2010).

Tosi, Arturo, *Language and the Grand Tour: Linguistic Experiences of Travelling in Early Modern Europe* (Cambridge: Cambridge University Press, 2020).

Towner, John, 'The Grand Tour: A Key Phase in the History of Tourism', *Annals of Tourism Research*, 12 (1985), 297–333.

Towner, John, 'The English Tourist and War, 1500–1800', in Richard Butler and Wantanee Suntikul, eds, *Tourism and War* (London: Routledge, 2013), pp. 49–63.

Turner, Katherine, *British Travel Writers in Europe, 1750–1800: Authorship, Gender and National Identity* (Aldershot: Ashgate, 2001).

Upton, Anthony F., *Sir Arthur Ingram, c. 1565–1642: A Study of the Origins of an English Landed Family* (London: Oxford University Press, 1961).

Van Strien, C. D., *British Travellers in Holland during the Stuart Period: Edward Browne and John Locke as Tourists in the United Provinces* (Leiden: Brill, 1993).

Verhoeven, Gerrit, 'Foreshadowing Tourism: Looking for Modern and Obsolete Features – or Some Missing Link – in Early Modern Travel Behaviour (1675–1750)', *Annals of Tourism Research*, 42 (2013), 262–83.

Verhoeven, Gerrit, *Europe within Reach: Netherlandish Travellers on the Grand Tour and Beyond (1585–1750)*, trans. Diane Webb (Leiden: Brill, 2015).

Viazzo, Pier Paolo, and Katherine A. Lynch, 'Anthropology, Family History, and the Concept of Strategy', *International Review of Social History*, 47:3 (2002), 423–52.

Vickery, Amanda, *The Gentleman's Daughter: Women's Lives in Georgian England* (New Haven, CT, and London: Yale University Press, 1998).

Villani, Stefano, '"Una piccola epitome di Inghilterra". La comunità inglese di Livorno negli anni di Ferdinando II: questioni religiose e politiche', *Cromohs*, 8 (2003), 1–23.

Walker, Claire, *Gender and Politics in Early Modern Europe: English Convents in France and the Low Countries* (Basingstoke: Palgrave Macmillan, 2003).

Wallis, Patrick, and Cliff Webb, 'The Education and Training of Gentry Sons in Early Modern England', *Social History*, 36:1 (2011), 36–53.

Walsh, Patrick, *The South Sea Bubble and Ireland: Money, Banking and Investment, 1690–1721* (Woodbridge: Boydell, 2014).

Walter, John, 'Gesturing at Authority: Deciphering the Gestural Code of Early Modern England', in Michael J. Braddick, ed., *The Politics of Gesture: Historical Perspectives* (Oxford: Oxford University Press, 2009), pp. 96–127.

Warneke, Sara, *Images of the Educational Traveller in Early Modern England* (Leiden: Brill, 1995).

Wasson, Ellis Archer, 'The House of Commons, 1660–1945: Parliamentary Families and the Political Elite', *English Historical Review*, 106:420 (1991), 635–51.

Williams, Mark R. F., *The King's Irishmen: The Irish in the Exiled Court of Charles II, 1649–1660* (Woodbridge: Boydell, 2014).
Williams, Mark R. F., 'The Inner Lives of Early Modern Travel', *Historical Journal*, 62:2 (2019), 349–73.
Wilson, Rachel, *Elite Women in Ascendancy Ireland, 1690–1745: Imitation and Innovation* (Martlesham: Boydell, 2015).
Wilton, Andrew, and Ilaria Bignamini, eds, *Grand Tour: The Lure of Italy in the Eighteenth Century* (London: Tate Gallery, 1996).
Wilton-Ely, John, '"Classic Ground": Britain, Italy, and the Grand Tour', *Eighteenth-Century Life*, 28:1 (2004), 136–65.
Woolf, D. R., *Reading History in Early Modern England* (Cambridge: Cambridge University Press, 2000).
Woolfson, Jonathan, *Padua and the Tudors: English Students in Italy, 1485–1603* (Cambridge: James Clarke, 1998).
Wrightson, Keith, 'The Family in Early Modern England: Continuity and Change', in Stephen Taylor, Richard Connors and Clyve Jones, eds, *Hanoverian Britain and Empire: Essays in Memory of Philip Lawson* (Woodbridge: Boydell, 1998), pp. 1–22.
Wrigley, E. A., and R. S. Schofield, *The Population History of England 1541–1871: A Reconstruction* (Cambridge: Cambridge University Press, 1989).

Index

Page numbers in *italics* refer to illustrations.

D

E

S

9 780197 267271

Printed and bound by CPI Group (UK) Ltd, Croydon, CR0 4YY
16/09/2026
14963460-0003